OXFORD READINGS IN PHILOSOPHY

EXPLANATION

D0139894

EXPLANATION

Edited by

DAVID-HILLEL RUBEN

OXFORD UNIVERSITY PRESS
1993

Oxford University Press, Walton Street, Oxford OX2 6DP
Oxford New York Toronto
Delhi Bombay Calcutta Madras Karachi
Kuala Lumpur Singapore Hong Kong Tokyo
Nairobi Dar es Salaam Cape Town
Melbourne Auckland Madrid
and associated companies in
Berlin Ibadan

Oxford is a trade mark of Oxford University Press

Published in the United States
by Oxford University Press Inc., New York

Introduction and Selection © Oxford University Press 1993

British Library Cataloguing in Publication Data
Data available

Library of Congress Cataloging in Publication Data
Explanation/edited by David-Hillel Ruben.
p. cm. — (Oxford readings in philosophy)
Includes bibliographical references and index.
1. Explanation. I. Ruben, David-Hillel. II. Series.
BD237.E86 1993 121'.4—dc20 92–33939
ISBN 0–19–875129–X
ISBN 0–19–875130–3 (Pbk.)

1 3 5 7 9 10 8 6 4 2

Typeset by Pure Tech Corporation, Pondicherry, India
Printed in Great Britain
on acid-free paper by
Biddles Ltd, Guildford and King's Lynn

CONTENTS

INTRODUCTION[*]

DAVID-HILLEL RUBEN

The concept of explanation should not be monopolized by the philosophy of the natural sciences. The concept of explanation, like the concept of knowledge to which it is closely related, is an epistemic concept, and therefore has a philosophical location within the theory of knowledge, widely conceived. The philosophy of science has a great contribution to make to a theory of explanation, just as it does to a theory of knowledge, but it is not the sole proprietor of either concept.

Many of the best contributions to the literature on explanation are written by philosophers of science, and the contents of this collection reflect that fact. Of course, even when an article has been written in a more technical style by a philosopher of science, it does not follow that the contribution only has relevance to explanation in the natural sciences. Many such contributions intend to address the topic of explanation in a general way, not confined to explanation in natural science. Carl Hempel's article, for example, is intended to apply to explanation in history and to rational explanation of action. However, I have tried to achieve what might be called a stylistic balance by including some articles not written by philosophers of natural science. The articles by Kim, Lewis, Lipton, and Matthews are clearly in this vein.

In deciding what to include in this volume, I have been careful to minimize overlap with Joseph Pitt's *Theories of Explanation*; only one article appears in both collections. Pitt's collection is, I believe, most suitable for the student of the philosophy of natural science. This collection is intended equally for the student of general philosophy, or anyway of metaphysics and epistemology. Taken together, the two collections cover most of the main approaches to the philosophy of explanation advanced in the last few decades.

I have said that the focus of the collection was on explanation generally, and not just explanation in the natural sciences. Although I believe

* I am grateful to those contributors to this collection who read over the Introduction, and especially to David Lewis and Wesley Salmon who provided me with written comments which allowed me to correct a number of points.

this to be so, there are some notable omissions in this collection. There is, for example, no contribution that addresses either intentional explanation of human action (beyond the brief discussion by Hempel), or functional explanation in biology and the social sciences. The articles I have included have a unity of focus that I thought should not be lost through dilution. For those interested in the topic of action explanation I can suggest no better place to start than Richard Taylor's book, now out of print, *Action and Purpose*.

Contemporary philosophical discussion about the concept of explanation began with Carl Hempel and Paul Oppenheim's 'Studies in the Logic of Explanation', 1948.[1] In that and subsequent articles, Hempel developed his philosophy of explanation. Many of the main ideas in those article can be found in John Stuart Mill's *A System of Logic*, 1843. But Hempel presents those ideas in a technically sophisticated, detailed, and far more convincing form than one finds in Mill. Indeed, almost all subsequent philosophical discussion about explanation, and most of the articles in this collection, take Hempel's account as their point of departure, even when they reject his main views.

Hempel's account of explanation, like that of Mill, can be placed squarely within the tradition of empiricist philosophy. Hempel attempts to account for explanation using only concepts which are acceptable to an empiricist philosopher, such as: entailment, inductive support, empirical content, truth, and laws as universally quantified generalizations. Some empirically minded philosophers, like Pierre Duhem, had come to believe that explanation was an experientially transcendent idea, and hence had no legitimate role to play within an empirically based, scientific outlook.[2] Hempel's intention is to demystify explanation, by explicating it in terms which are beyond empirical suspicion.

Hempel's 'Explanation in Science and in History' was first published in 1962. It presents the main outlines of his account in an admirably clear and succinct way. There are, according to Hempel, two main 'models' of explanation: deductive-nomological (hereafter, D-N) explanation, and probabilistic-statistical (or, inductive-statistical) explanation (hereafter, I-S explanation). The idea of I-S explanation goes beyond anything to be found in Mill. Some have claimed C. S. Peirce as an anticipator of Hempel on the topic of I-S explanation.

---

[1] Carl Hempel and Paul Oppenheim, 'Studies in the Logic of Explanation', *Philosophy of Science*, 15 (1948), 135–75, repr. in Hempel, *Aspects of Scientific Explanation* (New York: Free Press, 1965), 245–95, with an additional postscript.

[2] Pierre Duhem, *The Aim and Structure of Physical Theory* (New York: Atheneum, 1977).

On Hempel's view, full or complete explanations of both types are arguments (hereafter, 'the argument thesis'). Full D-N explanations are a proper subset of those deductively sound arguments whose conclusions are that the event to be explained has occurred; full I-S explanations are a proper subset of those inductively good arguments which show that the event to be explained was highly probable. Nothing Hempel says commits him to the view that there can be at most only one full explanation for some occurrence. Not only might there be both a full D-N and a full I-S explanation for something, but there may even be more than one full D-N explanation for it.

It is no objection to this account to point out that explanations in science and in common affairs sometimes, or even always, fall short of being sound or good arguments, since Hempel accepts that explanations as they are actually given can be incomplete in various ways. These models of full explanation are 'ideal types' or 'idealizations' to which actual practice may more or less correspond. Typically, it will be justified in the circumstances to give only a partial or incomplete explanation (suppose, for example, that giving a full explanation always put the hearer to sleep before it could be completed).

One might make Hempel's point by using a Kantian distinction between the constitutive and the regulative. Hempel's models of explanation provide ideals in the sense that they constitute for us what an ideal complete explanation would be like. They are not regulative ideals which we ought to aspire, as best we can, to provide in all, or perhaps even in any, circumstances. Peter Railton's idea of an ideal explanatory text, which he develops in 'Probability, Explanation, and Information', and David Lewis's idea of the complete causal history of an event, which he sets out in 'Causal Explanation', attempt to develop the distinction between an ideally full explanation and the incomplete sorts of explanation we actually give.

Laws play a particularly central role in Hempel's account of full explanation. The concept of a law of nature is itself not above empirical suspicion, since on some accounts at least it seems to require the idea of non-logical, or physical, necessity. Hempel has expressed the hope that a full reductive analysis of natural law can be offered in terms of a certain sort of universally quantified generalization ('a lawlike generalization'), although he does not himself seek to provide that analysis.[3] If such an analysis could be given, no commitment to non-logical necessity would be required in order to understand the idea of a law of nature.

---

[3] Carl Hempel, 'Aspects of Scientific Explanation', *Aspects*, 338–43.

Of course, there can be sound deductive arguments that include no law or generalization. But given some further reasonable restrictions on which sound arguments are explanatory (e.g., that no statement can explain itself), the premisses of every full explanatory argument must essentially include at least one law or lawlike generalization (hereafter, 'the law thesis'). Sometimes Hempel's account of explanation is called 'the covering law theory', to mark this requirement. James Woodward, in his 'A Theory of Singular Causal Explanation', presents an account of the explanation of singular events which rejects Hempel's law thesis. He makes use of the alternative idea of counterfactuals; all explanations, including singular causal explanation, explain in virtue of answering the question, 'What if things had been different?' Whether this line of thought is, at the end of the day, a genuine alternative to the law thesis will depend on what one takes the truth conditions for counterfactuals to be.

Hempel does not claim to be providing an account of explanation *sans phrase*. He claims that he is providing an account of scientific explanation, and is prepared to admit that his account will not cover such examples as explaining the meaning of a poem, explaining *how* to bake a Sacher torte, etc. So his models do not provide a constitutive ideal for explanation in all areas of life, but only for those cases in which we explain *why* something is the case. If we leave aside cases in which we do not explain *why* at all, what does the adjective 'scientific' add to 'explanation'? I have already noted that Hempel intends his models of explanation to apply to explanation in history, and to action explanation. He also intends it to apply to functional explanation, in biology and in the social sciences (although he does not think that the social sciences actually provide us with many successful explanations of that type). In my view, 'scientific' can only mean for Hempel 'full', or 'complete'. Hempel's view, thus construed, is that his models correctly describe full explanations whenever we seek to explain *why* some contingent fact obtains, whether in the natural sciences or elsewhere.[4]

Hempel's account of the explanation of token, particular events is based on the view that explanation can be identified as nomic expectability; we have explained an event when we can show that it was, in the circumstances, to be expected. In Hempel's words, 'The explanation here outlined may be regarded as an argument to the effect that the phenomenon to be explained . . . was to be expected in virtue of certain explanatory facts.'[5] According to Hempel, there is a symmetry between explanation

---

[4] I have argued this point more fully in *Explaining Explanation* (London: Routledge, 1990), 16–19.

[5] Hempel, 'Aspects', 336.

and prediction; every full explanation is a potential prediction, and conversely. The difference between them is merely pragmatic, depending on whether or not the event mentioned in the argument's conclusion has already occurred by the time at which the argument is produced.

I claimed at the beginning of the Introduction that the concept of explanation belongs basically to epistemology. One might summarize Hempel's view in the preceding paragraph by saying that explanation for Hempel is only an epistemic idea. It concerns the provision of reasons to believe that something has occurred or will occur. But the question arises of whether one can account for explanation wholly within such terms, or whether there are additionally some sort of non-epistemic, worldly constraints on explanation. Wesley Salmon, in the first part of the selection from his *Scientific Explanation and the Causal Structure of the World* describes three different conceptions of explanation: the epistemic, the modal, and the ontic. Salmon embraces an alternative, ontic conception of explanation: 'to explain an event is to exhibit it as occupying its (nomologically necessary) place in the discernible pattern of the world' (p. 81). Elsewhere, he says that this is accomplished by finding the causal mechanisms responsible for the production of the event to be explained. Salmon contrasts his ontic conception with an epistemic one like Hempel's. An account of explanation is ontic, let us say, if it requires of an explanation that it be about some real, worldly feature, relation, or whatever. Since I will not introduce what Salmon calls 'modal' conceptions as a separate category, I will use 'ontic' and 'non-epistemic' interchangeably in what follows.

Unlike Salmon, Hempel introduces no requirement concerning what sorts of entities or events in the world the explanatory premises must be about; the premises need only be true and have some empirical content. In particular, there is no requirement by Hempel that the premisses must tell us anything about the cause of the event to be explained. Hempel is particularly insistent on this point. Even if we restrict ourselves to the explanation of singular events rather than regularities, we can, he says, explain an event in terms of a simultaneous event, and no pair of simultaneous events are related as cause and effect. For instance, we can explain the period of a pendulum at time t by its length at t, and we can explain a gas's pressure at t by its volume and temperature at t. Indeed, Hempel nowhere even rules out the explanation of an event by subsequent events.

Which proper subset of good I-S *arguments* that show that the event mentioned in the conclusion was highly probable are I-S *explanations* of that event? I have included a brief excerpt from Hempel's 'Aspects of

Scientific Explanation', which deals with one of the main difficulties encountered by his account of I-S explanation, the problem of the epistemic ambiguity of I-S explanation. J. Alberto Coffa's 'Hempel's Ambiguity' discusses Hempel's views on the epistemic ambiguity of I-S explanation, and Hempel's relativization of I-S explanations to a specific knowledge situation. Coffa connects this problem with the difference between an epistemic and an ontic conception of explanation. The problem is this. There can be two equally good inductive arguments, both of which contain only true premises which make their conclusions highly likely or probable, but which have contradictory conclusions. However, the two equally good I-S arguments are not equally good I-S explanations. One might inductively infer that Jones, who is a Texan philosopher, is rich from the fact that he is a Texan, and might inductively infer that he is poor from the fact that he is a philosopher. Hempel, in 'Aspects of Scientific Explanation', accepts that I-S explanation, quite unlike D-N explanation, must be relativized to a knowledge situation ('the requirement of maximal specificity'). The inductive argument that is explanatory must include amongst its premises all known relevant facts. Thus, what might be a good I-S explanation in one situation might not be so in another, as the current state of knowledge changes.

Coffa points out some of the difficulties with Hempel's proposed solution to the problem of epistemic ambiguity. The main point is that it makes the idea of an I-S explanation dependent on the state of knowledge at the time. On this view, there is a vast difference between D-N and I-S explanation, since the former requires no similar relativization. Coffa says that the effect of Hempel's approach is to deny the existence of inductive explanation in any real or objective sense. Hempel's proposal subjectivizes explanation in an unacceptable way. Coffa agrees with Hempel that an account of I-S explanation must include a resolution of what he calls 'the reference class problem'. Is the right reference class for Jones the class of Texans, the class of philosophers, or the class of Texan philosophers? Moving beyond statistical relevance, which he believes cannot provide an acceptable solution to this problem, Coffa argues that there are objective ways in which one might try to answer this question (in terms of causation and nomologicity), and the 'right' I-S explanation is the one that gets these causal and nomological facts right, whether or not our current state of knowledge permits us to formulate this I-S explanation.

In the second part of the selection from *Scientific Explanation and the Causal Structure of the World*, Wesley Salmon sketches an account of the statistical relevance relations appropriate for explaining an explanandum event. Previously, Salmon had argued for the view that an account of explanation could be given in terms of statistical-relevance (hereafter, S-R) relations alone. In this selection, he presents the main outlines of an S-R account. Following Coffa, Salmon wants his reference classes to be objectively homogeneous. No epistemic relativization is admitted. Following on from the remarks in the first part of his selection, this is part and parcel of Salmon's attempt to present an ontic model of explanation, one based on how things really are. Also like Coffa, Salmon does not now believe that S-R relations can do the whole job in accounting for explanation. In the selection from his book, Salmon says that explanation is a two-stage affair. First, he says that the event to be explained is subsumed under the appropriate set of S-R relations. These latter are then to be explained by means of causal relations. A full explanation can be achieved only when both stages have been completed. Salmon has an interesting and original account of causation, which I have not included in this selection.[6]

Which proper subset of sound D-N *arguments*, the conclusions of which are that the event to be explained has occurred, are D-N *explanations* of that event? Hempel's D-N model of explanation as he stated it cannot provide sufficient conditions for an argument's being an explanation, since that model is open to a number of well-known counter-examples. One such is the problem of asymmetry. Suppose that something is an *F* if and only if it is a *G*. One can deductively infer that some particular object is an *F* if it is a *G*, and infer that it is a *G* if it is an *F*. Typically, only one of the inferences will be explanatory. Aristotle provided an example of this. Suppose that a celestial object twinkles if and only if it is far away. One can infer that a planet is near from the fact that it does not twinkle, and that it does not twinkle from the fact that it is near, but presumably, even though the fact that it is near would explain why it does not twinkle, the fact that it does not twinkle could not explain why it is near. Another example, from Hempel, is the case of the falling barometer. Suppose a barometer falls if and only if a storm is coming. One can infer the coming of the storm from the falling barometer, and the falling barometer from the coming of the storm, but only the later inference is in any way explanatory.

A plausible thought is that some, at any rate, of these counter-examples arise only on an epistemic conception of explanation, since in general one may have grounds for expecting something even though those

---

[6] The interested reader might consult chs. 5–7 of his book, or, for a briefer account, his 'Why Ask, "Why?"?': An Inquiry Concerning Scientific Explanation', Presidential Address, *Proceedings and Addresses of the American Philosophical Association*, 1978.

grounds may fail to explain what they reasonably lead one to expect. A falling barometer gives one a reason to expect the coming of a storm, although the former does not explain the latter. If so, one might try to deal with some of these counter-examples by the introduction of some further ontic requirements which constrain what the premises of an explanatory argument must be about. Aristotle, for example, introduced a causal requirement to cure the asymmetry problem, and Brody, in 'Towards an Aristotelian Theory of Scientific Explanation', follows Aristotle and proposes to add the ideas of causation and essence to the Hempelian models, to deal with what would otherwise be counter-examples to it. For example, causation is unidirectional, causes cause effects and not vice versa, and Brody hopes to use this to account for the fact that explanation will go in only one direction, even when inference goes in both directions.

McCarthy's reply to Brody, 'On an Aristotelian Model of Scientific Explanation', shows that similar counter-examples can still be constructed, even on Brody's ontic modification of Hempel's models. That is, there are counter-examples to the thesis that full explanations are arguments which meet all Hempel's requirements, plus the further, ontic requirement that one of the argument's premises mention the cause or essence of the event to be explained. McCarthy's discussion suggests further criticisms of Hempel's models, and many models which are sophisticated variants of Hempel's models, criticisms which are further developed by Peter Achinstein in his 'Can There be A Model of Explanation?'

The lesson seems to be that it is insufficient for an explanatory argument to merely mention the cause, c, of what is to be explained, e. To defeat the possibility of McCarthy-style counter-examples, such a premiss would have not only to mention c, but also actually to say that c was the cause of e. However, if this were so, the conclusion that e occurred could be inferred from the single premise that says that c caused e. So no law or lawlike premise would need to be included in a full explanation, since its inclusion would be redundant. If this is so, then Hempel's law requirement would be otiose. Moreover, there seems to be nothing gained in continuing to think of an explanation as an argument with the form: 'c causes e, therefore e'. The explanation seems merely to be a proposition (or sentence, statement, or whatever): 'c causes e'. So the thrust behind McCarthy's and Achinstein's articles questions not only whether Hempel's law thesis, but also whether his argument thesis, provides necessary conditions for full explanation.

Peter Railton, in 'Probability, Explanation, and Information', provides an account of I-S explanation (he calls it 'D-N-P explanation') which allows for the explanation of events the occurrence of which is objectively improbable. It has seemed to many that Hempel's view that only highly probable events could be explained is arbitrary. Richard Jeffrey seems to be first to have made this point clearly, in his 'Statistical Explanation *vs.* Statistical Relevance'.[7] As Wesley Salmon has frequently pointed out in his writings, if there are two outcomes possible to a stochastic process, namely, e and f, e being highly probable and f highly improbable, and precisely the same information ('the process is stochastic, e is highly probable, f highly improbable') is relevant to both, it seems arbitrary to hold that we can explain the one if it occurs but not the other were it to occur. The improbable, when it occurs, does not seem to be inexplicable. If there can be explanations for improbable occurrences, this would provide a further reason to reject Hempel's argument thesis, since there is no such thing as a valid or good argument such that the conclusion is improbable, given the premises.

Ontic accounts (like Salmon's) have, thus far, been entirely concerned with causation as being the appropriate underpinning for the explanation of singular occurrences. David Lewis, in 'Causal Explanation', argues that this is the only such underpinning for singular explanation. According to Lewis, the whole explanation of a singular event is given by the complete causal history of that event, no doubt stretching back to the Big Bang. Actual explanation-giving is governed by pragmatic criteria concerning what more limited information extracted from that total history is of interest to us in the circumstances, and at what level of generality of description. On Lewis's view, the only kind of explanation it is possible to give for a singular event is a causal explanation, in the sense that the explanation will be some information taken from the complete causal history of that event; any explanation which is genuinely non-causal cannot be an explanation of a singular event or occurrence at all.

Peter Lipton's 'Contrastive Explanation' accepts the basic ingredients of a causal account of explanation, and addresses the question of how one selects, out of the total causal history leading up to the event to be explained, that part of the history which is of explanatory interest. To answer this question, Lipton makes use of the idea of explanatory contrasts (see also Lewis's, Woodward's, and van Fraassen's papers for this topic). On Hempel's theory, what gets explained by an explanation is the fact that some event e has some feature or property P. On the contrastive account, what gets explained is typically that some event e has a property or feature P rather than ... (where many different things can fill the blank; for example, that e has a different feature Q, or that some different

[7] In Wesley Salmon (ed.), *Statistical Explanation and Statistical Relevance* (Pittsburgh: University of Pittsburgh Press, 1971), 19–28.

event *f* has feature *P*). How the blank should be filled will vary, depending on shifts in explanatory interest and focus. Contrastive approaches to full explanation have often been associated with accounts of explanation assumed not to be within the ontic tradition; Lipton shows that this approach can be included within an ontic framework.

Many, perhaps even most, other writers have disagreed with the claim that ties singular explanation so intimately with causal explanation, and have produced lists of apparent counter-examples to the thesis. Let me give an *ad hoc* sample of alleged examples, which are not discussed in this collection. Philip Kitcher's non-causal cases are the explanation of why neon is chemically inert by quantum chemistry and various explanations in formal linguistics.[8] Nancy Cartwright mentions explanations invoking laws of association as non-causal: 'the equations of physics . . . [for instance] whenever the force on a classical particle of mass *m* is *f* the acceleration is *f/m* . . . [and] the probabilistic laws of Mendelian genetics . . .'[9] Clark Glymour argues that there 'remains, however, a considerable bit of science that sounds very much like explaining, and which perhaps has causal implications, but which does not seem to derive its point, its force, or its interest from the fact that it has something to do with causal relations (or their absence)'.[10] Glymour's examples are all concerned with explaining gravitation and electro-dynamics on the basis of some variational principle, and he gives three examples of this. Peter Railton says that 'some particular facts may be explained non-causally, e.g., by subsumption under structural laws such as the Pauli exclusion principle'.[11] Elliott Sober says that 'equilibrium explanations . . . present a distinct set of counterexamples to the causal requirement'.[12] John Forge reminds us that '. . . laws of co-existence are not causal laws . . . laws of co-existence do in fact appear in scientific explanations. Some of these explanations are of considerable significance, such as those involving applications of classical thermodynamics in chemistry.'[13]

[8] Philip Kitcher, 'Salmon on Explanation and Causality: Two Approaches to Explanation', *The Journal of Philosophy*, 82 (1985), 632–9; 636–7.

[9] Nancy Cartwright, *How the Laws of Physics Lie* (Oxford: Oxford University Press, 1986), 21.

[10] Clark Glymour, 'Causal Inference and Causal Explanation', in Robert McLaughlin (ed.), *What? Where? When? Why?* (Dordrecht: Reidel, 1982), 179–91; 184. His examples are from pp. 184–6.

[11] Peter Railton, 'A Deductive-Nomological Model of Probabilistic Explanation', *Philosophy of Science*, 45 (1978), 206–26; 207.

[12] Elliott Sober, 'Equilibrium Explanations', *Philosophical Studies*, 43 (1983), 201–10.

[13] John Forge, 'The Instance Theory of Explanation', *Australasian Journal of Philosophy*, 64 (1986), 132.

What sorts of cases, if any, should convince us that there are non-causal singular explanations? In particular, what is the concept of causation that is being used, either in the assertion or in the denial of the causal theory of singular explanation? I assume that at least two features of (ordinary, empirical) causation, are uncontroversial.[14] The two features are these: (1) nothing can directly cause itself; (2) the causal relation is contingent, in the sense that it is always logically possible for the event which is the cause to occur without having that effect, and conversely. I do not claim that (1) and (2) are logically independent. Of course, (1) has been denied for the case of allegedly necessary beings such as God, or Nature-as-a-Whole, and what we might call 'metaphysical explanation'. It is, however, uncontroversial in its application to contingent beings and empirical explanation, scientific and ordinary, which is what is under discussion in this collection. Lewis has argued that an event could indirectly cause itself via a closed causal loop.[15] Therefore, (1) only outlaws direct self-causation, theology apart.

No one, as far as I know, has ever disputed the claim that no (contingent) thing or event directly causes itself, (1) above. Causation in such cases must be a relation between two distinct existences. Since there are cases of direct, singular empirical explanation in which there are not two distinct (or even different) existences that figure in the explanans and the explanandum, it follows that there are some cases of non-causal singular explanation.[16] These cases provide, to my mind, the least controversial examples of non-causal singular explanation.

Peter Achinstein has discussed cases of this sort in some of his writings not included in this collection.[17] Achinstein's examples of this type of explanation include: explaining why the pH value of some solution is changing on the grounds that the concentration of hydrogen ions which that solution contains is changing; explaining why ice is water on the grounds that it is H_2O; explaining why some gas sample has temperature t on the grounds that its constituent molecules have a mean kinetic energy k. These at least appear to be explanations in the sense in which Hempel was interested. They are explanations of why some contingent fact obtains: explanations of why some change occurs (a change in the pH value

[14] My indebtedness in this and the following sections to Peter Achinstein's work will be obvious to anyone who knows his writings.

[15] David Lewis, 'The Paradoxes of Time Travel', in *Philosophical Papers* (New York: Oxford University Press, 1986), ii. 74.

[16] I do not use the contingency claim (2) here, because I want to leave open whether the identities are contingent or necessary.

[17] Peter Achinstein, *The Nature of Explanation* (New York: Oxford University Press, 1983), 233–7.

of a solution), or of why some substance is in a particular state (a gas having temperature t or ice being water).

Could they be causal explanations? In its simplest form, we can sometimes explain why some particular, a, has property P by identifying P with a property, Q, which a also has. In a somewhat less simple form, we can sometimes explain why a is P, by identifying a with the sum of its parts, [$b\&c\&d$], and identifying P with some property Q, true of the sum itself, or sometimes, with a property Q had individually by each member of the sum. Achinstein argues that identity explanations cannot be a species of causal explanation, since the having or acquiring of property P cannot cause the having or acquiring of property Q, if $P = Q$. It makes no difference to the argument whether these identities are metaphysically necessary or contingent.

Temperature = mean kinetic energy (for some temperature t and some mke m, having temperature t = having constituent molecules with mke m). I can explain a gas having a certain temperature t by its constituent molecules having mean kinetic energy m, and I can explain a change in a gas's temperature by a change in the mean kinetic energy of its constituent molecules. We explain in these cases, not just by laws of the coexistence of two types of phenomena, but by property or type-type identities. This kind of singular explanation, relying as it does on identities, cannot be assimilated to causal explanation.

If these cases are genuinely cases of non-causal explanation, it would raise the question of whether there is some fault intrinsic to the ontic conception of explanation, or whether the ontic conception may merely have been presented in too restricted a form, tied solely to causation. Jaegwon Kim's 'Explanatory Realism, Causal Realism, and Explanatory Exclusion', distinguishes between explanatory realism and explanatory irrealism. The distinction is close to that between the ontic and epistemic conceptions of explanation. Kim is prepared at least to consider real, worldly relations other than causation which might underpin explanation on a realist or ontic view. Some alternative possibilities are suggested by him in 'Noncausal Connections' (which is not reprinted in this collection).[18] He there argues that:

Events in this world are interrelated in a variety of ways. Among them, the ones we have called dependency or determination relations are of great importance. Broadly speaking, it is these relations, along with temporal and spatial ones, that give a significant structure to the world of events. The chief aim of the present paper has been to show that causation, though important and in many ways

[18] Jaegwon Kim, 'Noncausal Connections', *Nous*, 8 (1974), 41–52.

fundamental, is not the only such relation, and that there are other such determinative relations that deserve recognition and careful scrutiny. (p. 52)

There appear to be dependency relations between events that are not causal,
and, as I shall argue, universal determinism may be true even if not every event
has a cause. These non-causal dependency relations are pervasively present in the
web of events, and it is important to understand their nature, their interrelations,
and their relation to the causal relation if we are to have a clear and complete
picture of the ways in which events hang together in this world. (p. 41)

Metaphysically, Kim's point is that the world is structured by various
determinative or dependency relations, of which causal relations are only
one kind. On an ontic account that moves beyond causation as the only
relevant relation to underpin explanation, it is the presence of some of
these non-causal 'structural' determinative (and dependency) relations
that might make explanation possible.[19]

Let us grant, then, that identity may provide us with some examples
of non-causal singular explanation. Are there any relations, other than
identity, which yield examples of non-causal singular explanation? When
a metaphysically austere theorist surveys the relations in which objects
or events stand in the world, he is happy with causation and identity, but
is sceptical about almost everything else. The metaphysically florid
theorist thinks that there are other determinative relations that lie somewhere between causation and identity; they are not as strict or tightly
binding as identity, but not as loose or contingent as causation. Cambridge dependency (the relation, for example, between Socrates' death
and Xantippe's becoming a widow),[20] supervenience, the by-relation
(that relates actions), the relation between a disposition and its structural
basis, are further suggestions advanced by various florid theorists. There
is a great deal of controversy about each such alleged case. If any of
these relations are distinct from both identity and causation, some may
provide the basis for additional non-causal singular explanations.

So, whether or not there are other cases of non-causal singular explanation will depend, I think, on whether or not there are determinative (or
dependency) metaphysical relations between objects, events, or states
other than causation and identity. Kim, certainly a florid theorist, mentions these three as examples of non-causal determinative relations: Cambridge dependency, one action being done by means of another, and event
composition. The third, event composition, is similar to the ordinary

---

[19] I have discussed these issues at some length in my *Explaining Explanation* (London:
Routledge, 1990), ch. 7.
[20] I have discussed Cambridge determination in my 'A Puzzle About Posthumous Predication',
The Philosophical Review, 97 (1988), 211–36.

mereological relation of a part to a whole, but is defined for events rather than objects, and therefore the parthood in question is temporal rather than spatial. Examples of the first two kinds rest on highly contentious (but not obviously false) theses about event identity. Understood in this way, the argument between Lewis and Kim about whether there are these non-causal explanations of singular events is, at bottom, a disagreement about what is to count as an event. Kim's fine-grained and generous conception, according to which Socrates' death and Xantippe's becoming a widow count as distinct events, contrasts with Lewis's rougher-grained and more parsimonious conception; this difference at least partly accounts for the differences between them over non-causal singular explanation.

It is not clear, therefore, whether there are insuperable objections to an ontic approach to explanation. The ontic theorist could agree that causation alone is too narrow a basis on which to account for explanation, since there are non-causal singular explanations. But whether a more general version of an ontic approach survives will depend crucially on the sorts of considerations I have mentioned above.

Meanwhile, proponents of epistemic accounts have not been idle. There are three basic epistemic approaches to explanation: the inferential (principally, Hempel), the information-theoretic (not represented in this collection), and the erotetic (for example, van Fraassen). A sophisticated, and perhaps the best-known, post-Hempelian epistemic account of explanation is provided by Bas van Fraassen. Van Fraassen conceives of his approach to explanation as 'pragmatic'.

Van Fraassen bases his account on work done in erotetic logic, the logic of questions. His view, represented by a selection from chapter 5 of his *The Scientific Image*, is that an explanation is neither an argument nor a proposition, but rather an answer to a why-question. A question (e.g., 'Why did *e* occur?'), on this account, can be identified as an ordered triple, composed of a topic (*e*'s occurrence), a contrast class (e.g., rather than *f* or *g* occurring), and a relevance relation *R*. Van Fraassen discusses the relevance relation, but leaves it relatively unconstrained. It is difficult to assess just how well his theory stands up without a fuller grasp of what sorts of replies to a question can count as 'relevant' and for what reasons, and it this omission in his account on which Philip Kitcher and Wesley Salmon focus in 'Van Fraassen on Explanation'. Their argument is that, with an unconstrained relevance relation, just about anything can count as an answer to any question. Van Fraassen would, presumably, resist constraining the relevance relation with ontic requirements; it would be unacceptable on an epistemic account to say that a relevant answer to a why-question must give the cause . . . etc. of the event which figures in the question's topic. Van Fraassen will have to reply to the challenge by constraining the relevance relation in ways open to an epistemic account of explanation. For example, certain additional syntactic requirements might be useful in further constraining the relevance relation in a way open to a non-ontic theorist.

Hitherto I have been somewhat vague about what it is that makes an account of explanation 'epistemic' or otherwise; similar vagueness infects the idea of a 'pragmatic' theory of explanation. Presumably, explanation is a pragmatic idea if what is an explanation in one context might not be an explanation in another relevantly similar context. Peter Achinstein, in the second article by him in this collection, 'The Pragmatic Character of Explanation', examines the question of what would have to be true of the concept of explanation in order to rightly classify it as pragmatic. Achinstein argues that, when this is clarified, van Fraassen's theory turns out not to be a pragmatic one after all, in any interesting sense.

Thus far, we have been taking it for granted that we know what a singular explanation looks like. James Woodward, in 'Singular Causal Explanation', describes some of the frequently neglected logical features of singular explanation. A contrast is often drawn between the extensionality of sentences which report singular causal connections ('token event c causes token event e') and the non-extensionality of sentences that provide singular causal explanations ('token event c causally explains token event e'). Causal explanation of one chunk of the world (an event, for instance) by another, it is often held, is relative to a description, whereas causation itself holds between chunks of the world (again, events, for example) *simpliciter*, however these chunks are described. Thus, on the standard view, what is to be explained is never an event *per se*, but a fact or an existential generalization or something with some sort of propositional structure. This view holds that the explanation relation relates relata which are not themselves entirely mind-independent. Notice that this standard view is consistent with an overall ontic approach to explanation. Such an ontic theory would have to accept that the explanation relation does not itself relate real chunks, but it can certainly maintain that, whatever fact-like entities the explanation relation does relate, such cases of explanation are only such at least partly in virtue of how the chunks really stand in the world, causally speaking or otherwise.

Woodward rejects this standard view. On Woodward's view, as on Lewis's, in a singular causal explanation, one event can explain another event *tout court*. The explanation relation, in the case of singular causal

explanation, relates the events themselves. Woodward draws the important distinction between what an explanation explains and what it merely
presupposes but does not explain. Woodward disputes the standard view
that, unlike causal sentences, sentences which give causal explanations
are non-extensional in character.

'Explanation', like many words ending in 'ion', suffers from a process–product ambiguity.[21] It may refer either to the activity of explaining
(the process), or to the information conveyed by means of such an activity (the product). Hempel, in keeping with the other classical accounts
of explanation, has taken explanation-as-information-product to be conceptually prior to explanation-as-activity. For Hempel, an explanatory
product can be characterized independently of an explaining act; presumably, an explaining act is merely one in which some explanatory
information is presented, at a time after whatever is being explained has
happened.

Achinstein's account of explanation, briefly characterized in his 'The
Pragmatic Character of Explanation', is that an explanation is an ordered
pair consisting of a proposition and the act type, explaining that so-and-
so. This type of approach, on which the idea of an explanation product
is made dependent on the idea of explanatory activity, is also represented
in this volume by Robert Matthews's 'Explaining and Explanation'. The
key notion for Matthews is that of an episode or act which plays a certain
sort of causal role that eventuates in a hearer's understanding something.

Are theories of explanation which make the activity of explaining
primary ontic or not? No general answer can be given to this question;
the answer will depend on the details of each account. For example,
Achinstein's account is, I think, not ontic, since he explicitly rejects any
attempt to place ontic restrictions on what the content of the propositions
expressed in an explaining speech act must be about. A view like Matthews's could in principle be either, depending on what is added about
further restrictions on those acts which are able to bring about understanding in a hearer (hitting the hearer over the head will hardly qualify,
regardless of the effect of so doing). However, making explaining activity primary at least holds open the possibility of an epistemic account of
explanation, and has proved attractive to philosophers sympathetic to a
non-ontic approach to explanation.

---

[21] Romane Clark and Paul Welsh, *Introduction to Logic* (Princeton, NJ: van Nostrand, 1962),
153–4.

I

EXPLANATION IN SCIENCE AND IN HISTORY

CARL G. HEMPEL

1. INTRODUCTION

Among the divers factors that have encouraged and sustained scientific inquiry through its long history are two pervasive human concerns which provide, I think, the basic motivation for all scientific research. One of these is man's persistent desire to improve his strategic position in the world by means of dependable methods for predicting and, whenever possible, controlling the events that occur in it. The extent to which science has been able to satisfy this urge is reflected impressively in the vast and steadily widening range of its technological applications. But besides this practical concern, there is a second basic motivation for the scientific quest, namely, man's insatiable intellectual curiosity, his deep concern to *know* the world he lives in, and to *explain*, and thus to *understand*, the unending flow of phenomena it presents to him.

In times past questions as to the *what* and the *why* of the empirical world were often answered by myths; and to some extent, this is so even in our time. But gradually, the myths are displaced by the concepts, hypotheses, and theories developed in the various branches of empirical science, including the natural sciences, psychology, and sociological as well as historical inquiry. What is the general character of the understanding attainable by these means, and what is its potential scope? In this paper I will try to shed some light on these questions by examining what seem to me the two basic types of explanation offered by the natural sciences, and then comparing them with some modes of explanation and understanding that are found in historical studies.

First, then, a look at explanation in the natural sciences.

Carl G. Hempel, 'Explanation in Science and in History', in R. G. Colodny (ed.), *Frontiers of Science and Philosophy* (London and Pittsburgh: Allen & Unwin and University of Pittsburgh Press, 1962), 7–33. Reprinted by permission of the author, Allen & Unwin Ltd., and University of Pittsburgh Press.

2. TWO BASIC TYPES OF SCIENTIFIC EXPLANATION

2.1. Deductive-nomological explanation

In his book, *How We Think*,[1] John Dewey describes an observation he made one day when, washing dishes, he took some glass tumblers out of the hot soap suds and put them upside down on a plate: he noticed that soap bubbles emerged from under the tumblers' rims, grew for a while, came to a standstill, and finally receded inside the tumblers. Why did this happen? The explanation Dewey outlines comes to this: In transferring a tumbler to the plate, cool air is caught in it; this air is gradually warmed by the glass, which initially has the temperature of the hot suds. The warming of the air is accompanied by an increase in its pressure, which in turn produces an expansion of the soap film between the plate and the rim. Gradually, the glass cools off, and so does the air inside, with the result that the soap bubbles recede.

This explanatory account may be regarded as an argument to the effect that the event to be explained (let me call it the explanandum-event) was to be expected by reason of certain explanatory facts. These may be divided into two groups: (i) particular facts and (ii) uniformities expressed by general laws. The first group includes facts such as these: the tumblers had been immersed, for some time, in soap suds of a temperature considerably higher than that of the surrounding air; they were put, upside down, on a plate on which a puddle of soapy water had formed, providing a connecting soap film, etc. The second group of items presupposed in the argument includes the gas laws and various other laws that have not been explicitly suggested concerning the exchange of heat between bodies of different temperature, the elastic behaviour of soap bubbles, etc. If we imagine these various presuppositions explicitly spelled out, the idea suggests itself of construing the explanation as a deductive argument of this form:

$$(D) \qquad \frac{\begin{array}{c} C_1, C_2, \ldots, C_k \\ L_1, L_2, \ldots, L_r \end{array}}{E}$$

Here, C_1, C_2, $\ldots$, C_k are statements describing the particular facts invoked; L_1, L_2, $\ldots$, L_r are general laws: jointly, these statements will be said to form the explanans. The conclusion E is a statement describing the explanandum-event; let me call it the explanandum-statement, and

---

[1] See John Dewey, *How We Think* (Boston, New York, Chicago, 1910), ch. 6.

let me use the word 'explanandum' to refer to either E or to the event described by it.

The kind of explanation thus characterized I will call *deductive-nomological explanation*; for it amounts to a deductive subsumption of the explanandum under principles which have the character of general laws: it answers the question '*Why* did the explanandum event occur?' by showing that the event resulted from the particular circumstances specified in $C_1, C_2, \ldots, C_k$ in accordance with the laws $L_1, L_2, \ldots, L_r$. This conception of explanation, as exhibited in schema (D), has therefore been referred to as the covering law model, or as the deductive model, of explanation.[2]

A good many scientific explanations can be regarded as deductive-nomological in character. Consider, for example, the explanation of mirror-images, of rainbows, or of the appearance that a spoon handle is bent at the point where it emerges from a glass of water: in all these cases, the explanandum is deductively subsumed under the laws of reflection and refraction. Similarly, certain aspects of free fall and of planetary motion can be accounted for by deductive subsumption under Galileo's or Kepler's laws.

In the illustrations given so far the explanatory laws had, by and large, the character of empirical generalizations connecting different observable aspects of the phenomena under scrutiny: angle of incidence with angle of reflection or refraction, distance covered with falling time, etc. But science raises the question 'Why?' also with respect to the uniformities expressed by such laws, and often answers it in basically the same manner, namely, by subsuming the uniformities under more inclusive laws, and eventually under comprehensive theories. For example, the question, 'Why do Galileo's and Kepler's laws hold?' is answered by showing that these laws are but special consequences of the Newtonian laws of motion and of gravitation; and these, in turn, may be explained by subsumption under the more comprehensive general theory of relativity. Such subsumption under broader laws or theories usually increases both the breadth and the depth of our scientific understanding. There is

[2] For a fuller presentation of the model and for further references, see e.g. C. G. Hempel and P. Oppenheim, 'Studies in the Logic of Explanation', *Philosophy of Science*, 15 (1948), 135–75. (Sects. 1–7 of this article, which contain all the fundamentals of the presentation, are reprinted in H. Feigl and M. Brodbeck (eds.), *Readings in the Philosophy of Science* (New York, 1953).) The suggestive term 'covering law model' is W. Dray's; cf. his *Laws and Explanation in History* (Oxford, 1957), ch. 1. Dray characterizes this type of explanation as 'subsuming what is to be explained under a general law' (p. 1), and then rightly urges, in the name of methodological realism, that 'the requirement of a *single* law be dropped' (p. 24, author's italics): it should be noted, however, that, like the schema (D) above, several earlier publications on the subject (among them the article mentioned at the beginning of this note) make explicit provision for the inclusion of more laws than one in the explanans.

an increase in breadth, or scope, because the new explanatory principles cover a broader range of phenomena; for example, Newton's principles govern free fall on the earth and on other celestial bodies, as well as the motions of planets, comets, and artificial satellites, the movements of pendulums, tidal changes, and various other phenomena. And the increase thus effected in the depth of our understanding is strikingly reflected in the fact that, in the light of more advanced explanatory principles, the original empirical laws are usually seen to hold only approximately, or within certain limits. For example, Newton's theory implies that the factor g in Galileo's law, $s = \frac{1}{2} gt^2$, is not strictly a constant for free fall near the surface of the earth; and that, since every planet undergoes gravitational attraction not only from the sun, but also from the other planets, the planetary orbits are not strictly ellipses, as stated in Kepler's laws.

One further point deserves brief mention here. An explanation of a particular event is often conceived as specifying its *cause*, or causes. Thus, the account outlined in our first illustration might be held to explain the growth and the recession of the soap bubbles by showing that the phenomenon was *caused* by a rise and a subsequent drop of the temperature of the air trapped in the tumblers. Clearly, however, these temperature changes provide the requisite explanation only in conjunction with certain other conditions, such as the presence of a soap film, practically constant pressure of the air surrounding the glasses, etc. Accordingly, in the context of explanation, a cause must be allowed to consist in a more or less complex set of particular circumstances; these might be described by a set of sentences: $C_1, C_2, \ldots, C_k$. And, as suggested by the principle 'Same cause, same effect', the assertion that those circumstances jointly caused a given event—described, let us say, by a sentence E—implies that whenever and wherever circumstances of the kind in question occur, an event of the kind to be explained comes about. Hence, the given causal explanation implicitly claims that there are general laws—such as $L_1, L_2, \ldots, L_r$, in schema (D)—by virtue of which the occurrence of the causal antecedents mentioned in $C_1, C_2, \ldots, C_k$ is a sufficient condition for the occurrence of the event to be explained. Thus, the relation between causal factors and effect is reflected in schema (D): causal explanation is deductive-nomological in character. (However, the customary formulations of causal and other explanations often do not explicitly specify all the relevant laws and particular facts: to this point, we will return later.)

The converse does not hold: there are deductive-nomological explanations which would not normally be counted as causal. For one thing, the subsumption of laws, such as Galileo's or Kepler's laws, under more comprehensive principles is clearly not causal in character: we speak of causes only in reference to *particular* facts or events, and not in reference to *universal facts* as expressed by general laws. But not even all deductive-nomological explanations of particular facts or events will qualify as causal; for in a causal explanation some of the explanatory circumstances will temporally precede the effect to be explained: and there are explanations of type (*D*) which lack this characteristic. For example, the pressure which a gas of specified mass possesses at a given time might be explained by reference to its temperature and its volume at the same time, in conjunction with the gas law which connects simultaneous values of the three parameters.[3]

In conclusion, let me stress once more the important role of laws in deductive-nomological explanation: the laws connect the explanandum event with the particular conditions cited in the explanans, and this is what confers upon the latter the status of explanatory (and, in some cases, causal) factors in regard to the phenomenon to be explained.

2.2. *Probabilistic explanation*

In deductive-nomological explanation as schematized in (*D*), the laws and theoretical principles involved are of *strictly universal form*: they assert that in *all* cases in which certain specified conditions are realized an occurrence of such and such a kind will result; the law that any metal, when heated under constant pressure, will increase in volume, is a typical example; Galileo's, Kepler's, Newton's, Boyle's, and Snell's laws, and many others, are of the same character.

Now let me turn next to a second basic type of scientific explanation. This kind of explanation, too, is nomological, i.e. it accounts for a given phenomenon by reference to general laws or theoretical principles; but some or all of these are of *probabilistic-statistical form*, i.e. they are, generally speaking, assertions to the effect that if certain specified conditions are realized, then an occurrence of such-and-such a kind will come about with such-and-such a statistical probability.

For example, the subsiding of a violent attack of hay fever in a given case might well be attributed to, and thus explained by reference to, the administration of 8 milligrams of chlor-trimeton. But if we wish to connect this antecedent event with the explanandum, and thus to establish

---

[3] The relevance of the covering-law model to causal explanation is examined more fully in sect. 4 of C. G. Hempel, 'Deductive-Nomological *vs.* Statistical Explanation'. In H. Feigl, *et al.* (eds.), *Minnesota Studies in the Philosophy of Science*, 3 (Minneapolis, 1962).

its explanatory significance for the latter, we cannot invoke a universal law to the effect that the administration of 8 milligrams of that antihistamine will invariably terminate a hay fever attack: this simply is not so. What can be asserted is only a generalization to the effect that administration of the drug will be followed by relief with high statistical probability, i.e., roughly speaking, with a high relative frequency in the long run. The resulting explanans will thus be of the following type:

John Doe had a hay fever attack and took 8 milligrams of chlor-trimeton.

The probability for subsidence of a hay fever attack upon administration of 8 milligrams of chlor-trimeton is high.

Clearly, this explanans does not deductively imply the explanandum, 'John Doe's hay fever attack subsided'; the truth of the explanans makes the truth of the explanandum not certain (as it does in a deductive-nomological explanation) but only more or less likely or, perhaps 'practically' certain.

Reduced to its simplest essentials, a probabilistic explanation thus takes the following form:

$$(P) \qquad \frac{\begin{array}{c} Fi \\ p(O,F) \text{ is very high} \end{array}}{Oi} \qquad \text{makes very likely}$$

The explanandum, expressed by the statement 'Oi', consists in the fact that in the particular instance under consideration, here called i (e.g. John Doe's allergic attack), an outcome of kind O (subsidence) occurred. This is explained by means of two explanans-statements. The first of these, 'Fi', corresponds to $C_1, C_2, \ldots, C_k$ in (D); it states that in case i, the factors F (which may be more or less complex) were realized. The second expresses a law of probabilistic form, to the effect that the statistical probability for outcome O to occur in cases where F is realized is very high (close to 1). The double line separating explanandum from explanans is to indicate that, in contrast to the case of deductive-nomological explanation, the explanans does not logically imply the explanandum, but only confers a high likelihood upon it. The concept of likelihood here referred to must be clearly distinguished from that of statistical probability, symbolized by 'p' in our schema. A statistical probability is, roughly speaking, the long-run relative frequency with which an occurrence of a given kind (say, F) is accompanied by an 'outcome' of a specified kind (say, O). Our likelihood, on the other hand, is a relation (capable of gradations) not between kinds of occurrences, but between statements. The likelihood referred to in (P) may be characterized as the strength of the inductive support, or the degree of rational credibility, which the explanans confers upon the explanandum; or, in Carnap's terminology, as the *logical*, or *inductive* (in contrast to statistical) *probability* which the explanandum possesses relative to the explanans.

Thus, probabilistic explanation, just like explanation in the manner of schema (*D*), is nomological in that it presupposes general laws; but because these laws are of statistical rather than of strictly universal form, the resulting explanatory arguments are inductive rather than deductive in character. An inductive argument of this kind *explains* a given phenomenon by showing that, in view of certain particular events and certain statistical laws, its occurrence was to be expected with high logical, or inductive, probability.

By reason of its inductive character, probabilistic explanation differs from its deductive-nomological counterpart in several other important respects; for example, its explanans may confer upon the explanandum a more or less high degree of inductive support; in this sense, probabilistic explanation admits of degrees, whereas deductive-nomological explanation appears as an either-or affair: a given set of universal laws and particular statements either does or does not imply a given explanandum statement. A fuller examination of these differences, however, would lead us far afield and is not required for the purposes of this paper.[4]

One final point: the distinction here suggested between deductive-nomological and probabilistic explanation might be questioned on the ground that, after all, the universal laws invoked in a deductive explanation can have been established only on the basis of a finite body of evidence, which surely affords no exhaustive verification, but only more or less strong probability for it; and that, therefore, all scientific laws have to be regarded as probabilistic. This argument, however, confounds a logical issue with an epistemological one: it fails to distinguish properly between the *claim* made by a given law-statement and the *degree* of *confirmation*, or *probability*, which it possesses on the available evidence. It is quite true that statements expressing laws of either kind can be only incompletely confirmed by any given finite set—however large—of data about particular facts; but law-statements of the two different types make claims of different kind, which are reflected in their logical forms: roughly, a universal law-statement of the simplest kind asserts that *all* elements of an indefinitely large reference class (e.g. copper objects) have a certain characteristic (e.g. that of being good conductors of electricity); while statistical law-statements assert that in

---

[4] The concept of probabilistic explanation, and some of the peculiar logical and methodological problems engendered by it, are examined in some detail in Part 2 of the essay cited in the previous note.

the long run, a specified proportion of the members of the reference class have some specified property. And our distinction of two types of law and, concomitantly, of two types of scientific explanation, is based on this difference in claim as reflected in the difference of form.

The great scientific importance of probabilistic explanation is eloquently attested to by the extensive and highly successful explanatory use that has been made of fundamental laws of statistical form in genetics, statistical mechanics, and quantum theory.

3. ELLIPTIC AND PARTIAL EXPLANATIONS: EXPLANATION SKETCHES

As I mentioned earlier, the conception of deductive-nomological explanation reflected in our schema (D) is often referred to as the covering law model, or the deductive model, of explanation: similarly, the conception underlying schema (P) might be called the probabilistic or the inductive-statistical, model of explanation. The term 'model' can serve as a useful reminder that the two types of explanation as characterized above constitute ideal types or theoretical idealizations and are not intended to reflect the manner in which working scientists actually formulate their explanatory accounts. Rather, they are meant to provide explications, or rational reconstructions, or theoretical models, of certain modes of scientific explanation.

In this respect our models might be compared to the concept of mathematical proof (within a given theory) as construed in meta-mathematics. This concept, too, may be regarded as a theoretical model: it is not intended to provide a descriptive account of how proofs are formulated in the writings of mathematicians: most of these actual formulations fall short of rigorous and, as it were, ideal, meta-mathematical standards. But the theoretical model has certain other functions: it exhibits the rationale of mathematical proofs by revealing the logical connections underlying the successive steps; it provides standards for a critical appraisal of any proposed proof constructed within the mathematical system to which the model refers; and it affords a basis for a precise and far-reaching theory of proof, provability, decidability, and related concepts. I think the two models of explanation can fulfil the same functions, if only on a much more modest scale. For example, the arguments presented in constructing the models give an indication of the sense in which the models exhibit the rationale and the logical structure of the explanations they are intended to represent.

I now want to add a few words concerning the second of the functions just mentioned; but I will have to forgo a discussion of the third.

When a mathematician proves a theorem, he will often omit mention of certain propositions which he presupposes in his argument and which he is in fact entitled to presuppose because, for example, they follow readily from the postulates of his system or from previously established theorems or perhaps from the hypothesis of his theorem, if the latter is in hypothetical form; he then simply assumes that his readers or listeners will be able to supply the missing items if they so desire. If judged by ideal standards, the given formulation of the proof is elliptic or incomplete; but the departure from the ideal is harmless: the gaps can readily be filled in. Similarly, explanations put forward in everyday discourse and also in scientific contexts are often *elliptically formulated*. When we explain, for example, that a lump of butter melted because it was put into a hot frying pan, or that a small rainbow appeared in the spray of the lawn sprinkler because the sunlight was reflected and refracted by the water droplets, we may be said to offer elliptic formulations of deductive-nomological explanations; an account of this kind omits mention of certain laws or particular facts which it tacitly takes for granted, and whose explicit citation would yield a complete deductive-nomological argument.

In addition to elliptic formulation, there is another, quite important, respect in which many explanatory arguments deviate from the theoretical model. It often happens that the statement actually included in the explanans, together with those which may reasonably be assumed to have been taken for granted in the context at hand, explain the given explanandum only *partially*, in a sense which I will try to indicate by an example. In his *Psychopathology of Everyday Life*, Freud offers the following explanation of a slip of the pen that occurred to him: 'On a sheet of paper containing principally short daily notes of business interest, I found, to my surprise, the incorrect date, "Thursday, October 20th", bracketed under the correct date of the month of September. It was not difficult to explain this anticipation as the expression of a wish. A few days before I had returned fresh from my vacation and felt ready for any amount of professional work, but as yet there were few patients. On my arrival I had found a letter from a patient announcing her arrival on the 20th of October. As I wrote the same date in September I may certainly have thought "X ought to be here already; what a pity about that whole month!", and with this thought I pushed the current date a month ahead.'[5]

5 S. Freud, *Psychopathology of Everyday Life*, trans. A. A. Brill (New York, 1951), 64.

Clearly, the formulation of the intended explanation is *at least incomplete* in the sense considered a moment ago. In particular, it fails to mention any laws or theoretical principles in virtue of which the subconscious wish, and the other antecedent circumstances referred to, could be held to explain Freud's slip of the pen. However, the general theoretical considerations Freud presents here and elsewhere in his writings suggests strongly that his explanatory account relies on a hypothesis to the effect that when a person has a strong, though perhaps unconscious, desire, then if he commits a slip of pen, tongue, memory, or the like, the slip will take a form in which it expresses, and perhaps symbolically fulfils, the given desire.

Even this rather vague hypothesis is probably more definite than what Freud would have been willing to assert. But for the sake of the argument let us accept it and include it in the explanans, together with the particular statements that Freud did have the subconscious wish he mentions, and that he was going to commit a slip of the pen. Even then, the resulting explanans permits us to deduce only that the slip made by Freud would, *in some way or other*, express and perhaps symbolically fulfil Freud's subconscious wish. But clearly, such expression and fulfilment might have been achieved by many other kinds of slip of the pen than the one actually committed.

In other words, the explanans does not imply, and thus fully explain, that the particular slip, say *s*, which Freud committed on this occasion, would fall within the narrow class, say *W*, of acts which consist in writing the words 'Thursday, October 20th'; rather, the explanans implies only that *s* would fall into a wider class, say *F*, which includes *W* as a proper subclass, and which consists of all acts which would express and symbolically fulfil Freud's subconscious wish in *some way or other*.

The argument under consideration might be called a *partial explanation*: it provides complete, or conclusive, grounds for expecting *s* to be a member of *F*, and since *W* is a subclass of *F*, it thus shows that the explanandum, i.e. *s* falling within *W*, accords with, or bears out, what is to be expected in consideration of the explanans. By contrast, a deductive-nomological explanation of the form (*D*) might then be called *complete* since the explanans here does imply the explanandum.

Clearly, the question whether a given explanatory argument is complete or partial can be significantly raised only if the explanandum sentence is fully specified; only then can we ask whether the explanandum does or does not follow from the explanans. Completeness of explanation, in this sense, is relative to our explanandum sentence. Now, it might seem much more important and interesting to consider instead the notion of a complete explanation of some *concrete event*, such as the destruction of Pompeii, or the death of Adolf Hitler, or the launching of the first artifical satellite: we might want to regard a particular event as completely explained only if an explanatory account of deductive or of inductive form had been provided for all of its aspects. This notion, however, is self-defeating; for any particular event may be regarded as having infinitely many different aspects or characteristics, which cannot all be accounted for by a finite set, however large, of explanatory statements.

In some cases, what is intended as an explanatory account will depart even further from the standards reflected in the model schemata (D) and (P) above. An explanatory account, for example, which is not explicit and specific enough to be reasonably qualified as an elliptically formulated explanation or as a partial one, can often be viewed as an *explanation sketch*: it may suggest, perhaps quite vividly and persuasively, the general outlines of what, it is hoped, can eventually be supplemented so as to yield a more closely reasoned argument based on explanatory hypotheses which are indicated more fully, and which more readily permit of critical appraisal by reference to empirical evidence.

The decision whether a proposed explanatory account is to be qualified as an elliptically formulated deductive or probabilistic explanation, as a partial explanation, as an explanation sketch, or perhaps as none of these is a matter of judicious interpretation; it calls for an appraisal of the intent of the given argument and of the background assumptions that may be assumed to have been tacitly taken for granted, or at least to be available, in the given context. Unequivocal decision rules cannot be set down for this purpose any more than for determining whether a given informally stated inference which is not deductively valid by reasonably strict standards is to count nevertheless as valid but enthymematically formulated, or as fallacious, or as an instance of sound inductive reasoning, or perhaps, for lack of clarity, as none of these.

4. NOMOLOGICAL EXPLANATION IN HISTORY

So far, we have examined nomological explanation, both deductive and inductive, as found in the natural sciences; and we have considered certain characteristic ways in which actual explanatory accounts often depart from the ideal standards of our two basic models. Now it is time to ask what light the preceding inquiries can shed on the explanatory procedures used in historical research.

In examining this question, we will consider a number of specific explanatory arguments offered by a variety of writers. It should be understood from the beginning that we are here concerned, not to appraise the factual adequacy of these explanations, but only to attempt an explication of the claims they make and of the assumptions they presuppose.

Let us note first, then, that some historical explanations are surely nomological in character: they aim to show that the explanandum phenomenon resulted from certain antecedent, and perhaps, concomitant, conditions; and in arguing these, they rely more or less explicitly on relevant generalizations. These may concern, for example, psychological or sociological tendencies and may best be conceived as broadly probabilistic in character. This point is illustrated by the following argument, which might be called an attempt to explain Parkinson's Law by subsumption under broader psychological principles:

As the activities of the government are enlarged, more people develop a vested interest in the continuation and expansion of governmental functions. People who have jobs do not like to lose them; those who are habituated to certain skills do not welcome change; those who have become accustomed to the exercise of a certain kind of power do not like to relinquish their control—if anything, they want to develop greater power and correspondingly greater prestige. . . . Thus, government offices and bureaus, once created, in turn institute drives, not only to fortify themselves against assault, but to enlarge the scope of their operations.[6]

The psychological generalizations here explicitly adduced will reasonably have to be understood as expressing, not strict uniformities, but strong *tendencies*, which might be formulated by means of rough probability statements; so that the explanation here suggested is probabilistic in character.

As a rule, however, the generalizations underlying a proposed historical explanation are largely left unspecified; and most concrete explanatory accounts have to be qualified as partial explanations or as explanation sketches. Consider, for example, F. J. Turner's essay 'The Significance of the Frontier in American History',[7] which amplifies and defends the view that

Up to our own day American history has been in a large degree the history of the colonization of the Great West. The existence of an area of free land, its continuous recession, and the advance of American settlement westward explain American development. . . . The peculiarity of American institutions is the fact that they have been compelled to adapt themselves . . . to the changes involved in crossing

---

[6] D. W. McConnell *et al.*, *Economic Behavior* (New York, 1939), 894–5.

[7] First published in 1893, and reprinted in several publications, among them: Everett E. Edwards (ed.), *The Early Writings of Frederick Jackson Turner* (Madison, Wis., 1938). Subsequent page-references are to this book.

a continent, in winning a wilderness, and in developing at each area of this progress, out of the primitive economic and political conditions of the frontier, the complexity of city life.[8]

One of the phenomena Turner considers in developing his thesis is the rapid westward advance of what he calls the Indian trader's frontier. 'Why was it', Turner asks, 'that the Indian trader passed so rapidly across the continent?'; and he answers,

The explanation of the rapidity of this advance is bound up with the effects of the trader on the Indian. The trading post left the unarmed tribes at the mercy of those that had purchased firearms—a truth which the Iroquois Indians wrote in blood, and so the remote and unvisited tribes gave eager welcome to the trader. . . . This accounts for the trader's power and the rapidity of his advance.[9]

There is no explicit mention here of any laws, but it is clear that this sketch of an explanation presupposes, first of all, various particular facts, such as that the remote and unvisited tribes had heard of the efficacy and availability of firearms, and that there were no culture patterns or institutions precluding their use by those tribes; but in addition, the account clearly rests also on certain assumptions as to how human beings will tend to behave in situations presenting the kinds of danger and of opportunity that Turner refers to.

Similar comments apply to Turner's account of the westward advance of what he calls the farmer's frontier:

Omitting those of the pioneer farmers who move from the love of adventure, the advance of the more steady farmer is easy to understand. Obviously the immigrant was attracted by the cheap lands of the frontier, and even the native farmer felt their influence strongly. Year by year the farmers who lived on soil, whose returns were diminished by unrotated crops, were offered the virgin soil of the frontier at nominal prices. Their growing families demanded more lands, and these were dear. The competition of the unexhausted, cheap, and easily tilled prairie lands compelled the farmer either to go West . . . or to adopt intensive culture.[10]

This passage is clearly intended to do more than describe a sequence of particular events: it is meant to afford an understanding of the farmers' westward advance by pointing to their interests and needs and by calling attention to the facts and the opportunities facing them. Again, this explanation takes it for granted that under such conditions normal human beings will tend to seize new opportunities in the manner in which the pioneer farmers did.

Examining the various consequences of this moving-frontier history, Turner states that 'the most important effect of the frontier has been in

[8] Ibid. 185–6.   [9] Ibid. 200–1.   [10] Ibid. 210.

the promotion of democracy here and in Europe',[11] and he begins his elaboration of this theme with the remark that 'the frontier is productive of individualism. . . . The tendency is anti-social. It produces antipathy to control, and particularly to any direct control':[12] and this is, of course, a sociological generalization in a nutshell.

Similarly, any explanation that accounts for a historical phenomenon by reference to economic factors or by means of general principles of social or cultural change are nomological in import, even if not in explicit formulation.

But if this be granted there still remains another question, to which we must now turn, namely, whether, in addition to explanations of a broadly nomological character, the historian also employs certain other distinctly historical ways of explaining and understanding whose import cannot be adequately characterized by means of our two models. The question has often been answered in the affirmative, and several kinds of historical explanation have been adduced in support of this affirmation. I will now consider what seem to me two especially interesting candidates for the role of specifically historical explanation; namely first, genetic explanation, and secondly, explanation of an action in terms of its underlying rationale.

5. GENETIC EXPLANATION IN HISTORY

In order to make the occurrence of a historical phenomenon intelligible, a historian will frequently offer a 'genetic explanation' aimed at exhibiting the principal stages in a sequence of events which led up to the given phenomenon.

Consider, for example, the practice of selling indulgences as it existed in Luther's time. H. Boehmer, in his work, *Luther and the Reformation*, points out that until about the end of the nineteenth century, 'the indulgence was in fact still a great unknown quantity, at sight of which the scholar would ask himself with a sigh: "Where did it come from?" '[13] An answer was provided by Adolf Gottlob,[14] who tackled the problem by asking himself what led the Popes and Bishops to offer indulgences. As a result, 'origin and development of the unknown quantity appeared clearly in the light, and doubts as to its original meaning came to an end.

---

[11] Ibid. 219.   [12] Ibid. 220.

[13] H. Boehmer, *Luther and the Reformation*, trans. E. S. G. Potter (London, 1930), 91.

[14] Gottlob's study, *Kreuzablass und Almosenablass*, was published in 1906; cf. the references to the work of Gottlob and other investigators in E. G. Schwiebert, *Luther and his Times* (St Louis, 1950), notes to ch. 10.

It revealed itself as a true descendant of the time of the great struggle between Christianity and Islam, and at the same time a highly characteristic product of Germanic Christianity.'[15]

In brief outline,[16] the origins of the indulgence appear to go back to the ninth century, when the popes were strongly concerned with the fight against Islam. The Mohammedan fighter was assured by the teachings of his religion that if he were to be killed in battle his soul would immediately go to heaven; but the defender of the Christian faith had to fear that he might still be lost if he had not done the regular penance for his sins. To allay these doubts, John VII, in 877, promised absolution for their sins to crusaders who should be killed in battle. 'Once the crusade was so highly thought of, it was an easy transition to regard participation in a crusade as equivalent to the performance of atonement . . . and to promise remission of these penances in return for expeditions against the Church's enemies.'[17] Thus, there was introduced the indulgence of the Cross, which granted complete remission of the penitential punishment to all those who participated in a religious war. 'If it is remembered what inconveniences, what ecclesiastical and civil disadvantages the ecclesiastical penances entailed, it is easy to understand that the penitents flocked to obtain this indulgence.'[18] A further strong incentive came from the belief that whoever obtained an indulgence secured liberation not only from the ecclesiastical penances, but also from the corresponding suffering in purgatory after death. The benefits of these indulgences were next extended to those who, being physically unfit to participate in a religious war, contributed the funds required to send a soldier on a crusade: in 1199, Pope Innocent III recognized the payment of money as adequate qualification for the benefits of a crusading indulgence.

When the crusades were on the decline, new ways were explored of raising funds through indulgences. Thus, there was instituted a 'jubilee indulgence', to be celebrated every hundred years, for the benefit of pilgrims coming to Rome on that occasion. The first of these indulgences, in 1300, brought in huge sums of money; and the time interval between successive jubilee indulgences was therefore reduced to 50, 33, and even 25 years. And from 1393 on the jubilee indulgence was made available, not only in Rome, for the benefit of pilgrims, but everywhere in Europe, through special agents who were empowered to absolve the penitent of their sins upon payment of an appropriate amount. The development went

[15] Boehmer, *Luther and the Reformation*, 10.
[16] This outline follows the accounts given by Boehmer, ibid., ch. 3, and by Schwiebert, *Luther and his Times*, ch. 10.
[17] Boehmer, *Luther and the Reformation*, 92.
[18] Ibid. 93.

even further: in 1477, a dogmatic declaration by Sixtus IV attributed to the indulgence the power of delivering even the dead from purgatory.

Undeniably, a genetic account of this kind can enhance our understanding of a historical phenomenon. But its explanatory role, far from being *sui generis*, seems to me basically nomological in character. For the successive stages singled out for consideration surely must be qualified for their function by more than the fact that they form a temporal sequence and that they all precede the final stage, which is to be explained: the mere enumeration in a yearbook of 'the year's important events' in the order of their occurrence clearly is not a genetic explanation of the final event or of anything else. In a genetic explanation each stage must be shown to 'lead to' the next, and thus to be linked to its successor by virtue of some general principle which makes the occurrence of the latter at least reasonably probable, given the former. But in this sense, even successive stages in a physical phenomenon such as the free fall of a stone may be regarded as forming a genetic sequence whose different stages—characterized, let us say, by the position and the velocity of the stone at different times—are interconnected by strictly universal laws; and the successive stages in the movement of a steel ball bouncing its zigzaggy way down a Galton pegboard may be regarded as forming a genetic sequence with probabilistic connections.

The genetic accounts given by historians are not, of course, of the purely nomological kind suggested by these examples from physics. Rather, they combine a certain measure of nomological interconnecting with more or less large amounts of straight description. For consider an intermediate stage mentioned in a genetic account: some aspects of it will be presented as having evolved from the preceding stages (in virtue of connecting laws, which often will be no more than hinted at); while other aspects, which are not accounted for by information about the preceding development, will be descriptively added because they are relevant to an understanding of subsequent stages in the genetic sequence. Thus, schematically speaking, a genetic explanation will begin with a pure description of an initial stage; thence, it will proceed to an account of a second stage, part of which is nomologically linked to, and explained by, the characteristic features of the initial stage; while the balance is simply described as relevant for a nomological account of some aspects of the third stage; and so forth.[19]

---

[19] The logic of genetic explanations in history is examined in some detail in E. Nagel's recent book, *The Structure of Science* (New York, 1961), 564–8. The conception outlined in the present paper, though arrived at without any knowledge of Nagel's work on this subject, accords well with the latter's results.

In our illustration the connecting laws are hinted at in the mention made of motivating factors: the explanatory claims made for the interest of the popes in securing a fighting force and in amassing ever larger funds clearly presuppose suitable psychological generalizations as to the manner in which an intelligent individual will act, in the light of his factual beliefs, when he seeks to attain a certain objective. Similarly, general assumptions underlie the reference to the fear of purgatory in explaining the eagerness with which indulgences were bought. And when, referring to the huge financial returns of the first jubilee indulgence, Schwiebert says 'This success only whetted the insatible appetite of the popes. The intervening period of time was variously reduced from 100 to 50, to 33, to 25 years . . . ',[20] the explanatory force here implied might be said to rest on some principle of reinforcement by rewards. As need hardly be added, even if such a principle were explicitly introduced, the resulting account would provide at most a partial explanation; it could not be expected to show, for example, why the intervening intervals should have the particular lengths here mentioned.

In the genetic account of the indulgences, those factors which are simply described (or tacitly presupposed) rather than explained include, for example, the doctrines, the organization, and the power of the Church; the occurrence of the crusades and their eventual decline; and innumerable other factors which are not even explicitly mentioned, but which have to be understood as background conditions if the genetic survey is to serve its explanatory purpose.

The general conception here outlined of the logic of genetic explanation could also be illustrated by reference to Turner's studies of the American frontier; this will be clear even from the brief remarks made earlier on Turner's ideas.

Some analysts of historical development put special emphasis on the importance of the laws underlying a historical explanation; thus, e.g., A. Gerschenkron maintains, 'Historical research consists essentially in application to empirical material of various sets of empirically derived hypothetical generalizations and in testing the closeness of the resulting fit, in the hope that in this way certain uniformities, certain typical situations, and certain typical relationships among individual factors in these situations can be ascertained',[21] and his subsequent substantive observations include a brief genetic survey of patterns of industrial

[20] Schwiebert, *Luther and his Times*, 304.

[21] A. Gerschenkron, 'Economic Backwardness in Historical Perspective', in B. F. Hoselitz (ed.), *The Progress of Underdeveloped Areas* (Chicago, 1952), 3–29.

development in nineteenth-century Europe, in which some of the presumably relevant uniformities are made reasonably explicit.

6. EXPLANATION BY MOTIVATING REASONS

Let us now turn to another kind of historical explanation that is often considered as *sui generis*, namely, the explanation of an action in terms of the underlying rationale, which will include, in particular, the ends the agent sought to attain, and the alternative courses of action he believed to be open to him. The following passage explaining the transition from the indulgence of the Cross to the institution of the jubilee indulgence illustrates this procedure:

in the course of the thirteenth century the idea of a crusade more and more lost its power over men's spirits. If the Popes would keep open the important source of income which the indulgence represented, they must invent new motives to attract people to the purchase of indulgences. It is the merit of Boniface VIII to have recognized this clearly. By creating the jubilee indulgence in 1300 he assured the species a further long development most welcome to the Papal finances.[22]

This passage clearly seeks to explain the establishment of the first jubilee indulgence by suggesting the reasons for which Boniface VIII took this step. If properly spelled out, these reasons would include not only Boniface's objective of ensuring a continuation of the income so far derived from the indulgence of the Cross, but also his estimate of the relevant empirical circumstances, including the different courses of action open to him, and their probable efficacy as well as potential difficulties in pursuing them and adverse consequences to which they might lead.

The kind of explanation achieved by specifying the rationale underlying a given action is widely held to be fundamentally different from nomological explanation as found in the natural sciences. Various reasons have been adduced in support of this view; but I will limit my discussion largely to the stimulating ideas on the subject that have been set forth by Dray.[23] According to Dray, there is an important type of historical explanation the features of which 'make the covering law model peculiarly inept'; he calls it 'rational explanation', i.e. 'explanation which displays the *rationale* of what was done', or, more fully, 'a reconstruction of the agent's *calculation* of means to be adopted toward

[22] Boehmer, *Luther and the Reformation*, 93–4.
[23] W. Dray, *Laws and Explanation in History* (Oxford, 1957), ch. 5. All quotations are from this chapter; italics in the quoted passages are Dray's.

his chosen end in the light of the circumstances in which he found himself'. The object of rational explanation is not to subsume the explanandum under general laws, but 'to show that what was done was the thing to have done for the reasons given, rather than merely the thing that is done on such occasions, perhaps in accordance with certain laws'. Hence, a rational explanation has 'an element of *appraisal*' in it: it 'must exhibit what was done as appropriate or justified'. Accordingly, Dray conceives a rational explanation as being based on a standard of appropriateness or of rationality of a special kind which he calls a '*principle of action*', i.e. 'a judgement of the form "When in a situation of type $C_1, C_2, \ldots, C_n$ the thing to do is X." '

Dray does not give a full account of the kind of 'situation' here referred to; but to do justice to his intentions, these situations must evidently be taken to include, at least, items of the following three types: (1) the end the agent was seeking to attain; (2) the empirical circumstances, as seen by the agent, in which he had to act; (3) the moral standards or principles of conduct to which the agent was committed. For while this brief list requires considerable further scrutiny and elaboration, it seems clear that only if at least these items are specified does it make sense to raise the question of the appropriateness of what the agent did in the given 'situation'.

It seems fair to say, then, that according to Dray's conception a rational explanation answers a question of the form 'Why did agent A do X?' by offering an explanans of the following type (our formulation replaces the notation $C_1, C_2, \ldots, C_n$ by the simpler 'C', without, of course, precluding that the kind of situation thus referred to may be extremely complex):

(R) A was in a situation of type C
In a situation of type C, the appropriate thing to do is X

But can an explanans of this type possibly serve to explain A's having in fact done X? It seems to me beyond dispute that in any adequate explanation of an empirical phenomenon the explanans must provide good grounds for believing or asserting that the explanandum phenomenon did in fact occur. Yet this requirement, which is necessary though not sufficient[24] for an adequate explanation, is not met by a rational explanation as conceived by Dray. For the two statements included in the contemplated explanans (R) provide good reasons for believing that the appropriate thing for A to do was X, but not for believing that A did in fact do X. Thus, a rational explanation in the sense in which Dray appears to understand it does not explain what it is meant to explain.

[24] Empirical evidence supporting a given hypothesis may afford strong grounds for believing the latter without providing an explanation for it.

Indeed, the expression 'the thing to do' in the standard formulation of a principle of action, 'functions as a value term', as Dray himself points out: but then, it is unclear, on purely logical grounds, how the valuational principle expressed by the second sentence in (R), in conjunctiton with the plainly empirical, non-valuational first sentence, should permit any inferences concerning empirical matters such as A's action, which could not be drawn from the first sentence alone.

To explain, in the general vein here under discussion, why A did in fact do X, we have to refer to the underlying rationale not by means of a normative principle of action, but by descriptive statements to the effect that, at the time in question A was a rational agent, or had the disposition to act rationally; and that a rational agent, when in circumstances of kind C, will always (or: with high probability) do X. Thus construed, the explanans takes on the following form:

(R') 1. A was in a situation of type C
   2. A was disposed to act rationally
   3. Any person who is disposed to act rationally will, when in a situation of type C, invariably (with high probability) do X

But by this explanans, A's having done X is accounted for in the manner of a deductive or of a probabilistic nomological explanation. Thus, in so far as reference to the rationale of an agent does explain his action, the explanation conforms to one of our nomological models.

An analogous diagnosis applies, incidentally, also to explanations which attribute an agent's behaviour in a given situation not to rationality and more or less explicit deliberation on his part, but to other dispositional features, such as his character and emotional make-up. The following comment on Luther illustrates this point: 'Even stranger to him than the sense of anxiety was the allied sense of fear. In 1527 and 1535, when the plague broke out in Wittenberg, he was the only professor besides Bugenhagen who remained calmly at his post to comfort and care for the sick and dying. . . . He had, indeed, so little sense as to take victims of the plague into his house and touch them with his own hand. Death, martyrdom, dishonor, contempt . . . he feared just as little as infectious disease.'[25] It may well be said that these observations give more than a description: that they shed some explanatory light on the particular occurrences mentioned. But in so far as they explain, they do so by presenting Luther's actions as manifestations of certain personality traits, such as fearlessness; thus, the particular acts are again subsumed under generalizations as to how a fearless person is likely to behave under certain circumstances.

[25] Boehmer, *Luther and the Reformation*, 234.

It might seem that both in this case and in rational explanation as construed in (R'), the statements which we took to express general laws—namely, (3) in (R'), and the statement about the probable behaviour of a fearless person in our last illustration—do not have the character of empirical laws at all, but rather that of analytic statements which simply express part of what is *meant* by a rational agent, a fearless person, or the like. Thus, in contrast to nomological explanations, these accounts in terms of certain dispositional characteristics of the agent appear to presuppose no general laws at all. Now, the idea of analyticity gives rise to considerable philosophical difficulties; but let us disregard these here and take the division of statements into analytic and synthetic to be reasonably clear. Even then, the objection just outlined cannot be upheld. For dispositional concepts of the kind invoked in our explanations have to be regarded as governed by entire clusters of general statements—we might call them symptom statements—which connect the given disposition with various specific manifestations, or symptoms, of its presence (each symptom will be a particular mode of 'responding', or acting, under specified 'stimulus' conditions); and the whole cluster of these symptom statements for a given disposition will have implications which are plainly not analytic (in the intuitive sense here assumed). Under these circumstances it would be arbitrary to attribute to some of the symptom statements the analytic character of partial definitions.

The logic of this situation has a precise representation in Carnap's theory of reduction sentences.[26] Here, the connections between a given disposition and its various manifest symptoms are assumed to be expressed by a set of so-called reduction sentences (these are characterized by their logical form). Some of these state, in terms of manifest characteristics, sufficient conditions for the presence of the given disposition; others similarly state necessary conditions. The reduction sentences for a given dispositional concept cannot, as a rule, all be qualified as analytic; for jointly they imply certain non-analytic consequences which have the status of general laws connecting exclusively the manifest characteristics; the strongest of the laws so implied is the so-called representative sentence, which 'represents, so to speak, the factual content of the set' of all the reduction sentences for the given disposition concept. This representative sentence asserts, in effect, that whenever at least one of the sufficient conditions specified by the given reduction sentences is

[26] See especially Carnap's classical essay, 'Testability and Meaning', *Philosophy of Science*, 3 (1936), 419–71, and 4 (1937), 1–40, reprinted, with some omissions, in Feigl and Brodbeck, *Readings in the Philosophy of Science*. On the point here under discussion, see sects. 9 and particularly 10 of the original essay or sect. 7 of the reprinted version.

satisfied, then so are all the necessary conditions laid down by the reduction sentences. And when *A* is one of the manifest criteria sufficient for the presence of a given disposition, and *B* is a necessary one, then the statement that whenever *A* is present so is *B* will normally turn out to be synthetic.

So far then, I have argued that Dray's construal of explanation by motivating reasons is untenable; that the normative principles of action envisaged by him have to be replaced by statements of a dispositional kind; and that, when this is done, explanations in terms of a motivating rationale, as well as those referring to other psychological factors, are seen to be basically nomological.

Let me add a few further remarks on the idea of rational explanation. First: in many cases of so-called purposive action, there is no conscious deliberation, no rational calculation that leads the agent to his decision. Dray is quite aware of this; but he holds that a rational explanation in his sense is still possible; for 'in so far as we say an action is purposive at all, no matter at what level of conscious deliberation, there is a calculation which could be constructed for it: the one the agent would have gone through if he had had time, if he had not seen what to do in a flash, if he had been called upon to account for what he did after the event, etc. And it is by eliciting some such calculation that we explain the action.[27] But the explanatory significance of reasons or 'calculations' which are 'reconstructed' in this manner is certainly puzzling. If, to take Dray's example, an agent arrives at his decision 'in a flash' rather than by deliberation, then it would seem to be simply false to say that the decision can be accounted for by some argument which the agent might have gone through under more propitious circumstances, or which he might produce later if called upon to account for his action; for, by hypothesis, no such argument was in fact gone through by the agent at the crucial time; considerations of appropriateness or rationality played no part in shaping his decision; the rationale that Dray assumes to be adduced and appraised in the corresponding rational explanation is simply fictitious.

But, in fairness to Dray, these remarks call for a qualifying observation: in at least some of the cases Dray has in mind it might not be fictitious to ascribe the action under study to a disposition which the agent acquired through a learning process whose initial stages did involve conscious ratiocination. Consider, for example, the various complex manoeuvres of accelerating, braking, signalling, dodging jaywalkers and animals, swerving into and out of traffic lanes, estimating the

---

[27] Dray, *Laws . . . History*, 123.

changes of traffic lights, etc., which are involved in driving a car through city traffic. A beginning driver will often perform these only upon some sort of conscious deliberation or even calculation; but gradually, he learns to do the appropriate thing automatically, 'in a flash', without giving them any conscious thought. The habit pattern he has thus acquired may be viewed as consisting in a set of dispositions to react in certain appropriate ways in various situations; and a particular performance of such an appropriate action would then be explained, not by a 'constructed' calculation which actually the agent did not perform but by reference to the disposition just mentioned and thus, again, in a nomological fashion.

The method of explaining a given action by 'constructing', in Dray's sense, the agent's calculation of means faces yet another though less fundamental, difficulty: it will frequently yield a rationalization rather than an explanation, especially when the reconstruction relies on the reasons the agent might produce when called upon to account for his action. As G. Watson remarks, 'Motivation, as presented in the perspective of history, is often too simple and straightforward, reflecting the psychology of the Age of Reason. . . . Psychology has come . . . to recognize the enormous weight of irrational and intimately personal impulses in conduct. In history, biography, and in autobiography, especially of public characters, the tendency is strong to present "good" reasons instead of "real" reasons.'[28] Accordingly, as Watson goes on to point out, it is important, in examining the motivation of historical figures, to take into account the significance of such psychological mechanisms as reaction formation, 'the dialectic dynamic by which stinginess cloaks itself in generosity, or rabid pacifism arises from the attempt to repress strong aggressive impulses'.[29]

These remarks have a bearing also on an idea set forth by P. Gardiner in his illuminating book on historical explanation.[30] Commenting on the notion of the 'real reason' for a man's action, Gardiner says: 'In general, it appears safe to say that by a man's "real reasons" we mean those reasons he would be prepared to give under circumstances where his confession would not entail adverse consequences to himself.' And he adds 'An exception to this is the psychoanalyst's usage of the expression where different criteria are adopted.'[31] This observation might be taken to imply that the explanation of human actions in terms of underlying

---

[28] G. Watson, 'Clio and Psyche: Some Interrelations of Psychology and History', in C. F. Ware (ed.), *The Cultural Approach to History* (New York, 1940), 34–47, 36.
[29] Ibid.   [30] P. Gardiner, *The Nature of Historical Explanation* (Oxford, 1952).
[31] Ibid. 136.

motives is properly aimed at exhibiting the agent's 'real reasons' in the ordinary sense of the phrase, as just described; and that, by implication, reasons in the psychoanalyst's sense require less or no consideration. But such a construal of explanation would give undue importance to considerations of ordinary language. Gardiner is entirely right when he reminds us that the 'language in which history is written is for the most part the language of ordinary speech';[32] but the historian in search of reasons that will correctly explain human actions will obviously have to give up his reliance on the everyday conception of 'real reasons' if psychological or other investigations show that real reasons, thus understood, do not yield as adequate an account of human actions as an analysis in terms of less familiar conceptions such as, perhaps, the idea of motivating factors which are kept out of the agent's normal awareness by processes of repression and reaction formation.

I would say, then, first of all, that historical explanation cannot be bound by conceptions that might be implicit in the way in which ordinary language deals with motivating reasons. But secondly, I would doubt that Gardiner's expressly tentative characterization does justice even to what we ordinarily mean when we speak of a man's 'real reasons'. For considerations of the kind that support the idea of subconscious motives are quite familiar in our time, and we are therefore prepared to say in ordinary, non-technical discourse that the reasons given by an agent may not be the 'real reasons' behind his action, even if his statement was subjectively honest, and he had no grounds to expect that it would lead to any adverse consequences for him. For no matter whether an explanation of human actions is attempted in the language of ordinary speech or in the technical terms of some theory, the overriding criterion for what-if-anything should count as a 'real', and thus explanatory, reason for a given action is surely not to be found by examining the way in which the term 'real reason' has thus far been used, but by investigating what conception of real reason would yield the most satisfactory explanation of human conduct; and ordinary usage gradually changes accordingly.

7. CONCLUDING REMARKS

We have surveyed some of the most prominent candidates for the role of characteristically historical mode of explanation; and we have found that they conform essentially to one or the other of our two basic types of scientific explanation.

---

[32] Ibid. 63.

This result and the arguments that led to it do not in any way imply a mechanistic view of man, of society, and of historical processes; nor, of course, do they deny the importance of ideas and ideals for human decision and action. What the preceding considerations do suggest is, rather, that the nature of understanding, in the sense in which explanation is meant to give us an understanding of empirical phenomena, is basically the same in all areas of scientific inquiry; and that the deductive and the probabilistic model of nomological explanation accommodate vastly more than just the explanatory arguments of, say, classical mechanics: in particular, they accord well also with the character of explanations that deal with the influence of rational deliberation, of conscious and subconscious motives, and of ideas and ideals on the shaping of historical events. In so doing, our schemata exhibit, I think, one important aspect of the methodological unity of all empirical science.

ASPECTS OF SCIENTIFIC EXPLANATION

CARL G. HEMPEL

3.3. *Inductive-Statistical Explanation*

As an explanation of why patient John Jones recovered from a streptococcus infection, we might be told that Jones had been given penicillin.
But if we try to amplify this explanatory claim by indicating a general
connection between penicillin treatment and the subsiding of a streptococcus infection we cannot justifiably invoke a general law to the effect
that in all cases of such infection, administration of penicillin will lead
to recovery. What can be asserted, and what surely is taken for granted
here, is only that penicillin will effect a cure in a high percentage of
cases, or with a high statistical probability. This statement has the general
character of a law of statistical form, and while the probability value is
not specified, the statement indicates that it is high. But in contrast to
the cases of deductive-nomological and deductive-statistical explanation,
the explanans consisting of this statistical law together with the statement
that the patient did receive penicillin obviously does not imply the
explanandum statement, 'the patient recovered', with deductive certainty, but only, as we might say, with high likelihood, or near certainty.
Briefly, then, the explanation amounts to this argument:

(3a) The particular case of illness of John Jones—let us call it j—was
an instance of severe streptococcal infection (Sj) which was treated with
large doses of penicillin (Pj); and the statistical probability $p(R, S \cdot P)$
of recovery in cases where S and P are present is close to 1; hence, the
case was practically certain to end in recovery (Rj).

This argument might invite the following schematization:

$$(3b) \qquad \qquad p(R, S \cdot P) \text{ is close to } 1$$
$$\underline{\qquad \qquad Sj \cdot Pj \qquad \qquad \qquad \qquad}$$

(Therefore:) It is practically certain (very likely) that Rj

Extract from *Aspects of Scientific Explanation and Other Essays in the Philosophy of Science* (New
York and London: Free Press and Collier Macmillan, 1965), 381–3, 394–403. Copyright © by The
Free Press, New York. Reprinted by permission of The Free Press, a division of Macmillan, Inc.

In the literature on inductive inference, arguments thus based on statistical hypotheses have often been construed as having this form or a similar one. On this construal, the conclusion characteristically contains a modal qualifier such as 'almost certainly', 'with high probability', 'very likely', etc. But the conception of arguments having this character is untenable. For phrases of the form 'it is practically certain that p' or 'It is very likely that p', where the place of 'p' is taken by some statement, are not complete self-contained sentences that can be qualified as either true or false. The statement that takes the place of 'p'—for example, 'Rj'—is either true or false, quite independently of whatever relevant evidence may be available, but it can be qualified as more or less likely, probable, certain, or the like only *relative to some body of evidence*. One and the same statement, such as 'Rj', will be certain, very likely, not very likely, highly unlikely, and so forth, depending upon what evidence is considered. The phrase 'it is almost certain that Rj' taken by itself is therefore neither true nor false; and it cannot be inferred from the premises specified in (3b) nor from any other statements.

The confusion underlying the schematization (3b) might be further illuminated by considering its analogue for the case of deductive arguments. The force of a deductive inference, such as that from 'all F are G' and 'a is F' to 'a is G', is sometimes indicated by saying that if the premises are true, then the conclusion is necessarily true or is certain to be true—a phrasing that might suggest the schematization

$$\text{All } F \text{ are } G$$
$$\underline{a \text{ is } F}$$
$$(\text{Therefore:}) \text{ It is necessary (certain) that } a \text{ is } G$$

But clearly the given premises—which might be, for example, 'all men are mortal' and 'Socrates is a man'—do not establish the sentence 'a is G' ('Socrates is mortal') as a necessary or certain truth. The certainty referred to in the informal paraphrase of the argument is relational: the statement 'a is G' is certain, or necessary, *relative to the specified premises*; i.e., their truth will guarantee its truth—which means nothing more than that 'a is G' is a logical consequence of those premises.

Analogously, to present our statistical explanation in the manner of schema (3b) is to misconstrue the function of the words 'almost certain' or 'very likely' as they occur in the formal wording of the explanation. Those words clearly must be taken to indicate that on the evidence provided by the explanans, or relative to that evidence, the explanandum is practically certain or very likely, i.e., that

(3c) 'Rj' is practically certain (very likely) relative to the explanans containing the sentences '$p (R, S \cdot P)$ is close to 1' and '$Sj \cdot Pj$'[33]

The explanatory argument misrepresented by (3b) might therefore suitably be schematized as follows:

(3d) $p (R, S \cdot P)$ is close to 1

$$\frac{Sj \cdot Pj}{Rj} \quad \text{[makes practically certain (very likely)]}$$

In this schema, the double line separating the 'premises' from the 'conclusion' is to signify that the relation of the former to the latter is not that of deductive implication but that of inductive support, the strength of which is indicated in square brackets.[34]

---

[33] Phrases such as 'It is almost certain (very likely) that j recovers', even when given the relational construal here suggested, are ostensibly concerned with relations between propositions, such as those expressed by the sentences forming the conclusion and the premises of an argument. For the purpose of the present discussion, however, involvement with propositions can be avoided by construing the phrases in question as expressing logical relations between corresponding *sentences*, e.g. the conclusion-sentence and the premise-sentence of an argument. This construal, which underlies the formulation of (3c), will be adopted in this essay, though for the sake of convenience we may occasionally use a paraphrase.

[34] In the familiar schematization of deductive arguments, with a single line separating the premises from the conclusion, no explicit distinction is made between a weaker and a stronger claim, either of which might be intended; namely (1) that the premises logically imply the conclusion and (2) that, in addition, the premises are true. In the case of our probabilistic argument, (3c) expresses a weaker claim, analogous to (1), whereas (3d) may be taken to express a 'proffered explanation' (the term is borrowed from I. Scheffler, 'Explanation, Prediction, and Abstraction', *British Journal for the Philosophy of Science* 7 (1957), sect. 1) in which, in addition, the explanatory premises are—however tentatively—asserted as true.

The considerations here outlined concerning the use of terms like 'probably' and 'certainly' as modal qualifiers of individual statements seem to me to militate also against the notion of categorical probability statement that C. I. Lewis sets forth in the following passage (italics the author's):

> Just as 'If D then (certainly) P, and D is the fact', leads to the categorical consequence, 'Therefore (certainly) P'; so too, 'If D then probably P, and D is the fact', leads to a categorical consequence expressed by 'It is probable that P'. And this conclusion is not merely the statement over again of the probability relation between 'P' and 'D'; any more than 'Therefore (certainly) P' is the statement over again of 'If D then (certainly) P'. 'If the barometer is high, tomorrow will probably be fair; and the barometer *is* high', categorically assures something expressed by 'Tomorrow will probably be fair'. This probability is still relative to the grounds of judgment; but if these grounds are actual, and contain all the available evidence which is pertinent, then it is not only categorical but may fairly be called *the* probability of the event in question (1946: 319).

This position seems to me to be open to just those objections suggested in the main text. If 'P' is a statement, then the expressions 'certainly P' and 'probably P' as envisaged in the quoted passage are not statements. If we ask how one would go about trying to ascertain whether they were true, we realize that we are entirely at a loss unless and until a reference set of statements or assumptions has been specified relative to which P may then be found to be certain, or to be highly probable, or neither. The expressions in question, then, are essentially incomplete; they are elliptic formulations of relational statements; neither of them can be the conclusion of an inference. However plausible Lewis's suggestion may seem, there is no analogue in inductive logic to *modus*

3.4. The Ambiguity of Inductive-Statistical Explanation and the Requirement of Maximal Specificity

3.4.1. The Problem of Explanatory Ambiguity

Consider once more the explanation (3d) of recovery in the particular case j of John Jones's illness. The statistical law there invoked claims recovery in response to penicillin only for a high percentage of streptococcal infections, but not for all of them; and in fact, certain streptococcus strains are resistant to penicillin. Let us say that an occurrence, e.g. a particular case of illness, has the property S^* (or belongs to the class S^*) if it is an instance of infection with a penicillin-resistant streptococcus strain. Then the probability of recovery among randomly chosen instances of S^* which are treated with penicillin will be quite small, i.e., $p(\underline{R}, S^* \cdot P)$ will be close to 0 and the probability of non-recovery, $p(\overline{R}, S^* \cdot P)$ will be close to 1. But suppose now that Jones's illness is in fact a streptococcal infection of the penicillin-resistant variety, and consider the following argument:

$$(3k) \qquad p(\overline{R}, S^* \cdot P) \text{ is close to 1}$$

$$\frac{S^*j \cdot Pj}{\overline{R}j} \qquad \text{[makes practically certain]}$$

This 'rival' argument has the same form as (3d), and on our assumptions, its premises are true, just like those of (3d). Yet its conclusion is the contradictory of the conclusion of (3d).

Or suppose that Jones is an octogenarian with a weak heart, and that in this group, S^{**}, the probability of recovery from a streptococcus infection in response to penicillin treatment, $p(R, S^{**} \cdot P)$, is quite small. Then, there is the following rival argument to (3d), which presents Jones's non-recovery as practically certain in the light of premises which are true:

ponens, or the 'rule of detachment', of deductive logic, which, given the information that 'D' and also 'if D then P', are true statements, authorizes us to detach the consequent 'P' in the conditional premiss and to assert it as a self-contained statement which must then be true as well.

At the end of the quoted passage, Lewis suggests the important idea that 'probably P' might be taken to mean that the total relevant evidence available at the time confers high probability upon P. But even this statement is relational in that it tacitly refers to some unspecified time, and, besides, his general notion of a categorical probability statement as a conclusion of an argument is not made dependent on the assumption that the premises of the argument include all the relevant evidence available.

It must be stressed, however, that elsewhere in his discussion, Lewis emphasizes the relativity of (logical) probability, and, thus, the very characteristic that rules out the conception of categorical probability statements.

Similar objections apply, I think, to Toulmin's construal of probabilistic arguments; cf. Toulmin (1958) and the discussion in Hempel (1960), sects. 1–3.

(3*l*) $p(\overline{R}, S^{**} \cdot P)$ is close to 1

$$\frac{S^{**}j \cdot Pj}{\overline{R}j}$$ [makes practically certain]

The peculiar logical phenomenon here illustrated will be called the *ambiguity of inductive-statistical explanation* or, briefly, of *statistical explanation*. This ambiguity derives from the fact that a given individual event (e.g. Jones's illness) will often be obtainable by random selection from any one of several 'reference classes' (such as $S \cdot P$, $S^* \cdot P$, $S^{**} \cdot P$), with respect to which the kind of occurrence (e.g. R) instantiated by the given event has very different statistical probabilities. Hence, for a proposed probabilistic explanation with true explanans which confers near certainty upon a particular event, there will often exist a rival argument of the same probabilistic form and with equally true premises which confers near certainty upon the non-occurrence of the same event. And any statistical explanation for the occurrence of an event must seem suspect if there is the possibility of a logically and empirically equally sound probabilistic account for its non-occurrence. *This predicament has no analogue in the case of deductive explanation*; for if the premises of a proposed deductive explanation are true then so is its conclusion; and its contradictory, being false, cannot be a logical consequence of a rival set of premises that are equally true.

Here is another example of the ambiguity of I-S explanation: Upon expressing surprise at finding the weather in Stanford warm and sunny on a date as autumnal as 27 November, I might be told, by way of explanation, that this was rather to be expected because the probability of warm and sunny weather (W) on a November day in Stanford (N) is, say, .95. Schematically, this account would take the following form, where 'n' stands for '27 November':

(3*m*) $p(W, N) = .95$

$$\frac{Nn}{Wn}$$ [.95]

But suppose it happens to be the case that the day before, 26 November, was cold and rainy, and that the probability for the immediate successors (S) of cold and rainy days in Stanford to be warm and sunny is .2; then the account (3*m*) has a rival in the following argument which, by reference to equally true premises, presents it as fairly certain that 27 November is not warm and sunny:

(3n) $$p(\overline{W}, S) = .8$$
$$\frac{Sn}{\overline{W}n} \quad [.8]$$

In this form, the problem of ambiguity concerns I-S arguments whose premises are in fact true, no matter whether we are aware of this or not. But, as will now be shown, the problem has a variant that concerns explanations whose explanans statements, no matter whether in fact true or not, are *asserted or accepted* by empirical science at the time when the explanation is proffered or contemplated. This variant will be called *the problem of the epistemic ambiguity of statistical explanation*, since it refers to what is presumed to be known in science rather than to what, perhaps unknown to anyone, is in fact the case.

Let K_t be the class of all statements asserted or accepted by empirical science at time t. This class then represents the total scientific information, or 'scientific knowledge' at time t. The word 'knowledge' is here used in the sense in which we commonly speak of the scientific knowledge at a given time. It is not meant to convey the claim that the elements of K_t are true, and hence neither that they are definitely known to be true. No such claim can justifiably be made for any of the statements established by empirical science; and the basic standards of scientific inquiry demand that an empirical statement, however well supported, be accepted and thus admitted to membership in K_t only tentatively, i.e. with the understanding that the privilege may be withdrawn if unfavourable evidence should be discovered. The membership of K_t therefore changes in the course of time; for as a result of continuing research, new statements are admitted into that class; others may come to be discredited and dropped. Henceforth, the class of accepted statements will be referred to simply as K when specific reference to the time in question is not required. We will assume that K is logically consistent and that it is closed under logical implication, i.e. that it contains every statement that is logically implied by any of its subsets.

The *epistemic ambiguity of I-S explanation* can now be characterized as follows: The total set K of accepted scientific statements contains different subsets of statements which can be used as premises in arguments of the probabilistic form just considered, and which confer high probabilities on logically contradictory 'conclusions'. Our earlier examples (3k), (3l) and (3m), (3n) illustrate this point if we assume that the premises of those arguments all belong to K rather than that they are all true. If one of two such rival arguments with premises in K is proposed as an explanation of an event considered, or acknowledged, in science to have occurred, then the conclusion of the argument, i.e. the explanandum statement, will accordingly belong to K as well. And since K is consistent, the conclusion of the rival argument will not belong to K. Nonetheless it is disquieting that we should be able to say: No matter whether we are informed that the event in question (e.g. warm and sunny weather on 27 November in Stanford) did occur or that it did not occur, we can produce an explanation of the reported outcome in either case; and an explanation, moreover, whose premises are scientifically established statements that confer a high logical probability upon the reported outcome.

This epistemic ambiguity, again, has no analogue for deductive explanation; for since K is logically consistent, it cannot contain premiss-sets that imply logically contradictory conclusions.

Epistemic ambiguity also bedevils the predictive use of statistical arguments. Here, it has the alarming aspect of presenting us with two rival arguments whose premises are scientifically well-established, but one of which characterizes a contemplated future occurrence as practically certain, whereas the other characterizes it as practically impossible. Which of such conflicting arguments, if any, are rationally to be relied on for explanation or for prediction?

3.4.2. The Requirement of Maximal Specificity and the Epistemic Relativity of Inductive-Statistical Explanation

Our illustrations of explanatory ambiguity suggest that a decision on the acceptability of a proposed probabilistic explanation or prediction will have to be made in the light of all the relevant information at our disposal. This is indicated also by a general principle whose importance for inductive reasoning has been acknowledged, if not always very explicitly, by many writers, and which has recently been strongly emphasized by Carnap, who calls it *the requirement of total evidence*. Carnap formulates it as follows: 'in the application of inductive logic to a given knowledge situation, the total evidence available must be taken as basis for determining the degree of confirmation.'[35] Using only a part of the total evidence is permissible if the balance of the evidence is irrelevant to the

---

[35] R. Carnap, *Logical Foundations of Probability* (Chicago, 1950), 211. The requirement is suggested, e.g., in the passage from Lewis quoted in n. 34. Similarly Williams speaks of 'the most fundamental of all rules of probability logic, that "the" probability of any proposition is its probability in relation to the known premisses and them only' (*The Ground of Induction* (Cambridge, Mass., 1947)).

I am greatly indebted to Professor Carnap for having pointed out to me in 1945, when I first noticed the ambiguity of probabilistic arguments, that this was but one of several apparent paradoxes of inductive logic that result from disregard of the requirement of total evidence.

S. F. Barker, *Induction and Hypothesis* (Ithaca, NY, 1957), 70–8, has given a lucid independent presentation of the basic ambiguity of probabilistic arguments, and a sceptical appraisal of the

inductive 'conclusion', i.e. if on the partial evidence alone, the conclusion has the same confirmation, or logical probability, as on the total evidence.[36]

The requirement of total evidence is not a postulate nor a theorem of inductive logic; it is not concerned with the formal validity of inductive arguments. Rather, as Carnap has stressed, it is a maxim for the *application* of inductive logic; we might say that it states a necessary condition of rationality of any such application in a given 'knowledge situation', which we will think of as represented by the set K of all statements accepted in the situation.

But in what manner should the basic idea of this requirement be brought to bear upon probabilistic explanation? Surely we should not insist that the explanans must contain all and only the empirical information available at the time. Not *all* the available information, because otherwise all probabilistic explanations acceptable at time t would have to have the same explanans, K_t; and not *only* the available information, because a proffered explanation may meet the intent of the requirement in not overlooking any relevant information available, and may nevertheless invoke some explanans statements which have not as yet been sufficiently tested to be included in K_t.

The extent to which the requirement of total evidence should be imposed upon statistical explanations is suggested by considerations such as the following. A proffered explanation of Jones's recovery based on the information that Jones had a streptococcal infection and was treated with penicillin, and that the statistical probability for recovery in such cases is very high is unacceptable if K includes the further information that Jones's streptococci were resistant to penicillin, or that Jones was an octogenarian with a weak heart, and that in these reference classes the probability of recovery is small. Indeed, one would want an acceptable explanation to be based on a statistical probability statement pertaining to the narrowest reference class of which, according to our total information, the particular occurrence under consideration is a member. Thus, if K tells us not only that Jones had a streptococcus infection and was treated with penicillin, but also that he was an octogenarian with a weak heart (and if K provides no information more specific than that) then we would require that an acceptable explanation of Jones's response to the

requirement of total evidence as a means of dealing with the problem. However, I will presently suggest a way of remedying the ambiguity of probabilistic explanation with the help of a rather severely modified version of the requirement of total evidence. It will be called the requirement of maximal specificity, and is not open to the same criticism.

[36] Cf. Carnap, *Logical Foundations*, 211 and 494.

treatment be based on a statistical law stating the probability of that response in the narrowest reference class to which our total information assigns Jones's illness, i.e. the class of streptococcal infections suffered by octogenarians with weak hearts.[37]

Let me amplify this suggestion by reference to an example concerning the use of the law that the half-life of radon is 3.82 days in accounting for the fact that the residual amount of radon to which a sample of 10 milligrams was reduced in 7.64 days was within the range from 2.4 to 2.6 milligrams. According to present scientific knowledge, the rate of decay of a radioactive element depends solely upon its atomic structure as characterized by its atomic number and its mass number, and it is thus unaffected by the age of the sample and by such factors as temperature, pressure, magnetic and electric forces, and chemical interactions. Thus, by specifying the half-life of radon as well as the initial mass of the sample and the time interval in question, the explanans takes into account all the available information that is relevant to appraising the probability of the given outcome by means of statistical laws. To state the point somewhat differently: Under the circumstances here assumed, our total information K assigns the case under study first of all to the reference class say F_1, of cases where a 10 milligram sample of radon is allowed to decay for 7.64 days; and the half-life law for radon assigns a very high probability, within F_1, to the 'outcome', say G, consisting in the fact that the residual mass of radon lies between 2.4 and 2.6 milligrams. Suppose now that K also contains information about the temperature of the given sample, the pressure and relative humidity under which it is kept, the surrounding electric and magnetic conditions, and so forth, so that K assigns the given case to a reference class much narrower than F_1, let us say, $F_1 F_2 F_3 \ldots F_n$. Now the theory of radioactive decay, which is equally included in K, tells us that the statistical probability of G within this narrower class is the same as within G. For this reason, it suffices in our explanation to rely on the probability $p(G, F_1)$.

Let us note, however, that 'knowledge situations' are conceivable in which the same argument would not be an acceptable explanation. Sup-

---

[37] This idea is closely related to one used by H. Reichenbach, (cf. *The Theory of Probability* (Berkeley, Calif., and Los Angeles, 1949), sect. 72) in an attempt to show that it is possible to assign probabilities to individual events within the framework of a strictly statistical conception of probability. Reichenbach proposed that the probability of a single event, such as the safe completion of a particular scheduled flight of a given commercial plane, be construed as the statistical probability which the *kind* of event considered (safe completion of a flight) possesses within the narrowest reference class to which the given case (the specified flight of the given plane) belongs, and for which reliable statistical information is available (e.g. the class of scheduled flights undertaken so far by planes of the line to which the given plane belongs, and under weather conditions similar to those prevailing at the time of the flight in question).

pose, for example, that in the case of the radon sample under study, the amount remaining one hour before the end of the 7.64 day period happens to have been measured and found to be 2.7 milligrams, and thus markedly in excess of 2.6 milligrams—an occurrence which, considering the decay law for radon, is highly improbable, but not impossible. That finding, which then forms part of the total evidence K, assigns the particular case at hand to a reference class, say F^*, within which, according to the decay law for radon, the outcome G is highly improbable since it would require a quite unusual spurt in the decay of the given sample to reduce the 2.7 milligrams, within the one final hour of the test, to an amount falling between 2.4 and 2.6 milligrams. Hence, the additional information here considered may not be disregarded, and an explanation of the observed outcome will be acceptable only if it takes account of the probability of G in the narrower reference class, i.e., $p(G, F_1F^*)$. (The theory of radioactive decay implies that this probability equals $p(G, F^*)$, so that as a consequence the membership of the given case in F_1 need not be explicitly taken into account.)

The requirement suggested by the preceding considerations can now be stated more explicitly; we will call it the *requirement of maximal specificity for inductive-statistical explanations*. Consider a proposed explanation of the basic statistical form

$$(3o) \qquad \begin{array}{c} p(G,\ F) = r \\ \hline Fb \\ \hline Gb \end{array} \quad [r]$$

Let s be the conjunction of the premises, and, if K is the set of all statements accepted at the given time, let k be a sentence that is logically equivalent to K (in the sense that k is implied by K and in turn implies every sentence in K). Then, to be rationally acceptable in the knowledge situation represented by K, the proposed explanation $(3o)$ must meet the following condition (the requirement of maximal specificity): If $s \cdot k$ implies[38] that b belongs to a class F_1, and that F_1 is a subclass of F, then $s \cdot k$ must also imply a statement specifying the statistical probability of G in F_1, say

$$p(G,\ F_1) = r_1$$

Here, r_1 must equal r unless the probability statement just cited is simply a theorem of mathematical probability theory.

---

[38] Reference to $s \cdot k$ rather than to k is called for because, as was noted earlier, we do not construe the condition here under discussion as requiring that all the explanans statements invoked be scientifically accepted at the time in question, and thus be included in the corresponding class K.

The qualifying unless-clause here appended is quite proper, and its omission would result in undesirable consequences. It is proper because theorems of pure mathematical probability theory cannot provide an explanation of empirical subject matter. They may therefore be discounted when we inquire whether $s \cdot k$ might not give us statistical laws specifying the probability of G in reference classes narrower than F. And the omission of the clause would prove troublesome, for if (3o) is proffered as an explanation, then it is presumably accepted as a fact that Gb; hence 'Gb' belongs to K. Thus K assigns b to the narrower class $F \cdot G$, and concerning the probability of G in that class, $s \cdot k$ trivially implies the statement that $p(G, F \cdot G) = 1$, which is simply a consequence of the measure-theoretical postulates for statistical probability. Since $s \cdot k$ thus implies a more specific probability statement for G than that invoked in (3o), the requirement of maximal specificity would be violated by (3o)—and analogously by any proffered statistical explanation of an event that we take to have occurred—were it not for the unless-clause, which, in effect, disqualifies the notion that the statement '$p(G, F \cdot G) = 1$' affords a more appropriate law to account for the presumed fact that Gb.

The requirement of maximal specificity, then, is here tentatively put forward as characterizing the extent to which the requirement of total evidence properly applies to inductive-statistical explanations. The general idea thus suggested comes to this: In formulating or appraising an I-S explanation, we should take into account all that information provided by K which is of potential *explanatory* relevance to the explanandum event; i.e. all pertinent statistical laws, and such particular facts as might be connected, by the statistical laws, with the explanandum event.[39]

The requirement of maximal specificity disposes of the problem of epistemic ambiguity; for it is readily seen that of two rival statistical arguments with high associated probabilities and with premises that all

---

[39] By its reliance on this general idea, and specifically on the requirement of maximal specificity, the method here suggested for eliminating the epistemic ambiguity of statistical explanation differs substantially from the way in which I attempted in an earlier study (Hempel, 'Deductive-Nomological *vs*. Statistical Explanation', esp. sect. 10) to deal with the same problem. In that study, which did not distinguish explicitly between the two types of explanatory ambiguity characterized earlier in this section, I applied the requirement of total evidence to statistical explanations in a manner which presupposed that the explanans of any acceptable explanation belongs to the class K, and which then demanded that the probability which the explanans confers upon the explanandum be equal to that which the total evidence, K, imparts to the explanandum. The reasons why this approach seems unsatisfactory to me are suggested by the arguments set forth in the present section. Note in particular that, if strictly enforced, the requirement of total evidence would preclude the possibility of any significant statistical explanation for events whose occurrence is regarded as an established fact in science; for any sentence describing such an occurrence is logically implied by K and thus trivially has the logical probability 1 relative to K.

belong to K, at least one violates the requirement of maximum specificity. Indeed, let

$$p(G, F) = r_1 \qquad\qquad p(\overline{G}, H) = r_2$$
$$\frac{Fb}{Gb} \quad [r_1] \qquad \text{and} \qquad \frac{Hb}{\overline{G}b} \quad [r_2]$$

be the arguments in question, with r_1 and r_2 close to 1. Then, since K contains the premisses of both arguments, it assigns b to both F and H and hence to $F \cdot H$. Hence if both arguments satisfy the requirement of maximal specificity, K must imply that

$$p(G, F \cdot H) = p(G, F) = r_1$$
$$p(\overline{G}, F \cdot H) = p(\overline{G}, H) = r_2$$
$$\text{But} \quad p(G, F \cdot H) + p(\overline{G}, F \cdot H) = 1$$
$$\text{Hence} \quad r_1 + r_2 = 1$$

and this is an arithmetic falsehood, since r_1 and r_2 are both close to 1; hence it cannot be implied by the consistent class K.

Thus, for I-S explanations that meet the requirement of maximal specificity the problem of epistemic ambiguity no longer arises. We are *never* in a position to say: No matter whether this particular event did or did not occur, we can produce an acceptable explanation of either outcome; and an explanation, moreover, whose premisses are scientifically accepted statements which confer a high logical probability upon the given outcome.

While the problem of epistemic ambiguity has thus been resolved, ambiguity in the first sense discussed in this section remains unaffected by our requirement; i.e. it remains the case that for a given statistical argument with true premisses and a high associated probability, there may exist a rival one with equally true premisses and with a high associated probability, whose conclusion contradicts that of the first argument. And though the set K of statements accepted at any time never includes all statements that are in fact true (and no doubt many that are false), it is perfectly possible that K should contain the premisses of two such conflicting arguments; but as we have seen, at least one of the latter will fail to be rationally acceptable because it violates the requirement of maximal specificity.

The preceding considerations show that *the concept of statistical explanation for particular events is essentially relative to a given knowledge situation as represented by a class K of accepted statements.* Indeed, the requirement of maximal specificity makes explicit and unavoidable reference to such a class, and it thus serves to characterize the concept of 'I-S explanation relative to the knowledge situation represented by K'. We will refer to this characteristic as the *epistemic relativity of statistical explanation.*

It might seem that the concept of deductive explanation possesses the same kind of relativity, since whether a proposed D-N or D-S (Deductive-Statistical) account is acceptable will depend not only on whether it is deductively valid and makes essential use of the proper type of general law, but also on whether its premises are well supported by the relevant evidence at hand. Quite so; and this condition of empirical confirmation applies equally to statistical explanations that are to be acceptable in a given knowledge situation. But the epistemic relativity that the requirement of maximal specificity implies for I-S explanations is of quite a different kind and has no analogue for D-N explanations. For the specificity requirement is not concerned with the evidential support that the total evidence K affords for the explanans statements: it does not demand that the latter be included in K, nor even that K supply supporting evidence for them. It rather concerns what may be called the concept of a *potential* statistical explanation. For it stipulates that no matter how much evidential support there may be for the explanans, a proposed I-S explanation is not acceptable if its potential explanatory force with respect to the specified explanandum is vitiated by statistical laws which are included in K but not in the explanans, and which might permit the production of rival statistical arguments. As we have seen, this danger never arises for deductive explanations. Hence, these are not subject to any such restrictive condition, and the notion of a potential deductive explanation (as contradistinguished from a deductive explanation with well-confirmed explanans) requires no relativization with respect to K.

As a consequence, we can significantly speak of true D-N and D-S explanations: they are those potential D-N and D-S explanations whose premises (and hence also conclusions) are true—no matter whether this happens to be known or believed, and thus no matter whether the premises are included in K. But this idea has no significant analogue for I-S explanation since, as we have seen, the concept of potential statistical explanation requires relativization with respect to K.

REFERENCES

Barker, S. F., *Induction and Hypothesis* (Ithaca, NY: Cornell University Press, 1957).
Boehmer, H., *Luther and the Reformation in the Light of Modern Research*, trans. E. S. G. Potter (New York: The Dial Press, 1930).

Carnap, R., 'Testability and Meaning', *Philosophy of Science* 3 (1936) and 4 (1937). Reprinted in part in Feigl and Brodbeck (eds.).

—— *Logical Foundations of Probability* (Chicago: University of Chicago Press 1950; second, rev., edn. 1962).

Dewey, John, *How We Think* (Boston: D. C. Heath & Co., 1910).

Dray, W., *Laws and Explanation in History* (Oxford: Oxford University Press, 1957).

Edwards, E. (ed.), *The Early Writings of Frederick Jackson Turner* (Madison, Wis.: University of Wisconsin Press, 1938).

Feigl, H., and Brodbeck, M. (eds.), *Readings in the Philosophy of Science* (New York: Appleton-Century-Crofts, 1953).

Feigl, H., and Maxwell, G. (eds.), *Minnesota Studies in the Philosophy of Science*, iii (Minneapolis: University of Minnesota Press, 1962).

Freud, S., *Psychopathology of Everyday Life*, trans. A. A. Brill (New York: The New American Library (Mentor Book Series), 1951).

Gardiner, P., *The Nature of Historical Explanation* (Oxford: Oxford University Press, 1952).

Gerschenkron, A., 'Economic Backwardness in Historical Perspective', in B. F. Hoselitz (ed.), *The Progress of Underdeveloped Areas* (Chicago, 1952).

Hempel, C. G., 'Inductive Inconsistencies', in *Synthese*, 12 (1960), 439–69, repr. in id., *Aspects of Scientific Explanation* (New York and London: Free Press and Collier Macmillan, 1965), 53–79.

—— 'Deductive-Nomological *vs.* Statistical Explanation', in Feigl and Maxwell (eds.), 98–169.

—— and Oppenheim, P., 'Studies in the Logic of Explanation', *Philosophy of Science* 15 (1948) 135–75.

Lewis, C. I., *An Analysis of Knowledge and Valuation* (La Salle, Ill.: Open Court, 1946).

Nagel, E., *The Structure of Science: Problems in the Logic of Scientific Explanation* (New York: Harcourt, Brace, & World, Inc., 1961).

Reichenbach, H., *The Theory of Probability* (Berkeley, Calif., and Los Angeles: The University of California Press, 1949).

Scheffler, I., 'Explanation, Prediction, and Abstraction', *The British Journal for the Philosophy of Science* 7 (1957) 293–309.

Schwiebert, E. G., *Luther and his Times* (St Louis: Concordia Publishing House, 1950).

Toulmin, S., *The Uses of Argument* (Cambridge: Cambridge University Press, 1958).

Ware, C. F. (ed.), *The Cultural Approach to History* (New York: Columbia University Press, 1940).

Watson, G., 'Clio and Psyche: Some Interrelations of Psychology and History', in Ware (ed.), 34–47.

Williams, D. C., *The Ground of Induction* (Cambridge, Mass.: Harvard University Press, 1947).

II

HEMPEL'S AMBIGUITY*

J. ALBERTO COFFA

Explanation theory abandoned its pre-theoretical stage and became a respectable branch of philosophical inquiry when, in the late forties, Hempel began to develop his model of deductive-nomological (D-N) explanation. In a sequence of now classic papers he succeeded in articulating an illuminating philosophical account of explanation which provided compelling evidence for the adequacy of the philosophical views captured by the D-N model.

In the early sixties Hempel turned his attention to the topic of inductive explanation. Until then, it had been generally believed that the inductive model had to be understood as a generalization—and a rather straightforward one at that—of the deductive model. Yet, already in the Hempel–Oppenheim paper a warning had been issued to the effect that such generalization raised 'a variety of new problems'. Indeed, when finally, in a sequence of illuminating papers on inductive explanation, Hempel decided to face one of these problems, he felt forced to propose a theory of inductive explanation which differed drastically from pre-analytic consensus on the nature of such explanations.

Not the least of these departures was Hempel's implicit rejection of the claim that the deductive model is a limiting instance of the inductive model. Yet, much more than this was involved. We should like to argue that, in spite of misleading appearances of continuity, the philosophical understanding of explanations implicit in the model of inductive-statistical (I-S) explanation which Hempel eventually produced is drastically different from, if not incompatible with, that which inspired his D-N model. One of the purposes of this paper is to draw attention to the nature and magnitude of the shift involved. Another is to explain why Hempel's views on inductive explanation ought not to be accepted.

J. Alberto Coffa, 'Hempel's Ambiguity'. First published in *Synthese*, 28 (1974), 141–63. Copyright © 1974 by D. Reidel Publishing Company, Dordrecht. Reprinted by permission of Kluwer Academic Publishers.

* This paper was read at the Lecture Series organized by the Center for the Philosophy of Science at the University of Pittsburgh. I gratefully acknowledge comments by Professors A. Grünbaum, W. C. Salmon, G. Massey, N. D. Belnap, Jr., and T. M. Simpson.

The evolution of Hempel's thought was causally related to his analysis of a problem which, pending more illuminating designations, we will refer to as 'Hempel's problem'. Due to it Hempel felt forced to propose an account of inductive explanation which contained a rather unexpected feature, a peculiar form of relativization to knowledge that, according to Hempel, is unavoidably present in every adequate theory of inductive explanation.

The question whether such epistemic relativization as Hempel has introduced is avoidable or not is interesting in its own right. But it becomes pressing if one believes—as I do—that no characterization of inductive explanation incorporating that feature can be backed by a coherent and intelligible philosophy of explanation. If I am right, it is of more than passing interest to give a closer look at Hempel's problem. For unless one can find a way to avoid the conclusion Hempel drew from it, one may well have to accept that the concept of inductive explanation is as much of a conceptual delusion as some have claimed modal concepts to be.

My strategy in this paper will be the following. I will first attempt to locate the nature of Hempel's problem. This will prove to be unexpectedly difficult. Next, I will examine the epistemic relativization that Hempel felt forced to introduce, as well as his reasons for introducing it. Having done this, we will be in a position to examine the question of the unavoidability of Hempel's epistemic relativization. Up to this point the argument will be reasonably inter-paradigmatic. It will become less so when I finally turn to explain why an epistemically relativized theory of explanation à la Hempel cannot be taken to be a theory of explanation. In a way, my argument will boil down to the claim that I cannot see how it could; but I will try to hide its unacademic form under the cloak of an analysis and rejection of those reasons that could conceivably be given in defence of such models. Here my remarks will be cautiously brief and somewhat cryptic. They will be even more so when, in the last paragraphs, I attempt to suggest the way in which an alternative theory of inductive explanation could be developed, based upon a solution of Hempel's problem. We turn, first, to Hempel's problem.

1. HEMPEL'S PROBLEM

There is a widespread tendency to view the simplest form of the D-N model (its 'basic form') as being that of

(1) $(x)\,(Fx \supset Gx)$

$$\frac{Fa}{Ga}$$

where 'Ga' is the explanandum, '$(x)\,(Fx \supset Gx)$' is a nomic statement and 'Fa' is a statement of initial (and boundary) conditions. When the premisses of (1) are true, one says that (1) is a D-N explanation, or (for emphasis) a true D-N explanation; when they are well confirmed on the available evidence one says that (1) is a well-confirmed D-N explanation, relative to the available evidence (or knowledge situation).

If one views the D-N model as in (1), attempts to generalize it into an inductive theory will naturally start fixing attention on what seem to be its only generalizable features: the kind of connection that the nomic premiss asserts to obtain between the attributes it mentions, and the kind of connection obtaining between the premisses and the conclusion of the argument. Both relationships can be generalized into probabilities, yet not into the same sort of probability. The deterministic connection asserted by the first premiss can be weakened or generalized into a frequentist or statistical correlation, whereas the deductive link between premisses and conclusion can be weakened or generalized into an inductive link. The deductive deterministic model can be thus generalized into an inductive-statistical model.

Thus, it seems most natural to conclude that inductive explanations of basic form ought to be understood as follows, in what we will refer to as *the naïve model of inductive explanation.* An I-S explanation (or true I-S explanation) of basic form will be an argument of the form

(2) $p(G,\ F) = r$

$$\frac{Fa}{Ga}$$

together with the number $c(Ga,\ p(G,\ F) = r\ \&\ Fa)$ which, we assume, is also r; where r is close to 1 and the premisses of (2) are true, the first one being a law of nature.

Although probably no one has ever offered this precise characterization of inductive explanations, one may conjecture that something essentially like this has been lurking in people's minds when they talked about inductive explanations. Yet, as soon as Hempel turned his attention to the inductive theory of explanation, he noticed that the naïve model was quite unacceptable.

Hempel observed that given that all of the premisses of (1) are true, it is still possible that, for some property H, 'Ha' is true and '$p(-G,\ H) = s$' is a true nomic statement where s is close to 1. But then, not only (2) but also

(3)
$$p(-G, H) = s$$
$$\frac{Ha}{-Ga}$$

would be an inductive argument with true premises, and a conclusion that is implied with high inductive probability by the premises. The existence of arguments like (2) and (3) is what Hempel called 'the phenomenon of ambiguity'.

Clearly, the phenomenon of ambiguity amounts to the fact that there can be naïve inductive explanations of mutually inconsistent statements. The assumptions that most Texans are millionaires and that most philosophers are not millionaires are compatible with the assumption that Jones is a Texan philosopher. If all of these assumptions are true and if the statistical correlations are nomic, then we have two naïve inductive explanations, one explaining that Jones is a millionaire and the other explaining that he is not.

Hempel felt that there was something very undesirable about this consequence; indeed, undesirable to the point of constituting a *reductio ad absurdum* of the naïve model. Thus, according to Hempel, the phenomenon of ambiguity shows that there is something hopelessly wrong about that model; it shows that it incorporates a certain feature that immediately implies the worthlessness of the model. If we could identify this 'bad feature' precisely, we would have a well-defined programme and a problem that every explanation theorist should attempt to solve: to produce a definition of inductive explanation for which it can be shown that it does not share the bad feature in question.

Most of Hempel's explicit statements on the topic suggest that there is an obvious way of identifying this 'bad feature': it would consist in the fact that certain (otherwise plausible) characterizations of inductive explanation have instances with mutually inconsistent conclusions, or rather, in Hempelian terminology, that such definitions have inductive inconsistencies as instances (an *inductive inconsistency* being a pair of inductive arguments with mutually inconsistent conclusions which are implied with high inductive probability by their respective premises).

Let us say that a definition of inductive explanation *suffers from ambiguity* when it has inductive inconsistencies as instances. (Note that since a definition of true inductive explanation must demand that its premises be true, such definitions may suffer from ambiguity only when the inductive inconsistencies that it has as instances have true premises. Hempel referred to this variety of ambiguity as *ontic ambiguity*. For definitions of inductive explanation demanding that the premises belong to a certain knowledge situation, the ambiguity in question was called by

Hempel *epistemic ambiguity*.) The seemingly obvious construal indicated above is that the 'bad feature' is to suffer from ambiguity. Hempel's problem would then be that of finding a definition of inductive explanation that does not suffer from ambiguity.

This construal receives further support from the fact that throughout a series of papers in which Hempel has attempted to solve the problem raised by ambiguity, his efforts have had a consistently unique form: in all cases, a definition of inductive explanation was first offered, and then an attempt was made to prove that it did not suffer from ambiguity.

Nevertheless, a closer look at the structure of these attempts engenders a quite different impression concerning what Hempel's problem really is.

Notice first that if Hempel thought that his problem was that of producing a definition of inductive explanation not suffering from ambiguity, there would have been a most trivial solution which would apply in all imaginable cases, a solution which surely couldn't have escaped Hempel's attention. Given an arbitrary definition of inductive explanation, one could make it comply with the requirement to avoid ambiguity by adding to it a clause to the effect that the explanandum should be true (or, in an epistemic characterization, *known*). This would imply in the most straightforward way that there are no (appropriately corrected) naïve inductive explanations of mutually inconsistent statements. Furthermore the additional clause seems easily justifiable on traditional Hempelian standards according to which one can only explain what is the case (respectively, what is known to be the case). Thus, had Hempel thought that the problem uncovered by the phenomenon of ambiguity was the one described above, this problem would have deserved no attention whatsoever.

Yet, not only did Hempel think that the problem had no easy solution; in fact, for the case of true inductive explanations he thought that it had no solution at all. And he claimed that this established the meaninglessness of the concept of true inductive explanation.

Anyone who interprets Hempel's problem as has been suggested above must also find it very difficult to comprehend why Hempel has set himself such narrow limits on what is to count as an allowable method to solve this problem. For, apart from a soon withdrawn appeal to the principle of total evidence in Hempel (1962), Hempel's way of dealing with his problem has always been to offer restrictions on the allowable reference classes in the nomic premises of inductive explanation. This was first done tentatively in Hempel (1962); a more precise requirement was offered in Hempel (1965), and he claimed there that he could prove that the enforcement of this requirement guaranteed that his definition did not suffer from ambiguity. When Grandy showed that this was not the case, Hempel (1968) offered a considerably more complex requirement of the same kind and a new proof that this new restriction on admissible reference classes guaranteed the desired effect. Yet, it turns out that the new proof also fails, in that it makes an unwarranted existential assumption.[1] The remarkable disproportion between the alleged aim of Hempel's seemingly unrealizable programme and its limited means is underscored by the ironic fact that Hempel's (1968) definition of inductive explanation requires that the explanandum belong to a knowledge situation; and this alone, under Hempel's demand that knowledge situations be consistent, immediately implies the result that Hempel has been at such great pains to prove on the basis of restrictions on admissible reference classes.

Are we to conclude that the Hempelian bird of happiness had always been in the philosopher's backyard? This seems unlikely. Rather, we would suggest, what seems called for is a revision of our understanding of what Hempel's problem is. Under our first construal, it appeared as a problem essentially concerned with the conclusions of inductive explanations, particularly, with the fact that they may contradict each other. But under this construal it proves impossible to make sense of Hempel's treatment of his problem. Hence, it seems necessary to provide a different account of what Hempel's problem is. The alternative that we would like to suggest is that Hempel is accurately, though maybe not clearly, perceiving a very real problem for inductive explanation, one that concerns not their conclusions but their premisses; more precisely, the nomic premiss, and, within it, its reference class. We would like to suggest that when Hempel turned his attention to the theory of inductive explanation what he stumbled upon was the fact that the problem of defining a model of inductive explanation for single events was the other side of the coin of the single case problem. He stumbled, that is, upon the reference-class problem.

There was in the theory of probability an old problem that had faded away, but not quite died. It was the problem of the reference class. This problem arises when, in the context of a frequency theory of probability, we attempt to answer the question, 'What is the probability of a single event?', for example, 'What is the probability that this particular airplane will fall in its next flight?' If one believes, as many frequentists did, that the only meaning probabilities can have is the frequency meaning, then in order to answer the question one must begin by identifying a reference

---

[1] It is assumed, without warrant, that for any K and any 'Gi', there is a maximally specific predicate related to 'Gi' in K. (It is possible to construct consistent and logically closed K's for which this assumption is false.)

class for the given event. But here is where the reference-class problem arises. One and the same event can be associated with different reference classes, different to the extent that the probabilities of the event in such reference classes differ. This airplane belongs to the reference class of all airplanes and in that reference class the probability of its falling in its next flight is not high, but it also belongs to the reference class of airplanes whose wings will fall during the next flight, and as such its probability of not quite succeeding in reaching safely its destination is rather high. For the frequentist the question, 'What is the probability of a single event?' could make sense only if we could find a 'natural', 'appropriate' reference class for each event. But frequentists have traditionally held a principle of reference-class democracy: in so far as the estimation of probabilities is concerned, all reference classes are created equal. Thus they would traditionally hold that it is strictly meaningless to assign a probability to a single event. The frequentist finds that, despite misleading appearances to the contrary, statements like 'the probability that this plane will fall on its next flight is high' are no more meaningful than guttural noises. Now, *if* one views statistical explanations as essentially concerned with placing the explanandum in an appropriate reference class, and *if* one countenances a frequentist interpretation for the nomic premiss, then one of the main problems one will have to face while developing a theory of inductive explanation is the reference-class problem. In effect, this is how Salmon saw the matter, since the essential ingredient of his model is the claim that the 'appropriate' reference class in an explanation is the maximal homogenous reference class. We want to suggest that, although less obviously, this is also Hempel's understanding of what the main problem for a theory of inductive explanation is.

2. EPISTEMIC RELATIVIZATION

Let us leave for a moment the question as to what Hempel's problem is, and let me turn to consider briefly the conclusion Hempel felt forced to draw from it: the thesis of the epistemic relativity of inductive explanation. Since it is crucial to have a full understanding of what this thesis amounts to, I will devote a few paragraphs to the explanation of certain conceptual distinctions which will help us to grasp its force.

It is an obvious fact that the meaning of some expressions or concepts can be given without referring to knowledge, whereas that of others cannot. Let me call the latter epistemic and the former non-epistemic expressions. Examples of non-epistemic expressions are easy to find.

'Table', 'chair', 'electron', according to many, 'truth', would be typical instances. Examples of epistemic notions are also readily available. The best known instance may be that of the concept of confirmation. Although the syntactic form of expressions like 'hypothesis h is well-confirmed' may mislead us into believing that confirmation is a property of sentences, closer inspection reveals the fact that it is a relation between sentences and knowledge situations and that the concept of confirmation cannot be properly defined (that is, its meaning cannot be given) without a reference to sentences intended to describe a knowledge situation. Just as there are clear-cut cases of each of these two kinds, there are concepts for which it is difficult to decide whether they are epistemic or not. Randomness is one such case. Some will argue that it describes possible properties of the world which may obtain or not obtain quite independently of the presence of knowledge. For these the concept is non-epistemic. Others argue that randomness means nothing if not relativized to knowledge, for the predication of randomness can only mean that the person who ascribes such a predicate lacks the (it is claimed, always potentially available) information that would allow him to uncover a certain order in the given system.

Having introduced the distinction between epistemic and non-epistemic concepts, we go on to notice that there is a further interesting distinction to be drawn within the class of epistemic notions based upon the kind of role knowledge plays in them. On the one hand there are those epistemic notions in which knowledge enters essentially as an argument in a confirmation function, or, equivalently, as an ingredient in a statement of rational belief. And then there is the obscure and largely unintelligible remainder.

In the first group we find a significant example provided by Hempel's theory of deductive explanation. After introducing his non-epistemic notion of D-N explanation Hempel went on to say that he could define now the concept of a well-confirmed D-N explanation, a well-confirmed D-N explanation in a tacitly assumed knowledge situation K being, in effect, an argument which in knowledge situation K it is rational to believe is a D-N explanation, i.e. a true D-N explanation. In precisely the same fashion we could correctly and uninterestingly define the concepts of well-confirmed table, well-confirmed chair, and well-confirmed electron, given that we started by having the concepts of table, chair, and electron. Since we can only have reason to believe meaningful sentences, a confirmational epistemic predicate is an articulation of independently meaningful components.

Of course we can understand what a well-confirmed chair is because we began by understanding what a chair is. If 'x is a chair' had not had a meaning, we would not even have been able to make sense of the statement of rational belief made about it. Similarly, we can understand, if not appreciate, the notion of well-confirmed D-N explanation, because we were told first what kind of thing a D-N explanation is.[2]

Now we are in a position to state Hempel's thesis of the epistemic relativity of inductive explanation. As a consequence of his analysis of the phenomenon of ambiguity, Hempel concludes that the concept of inductive explanation, unlike its deductive counterpart, is epistemic; and he goes on to add that it is not epistemic in the sense in which well-confirmed deductive explanations are. The concept of inductive explanation is a non-confirmational epistemic concept. Such is the thesis of the epistemic relativity of inductive explanation.

As Hempel is careful to point out, this means that there is no concept that stands to his epistemically relativized notion of inductive explanation as the concept of true D-N explanations stands to that of well-confirmed D-N explanation. According to the thesis of epistemic relativity there is no meaningful notion of true inductive explanation. Hence, we could not possibly have reasons to believe that anything is a true inductive explanation. Thus, it would be sheer confusion to see inductive explanations relative to K in Hempel's sense as those inductive arguments which in knowledge situation K it is rational to believe are inductive explanations.

It is then clear that, according to Hempel, there is a remarkable and surprising disanalogy between deductive and inductive explanations. When somebody asks us to give an account of deductive explanations, we can do so without referring to anybody's knowledge. If asked, for instance, what sort of thing would it be to explain deductively the present position of a planet, we would refer to descriptions of certain nomic and non-nomic facts but never to our or to anybody else's knowledge. This is a desirable feature in a non-psychologistic account of explanation. Yet, according to Hempel, when we ask what an inductive explanation of the same event would look like, there is no way in which an appropriate answer can be given without talking about knowledge; presumably, the knowledge available at the time of the explanation. Even more surprisingly, this reference to knowledge does not play the standard role that such references usually play, to wit, that of providing the epistemic platform for a judgement of rational belief. What role such reference

---

[2] Prof. Gerald Massey has drawn my attention towards the apparent opacity of epistemically relativized predicates. He has pointed out that this raises serious doubts concerning the possibility of viewing them as expressing properties.

plays is a question which deserves serious attention, since here we find the Achilles' heel of Hempel's whole construction.

Let me briefly survey what we have done so far. In effect, we have only done two things. First, we have inspected Hempel's problem, arguing that it is not clear what precisely the problem is. We contended that in most of the relevant texts Hempel seems to imply that to show that a definition of explanation avoids the problem in question is to show something about the conclusions of such explanations, to wit, that they never contradict each other. But we also argued that some of Hempel's remarks, his blatantly ignoring a most trivial solution to this problem, and the otherwise unintelligible decision to circumscribe solutions to his problem to restrictions to the reference class, suggest that Hempel perceives that the real problem behind ambiguity is a problem having to do with the premisses of explanations rather than with their conclusions, and that his problem is in fact, a new variant of the old reference-class problem. After arguing for this claim we moved on to inspect the meaning of a very peculiar conclusion that Hempel feels forced to draw from ambiguity. This is what we have just done. There is now a very natural gap to fill. We should explain what reasons Hempel has to argue that, given the problem of ambiguity, the thesis of the epistemic relativity of inductive explanation follows.

Here we find ourselves in an awkward position. For Hempel has said next to nothing explicitly on the connection between ambiguity and epistemic relativity, and the little he has said does not carry much weight. Rather than engaging in a frustrating exercise of Hempelian exegesis, I will attempt to reconstruct conjecturally the train of thought that may have led Hempel to his remarkable conclusion.

3. THE ARGUMENT FOR EPISTEMIC RELATIVIZATION

We have seen that Hempel claims that there is no acceptable definition of inductive explanation which is not relativized to K. More precisely he seems to hold that not only is it the case that a definition of true I-S explanation may instantiate ontic ambiguity; he holds that it *must* instantiate it. And in the context of his assumption that a necessary condition for the acceptability of a concept of explanation is that it should not suffer from ambiguity, this implies the non-existence of non-relativized inductive explanations. Let us trace more carefully the form of Hempel's argument. The following assumptions seem to be involved.

Assumption 1. A definition of inductive explanation is inadmissible if it suffers from ambiguity.

Assumption 2. The only way to improve upon the naïve definition of true I-S explanation is by introducing in the definition a new clause restricting those reference classes which are admissible in the nomic premiss of the explanation.

Assumption 3. There is no clause as the one described in Assumption 2 such that (*a*) it makes no reference to knowledge, and (*b*) when added to the definition of true I-S explanation it guarantees that the resulting definition does not suffer from ambiguity.

Before I try to inspect the sources of these assumptions, let me make a somewhat marginal remark. It seems clear that Hempel believes in the following two further assumptions:

Assumption 4. Inductive explanations relative to K ought to be defined as inductive arguments of form (2) with premises in K (rather than in the class T of true sentences) which imply the conclusion with high inductive probability, and which verify a requirement delimiting the class of allowable reference classes in the nomic premise. Thus, in purely syntactic terms, a proposed definition of I-S explanation relative to K will differ from a proposed definition of true I-S explanation only in that the class of true sentences T plays in the latter the role that the class of known sentences plays in the former.

Finally, Hempel obviously believes that

Assumption 5. There is a definition of I-S explanation relative to K for which it can be proved that it does not suffer from ambiguity.

Now, the minor point I want to make is that these five assumptions are inconsistent with the claim that the class of true sentences is a possible knowledge situation, or, in other words, with the assumption that it is logically possible that someone could know precisely what is true. The contradiction is obvious since, if one admits that T is one of the classes over which the variable K ranges, Assumption 3 affirms a universal of which Assumption 5 denies an instance. For Assumption 5 says that a certain definition of I-S explanation relative to K (where K is a free variable ranging over knowledge situations) does not suffer from ambiguity, in the sense of Assumption 1. But when we replace the variable K by T in the alleged definition of I-S explanation relative to K, what we obtain, in view of Assumptions 4 and 2, is a definition of true I-S explanation. Moreover, this definition of true I-S explanation does not suffer from (ontic) ambiguity, for there is, by Assumption 5, a proof that, for all K (hence also for T), no instances of inductive inconsistencies obtain. Since, when K becomes T epistemic ambiguity becomes ontic ambiguity, the proof of the avoidance of epistemic ambiguity for some definition of I-S explanation relative to K is *ipso facto* the proof of the avoidance of ontic ambiguity for the corresponding definition of true I-S explanation.

One might retort that this would hardly affect Hempel since the class of true sentences is not a possible knowledge situation. This would mean that the conception of an omniscient being is logically incoherent and that it could not, it *logically* could not be the case that someone might happen to know all and only true statements, so that, as a matter of meaning, every knowledge situation should contain at least one falsehood or lack one truth. This seems unlikely, but even if granted, it should be observed that a proof that a definition of I-S explanation relative to K does not suffer from epistemic ambiguity must use some assumption about K that is not true about T, for otherwise the same proof would establish for the associated concept of true inductive explanation that it avoids ontic ambiguity. Since none of the proofs of the alleged avoidance of epistemic ambiguity offered by Hempel so far make use of any features of K that are not true of T, had any of them been successful, it would have implied the falsehood of the epistemic relativity thesis.

This is the minor point I wanted to make. Let me now return to the main line of my argument. I was trying to understand the reasons which led Hempel to his claim that there are no true inductive explanations, and I said that the best sense I could make of such a claim was in terms of Assumptions 1–3. Now, these assumptions are not implausible. The first one constitutes Hempel's 'official' view on what is wrong about ambiguity, and, no doubt, this claim has a strong intuitive appeal. The second assumption looks similarly plausible, but I have been arguing that its main contention concerning the limits imposed to particular solutions to ambiguity is based upon an understanding of Hempel's problem in conflict with that which inspires the first assumption. For, as I have pointed out, were one to take seriously the idea that all that is wrong with ambiguity is that some definitions of inductive explanation instantiate inconsistencies, then the restriction imposed by this assumption on possible solutions to that problem would be intolerably arbitrary due to the existence of the above-mentioned trivial solution.

Let us turn briefly to the crucial third assumption. It is here that, in spite of his explicit espousal of a propensity interpretation, Hempel's tacit appeal to a frequency conception of probability—with its reference-class puzzlements—becomes apparent. Presumably, what could lead to this assumption is an argument like the following.

The source of ontic ambiguity lies in the fact that inductively inconsistent arguments with true premises fix attention upon what is only a partial aspect of the object of the explanation. The problem with our alternative explanations of Jones's financial condition is that one of them pays attention only to his being a Texan whereas the other pays attention only to his being a philosopher. Each alleged explanation is to be blamed on the grounds that it ignores a relevant aspect that the other takes into account. This is the problem. In order to have a notion of true inductive explanation one must be able to solve it. But, Hempel says, one cannot solve it.

That we cannot solve it seems suggested by the fact that although the 'partial aspects' approach seems excellent to indicate which reference classes will not do, it seems unusable to decide which ones will do. Prima facie, it would seem that what needs to be enforced is a principle demanding that all relevant aspects of the explanandum object be taken into account. The frequentist will explicate this intuitive idea of relevance in terms of the long-run convergence of the relative frequencies of the appropriate classes. When so understood, the principle seems to be as strong as to imply that no reference class is admissible, for every reference class seems condemned to be a partial relevant aspect of the explanandum object. The principle tells us that it is not enough to refer Jones to the class of Texans in order to explain his financial condition. But is it enough to refer him to the class of Texan philosophers? In all likelihood the demand to identify all relevant aspects implies that it is not. For it is most likely that there will be some other property of Jones, different from the one to be explained, which when conjoined to that of his being a Texan philosopher will determine a reference class in which the long-run frequency of the outcome differs from the one it has in the original class. Unknown to everyone, Jones may have been born at the same time as a Chinese mandarin sneezed, and if the long-run frequency of richness in the reference class of Texan philosophers born while a Chinese mandarin sneezes differs (as it well may) from that in the class of Texan philosophers, the demand to refer to all relevant aspects in an explanation rules out the class of Texan philosophers as an admissible reference class. Indeed, every reference class would be ruled out by some other reference class if a certain not unlikely assumption of the denseness of relevant reference classes is accepted: the assumption that given a reference class F and an attribute (explanandum) class G, there is a subclass of F (i.e. a class more specific than F) other than $F \& G$, a class to which the object of the explanation belongs, and in which the long-run frequency of the explanandum property G is different from the one it has in F.

As far as I can see, it is something like the above considerations that may have led Hempel to act in agreement with Assumption 3. And it is this assumption, together with the rather natural Assumption 2, which is to be held responsible for the thesis of epistemic relativity.

Only one more thing remains to be done before I can rest my case. I have argued that from a certain problem Hempel felt forced to draw the conclusion that the notion of inductive explanation is epistemically relativized. We have just seen what kinds of assumptions may have been involved in the argument leading to this conclusion. Now, I would like to explain why I find Hempel's conclusion worth avoiding. I will try to convince you that to accept Hempel's thesis of epistemic relativity amounts to accepting the claim that there are no inductive explanations, the concept of I-S explanation relative to K functioning as a placebo which can only calm the intellectual anxieties of the uncautious user. If I am right, anyone willing to hold that there are inductive explanations will have to begin by spotting a flaw in Hempel's argument. I will close this paper with a few remarks in which I will attempt to indicate in the barest outline how, by denying some of the assumptions on which Hempel's conclusion seems to rest, one can conceivably avoid epistemic relativity and introduce a satisfactory characterization of inductive explanation.

4. EPISTEMIC RELATIVITY REVISITED

Maybe the best way in which I can briefly convey my feelings about the oddity implicit in Hempel's theory of inductive explanation, is by noting, that in my view, Hempel's decision to develop a theory of I-S explanation relative to K after having argued that the notion of true inductive explanation makes no sense, seems comparable to that of a man who establishes conclusively that Hegel's philosophy is strict nonsense, and then proceeds to devote the rest of his life to produce the definitive edition of Hegel's writings. For I would like to suggest that the only purpose that could be served by the predicate 'being an inductive explanation relative to K' is that of identifying a class of inductive arguments, the respectability of which has been seriously undermined by Hempel's analysis of ambiguity.

Let me remind you of the general form of Hempel's characterization of I-S explanation relative to K. As our fourth assumption indicates, Hempel views I-S explanations relative to a knowledge situation K as arguments of form (2) with premises in K, which imply the conclusion with high inductive probability; furthermore, the probabilistic premiss is nomic and its reference class complies with a certain requirement of maximal specificity. The heart of the definition is this last requirement which is supposed to work roughly in the following way. Pick a knowledge situation K; now, a class to which the object of the explanation is known (in K) to belong will be an admissible reference class (relative to that knowledge situation) if it is the most specific class to which the object is known to belong (disregarding the explanandum property), or if it is a wider class but the probability of the explanandum property in all more specific classes to which the object of the explanation is known to belong, are known to be equal to that in the wider class.

Let me exemplify with our Texan philosopher, Jones. Suppose we know that he is a Texan and a philosopher, and suppose we know nothing other than the logical consequences of these facts. Then Hempel's principle says that we can explain his financial condition by referring him to the most specific class to which he is known to belong, that of Texan philosophers. And the principle adds that we could also refer him to the class of Texans, or to that of philosophers, provided that the probability of richness in the given class were equal to that in the class of Texan philosophers.

No doubt, you will have recognized in this description of the maximal specificity principle an epistemologized version of the demand to refer to all relevant aspects of the explanandum. By relativizing to knowledge that 'ontic' demand, the requirement becomes tractable, due to the fact that knowledge, or at any rate the instances of it that we are familiar with, is notable by how few traits of reality it captures. In effect, then, it is ignorance, rather than knowledge, that makes the maximal specificity principle look like a workable demand. It is because we do not know that Jones was born while that Chinese mandarin sneezed that we can confidently explain his financial condition by referring him to the class of Texan philosophers. Somewhat annoyingly, an increase in knowledge could leave us with no explanation at all. Even worse, if that denseness assumption of relevant classes which we seem to see behind Hempel's rejection of the demand for total relevant aspects were true, then as our knowledge increases, the principle becomes unusable. In the limit, God would find no inductive explanations relative to his knowledge situation: He knows too much.

Now, the question I would like to put to Hempel is the following. Take any I-S explanation relative to K, for some given K. It will be a sequence of formulas like (2). Assume, if you will, that the class K describes our knowledge situation. Now what is there about this inductive argument that makes it an explanation of its last formula? What reason could anyone have to say that it is an explanation of its conclusion?

It is not difficult to answer this question when we pose it, not for the inductive, but for the deductive case. If one asks, for example, what reason do we have to believe that a causal deductive explanation explains its explanandum, the answer is that its premises identify certain features of the world that are nomically responsible for the occurrence of the explanandum event.

Could we say, as in the deductive case, that I-S explanations relative to K explain because their premises somehow identify features of the world that are nomically responsible for the explanandum event? Certainly not. This is what we vaguely conceived to be possible while tacitly espousing the naïve model, until Hempel shattered our illusions to pieces by focusing the reference-class problem on the theory of explanation. Indeed, if there is no characterization of true inductive explanation, then it must be because there are no things which go on in the non-epistemic world of facts that can inductively explain the event. For if there were such non-epistemic goings-on, their characterization would be a characterization of true inductive explanation. Thus, the possibility of a notion of true explanation, inductive or otherwise, is not just a desirable but ultimately indispensable feature of a model of explanation: it is the *sine qua non* of its realistic, non-psychologistic inspiration. It is because certain features of the world can be deterministically responsible for others that we can describe a concept of true deductive explanation by simply describing the form of such features. If there are features of the world which can be non-deterministically responsible for others, then we should be able to define a model of true inductive explanation. And, conversely, if we could define a model of true inductive explanation, there could be features of the world non-deterministically responsible for others. The thesis of epistemic relativity implies that, for Hempel, there are no such features, What, then, is the interest of I-S explanations relative to K? Surely not, as we seen above, that in knowledge situation K we have reason to believe that they are inductive explanations. Then what? We detect in Hempel's writings not even a hint as to what an answer to this question might be.

As it often happens when someone finds obviously absurd what someone else finds too obvious for defence, it is likely that we are approaching here a zone in which some of my deepest philosophical prejudices interfere with some of Hempel's deepest philosophical principles. Although I believe there is no way in which, consistently with his assumptions,

Hempel can satisfy the urge to know why his I-S explanations relative to K are supposed to explain, there is a way in which he could overcome my criticism: by simply not taking it as such. He could agree, that is, that no answer can be given to my question, and he could then add that this is precisely the way it should be, for my question is, he might say, a pseudoquestion, the product of a serious philosophical confusion. I-S explanations relative to K are those arguments which, in all likelihood, educated people in knowledge situation K would take to be inductive explanations. Thus, one might rejoin that my question amounts to a challenge to the right of the scientific community to decide what things are explanations; a challenge raised from a suprascientific standpoint. But, the rejoinder would go on, there is no such standpoint since there are no standards to decide what is an explanation over and above the standards set by the scientific community.

Does it really make no sense to raise the question of whether those arguments that the scientific community identifies as, believes to be or calls explanations, really are explanations? Does agreement at this (or any other) level imply truth? It is obvious that space (and other) limitations do not allow me to even start to deal seriously with this very deep question. But I would like to sketch very briefly the reasons I have to suspect it deserves a negative answer.

It is said that Lincoln once raised the question 'If we call a dog's tail a leg, how many legs does it have?', to which he immediately answered 'Four, because calling the tail a leg does not make it a leg.' If we call an inductive argument 'inductive explanation', does that make it an inductive explanation?

Some people tend to believe that when the calling is universal, or almost universal, for users of a language, this quantitative change somehow becomes qualitative and calling, then, does imply being. For them, what the ordinary man says will be philosophically important because what the ordinary man calls things is what everyone, or almost everyone, calls them; hence (it is claimed) what they ought to be called in the given language; hence, what they are.

Other people tend to apply a more élitist version of the same argument: when the calling is universal, or almost universal in a community or by specialists, then, once again, the quantitative change becomes qualitative and the fact that something is generally called an X in that community is sufficient ground to accept that this is what such things ought to be called; hence, what they are.

In so far as an argument can be detected for these two views, it would seem that they rely on a certain hypothesis concerning the way in which words receive their meaning. It is sometimes claimed that it would be an impossibility of some kind that there were no material objects because of the way in which the expression 'material object' used in that claim acquires its meaning; presumably, by ostension of objects that are supposed definitionally to qualify as entities to which the expression 'material object' is meant to apply (or in the presence of circumstances in which as a matter of definition, it is correct to apply the expression). Similarly, it is said, there must be things that we know, instances of knowledge, because of the way in which the word 'knowledge' is taught—which would be, at least partly, by the ostension of some paradigmatic cases of what is to count as knowledge.

One problem with this line of argument is that the meaning of expressions is not always given exclusively by ostension or by the application of the expression to actually given circumstances. At least sometimes, one must appeal to abstract conditions for the correct application of the term. We learn (and teach) the meaning of 'knowledge', for example, not only by learning (and teaching) to recognize alleged case of knowledge, but also by relating to that expression additional conceptual information, as, e.g., the information that a sentence is not a case of knowledge if it is not true (if a strong sense of 'knowledge' is meant) or well-confirmed on the given evidence (if a weaker sense is intended).

Now, whenever alternative partial characterizations of the meaning of an expression are available, the question of their compatibility arises. If different instructions for the identification of instances of a concept are not compatible, it is not admissible to rely only on the verdicts of one, unless it has been made clear in the process of meaning ascription that such an instruction is to have overriding force.

It follows that the above-mentioned paradigm case argument can only be applied to expressions that receive their meaning exclusively or with overriding force via ostension. But for many philosophically important notions, meanings are assigned via the introduction of non-ostensive characterizations in the form of conditions to be satisfied by instances of the notion. Furthermore, it is often the case that such non-ostensional conditions tend to have overriding force, since we are taught to correct our judgements whenever they are not satisfied. We are taught to correct a knowledge claim, i.e. to say that what we called knowledge was not really such, when we find out that the statement claimed to be known was not true (in the strong sense) or was poorly confirmed on the available evidence (now in the weak sense). Hence, from the fact that we are taught (or teach) to call certain objects instances of knowledge, it does not follow that there is anything that we know.

Now, I would argue that the sense or senses of 'explanation' that are of interest in the philosophy of science do not receive their meaning exclusively or primarily by ostension but also by the specification of conditions that instances of such concepts ought to verify. Thus, from the fact that ordinary men or ordinary scientists are willing to call certain arguments 'explanations' in that sense, it will not follow that the things so-called happen to be explanations in that sense; nor even that there are any such explanations.

Thus, I remain convinced of the legitimacy of the question I have posed, and I conclude that the lack of an answer to it counts heavily against the significance of Hempel's theory of inductive explanation. I have just argued that Hempel's epistemic relativization is, if possible, worth avoiding. My search for the Hempelian assumptions that led to this thesis may now be seen as the search for the 'causes' of this intellectual malady. Maybe by removing some of them, we can also remove their 'effects'. In the few remaining paragraphs I would like to give a very rough idea of what I conceive to be an appropriate theory of inductive explanation.

5. SPECULATION

I side with Hempel's somewhat tacit belief that a good theory of inductive explanation ought to begin with an analysis of the reference-class problem. Thus, I accept the contention that, at present, the problem of defining inductive explanation is, essentially, the problem of identifying an appropriate requirement on admissible reference classes. I further agree with the demand (implicit in Hempel, explicit in Salmon[3]) that such requirements ought to be an explication of the demand to identify all *relevant* aspects of the explanandum. But I want to question Hempel's conclusion that this requirement is undefinable outside the limited frame-work provided by a human knowledge situation. Such conclusion seemed to rely upon the assumption that the only way to determine the relevance of a predicate is by determining its actual frequentist correlation with the explanandum predicate. But one might try to characterize the 'relevance' in those relevant aspects in a different way: not as statistical relevance but as nomic relevance, a predicate being nomically relevant to another when a law of nature determines that changes in the first one generate changes in the second one.

[3] Particularly in Salmon (1971).

Perhaps I could make my plea for nomic relevance more appealing if I could explain why I feel that statistical relevance is neither necessary nor sufficient to determine whether a reference class is the 'appropriate' one.

Consider first the famous bellic episode described by Scriven in a somewhat different context: an atom bomb falls over a bridge and the bridge is destroyed; but it so happens that the cause of the destruction is not the explosion of the atom bomb for, a fraction of a second before the atom bomb explodes, 1,000 kg of dynamite is detonated on the main span of the bridge, causing its collapse. It would now seem that a deductive, indeed, causal explanation can be offered for the fact that the bridge is destroyed at a certain appropriately chosen time t, by appealing to the nomic premiss that whenever an atom bomb explodes over a bridge at time t' (shortly before t), at time t the bridge is destroyed. In effect, we would be attempting to explain our explanandum deductively by referring the bridge to the class of bridges on which atom bombs fall a fraction of a second before their destruction. But it seems quite natural to contend that this alleged explanation is no explanation at all. The right explanation is provided by referring the bridge in question to the class of bridges on which 1,000 kg of dynamite is detonated. And this is so even though this reference class provides a statistically irrelevant partition for the property 'being destroyed a few seconds later' in the reference class defined in terms of the atom-bomb explosion. Notice that the reason we give for this choice is that the dynamite explosion was the nomically operant feature in the envisaged situation.

It is also possible to argue that statistical relevance is not sufficient in order that a reference class be preferred over another one. Thus, we should not care whether Jones was born while our Chinese mandarin sneezed because, even though it may well be that the class of persons born while a Chinese mandarin sneezes is statistically relevant to richness in the class of Texan philosophers, there is, in all likelihood, no natural law that correlates Texan philosophers born in such circumstances with the amount of money they possess. The only kind of property that could 'screen-off' that of being a Texan philosopher as inappropriate to explain Jones's financial condition is a property true of Jones (at the appropriate time) and such that a law of nature determines the nomic relevance of its conjunction with the screened-off property to Jones's financial condition.

Thus, when nomicity and statistical relevance enter into conflict, it seems clear that nomicity always wins the day. They seldom enter into conflict, for statistical relevance is the evidence that we may have for the presence of nomicity. But a model of explanation, i.e. an account of what explanations are—rather than an account of what counts as evidence for their presence—should contain a reference to the explanatory features rather than to the symptoms of their presence. In this way, one may be led to speculate that to explain a single event is to refer the object of the explanation to its most specific property relative to the explanandum outcome. A property P will be the most specific property of an object relative to an outcome property Q whenever there is no other property R instantiated by the object of the explanation (during the appropriate time interval) such that the property $P \& R$ is nomically related to Q. There is some reason to suspect that the obscure notion of nomic relation can be clarified in a non-*ad-hoc*ish way, consistently with the above speculation. But this is a subject for some other occasion.

Let me conclude with a summary of what I have attempted to do. I have first tried to explain the nature of what I take to be a most remarkable and unexpected development of Hempel's theory of explanation, his thesis of the epistemic relativity of inductive explanation. I have tried to understand Hempel's reasons for this claim, which seem to stem from his discovery of the problem of ambiguity. But, as we saw, Hempel's identification of the nature of this problem was, itself, ambiguous. I have argued that the real difficulty is just the old reference-class problem in a new guise. Then I contended that to accept Hempel's thesis is, in effect, to deny the existence of inductive explanations. If such consequence is to be avoided, some of Hempel's assumptions ought to be rejected. In agreement with my contention that Hempel's problem is the reference-class problem, I held that one must revise his requirement of maximal specificity, reformulating it in ontic, rather than in epistemic terms. It is not obvious that this can be done; but it can be done if sense can be made of a certain appeal to nomicity related to that contemporary *inintelligibile* the propensity interpretation of probability. I cannot possibly expect to have made this last claim plausible; but I would be satisfied if I had convinced you that there is a problem, maybe an interesting problem, where you thought there was none.

REFERENCES

Hempel, Carl G., 'Deductive Nomological *vs.* Statistical Explanation', in H. Feigl and G. Maxwell (eds.), *Minnesota Studies in the Philosophy of Science* (Minneapolis: University of Minnesota Press, 1962), 98–169.
—— *Aspects of Scientific Explanation* (New York: The Free Press, 1965).

—— 'Maximal Specificity and Lawlikeness in Probabilistic Explanation', *Philosophy of Science*, 35 (1968), 116–34.

Salmon, Wesley C., 'Statistical Explanation', in Salmon (ed.), *Statistical Explanation and Statistical Relevance* (Pittsburgh: University of Pittsburgh Press, 1971), 29–88.

III

SCIENTIFIC EXPLANATION AND THE CAUSAL STRUCTURE OF THE WORLD

WESLEY C. SALMON

THREE BASIC CONCEPTIONS

Laplace attributed our ability to explain comets to our knowledge of the laws of nature. Twentieth-century philosophers have echoed that view by maintaining that, *with the aid of suitable initial conditions, an event is explained by subsuming it under one or more laws of nature.* If these laws are regarded as deterministic, this formulation becomes hardly more than a translation into more up-to-date and less colourful terminology of Laplace's famous statement:

Given for one instant an intelligence which could comprehend all of the forces by which nature is animated and the respective situation of the beings who compose it—an intelligence sufficiently vast to submit all these data to analysis—it would embrace in the same formula the movements of the greatest bodies of the universe and those of the lightest atom; for it, nothing would be uncertain and the future, as the past, would be present to its eyes (1951: 4).

Such an intelligence would exemplify the highest degree of scientific knowledge; it would, on Laplace's view, be able to provide a complete scientific explanation of any occurrence whatsoever.

There are, it seems to me, at least three distinct ways in which such Laplacian explanations can be construed. In order to relate them to the modern context, we will need to introduce a bit of technical terminology. It is customary, nowadays, to refer to the event-to-be-explained as the *explanandum-event*, and to the statement that such an event has occurred as the *explanandum-statement*. Those facts—both particular and general—that are invoked to provide the explanation are known as the *explanans*. If we want to refer specifically to statements that express such facts, we may speak of the *explanans-statements*. The explanans and the explanandum taken together constitute the explanation. Let us now look at the three conceptions.

1. Epistemic Conception

Suppose that we attempt to explain some occurrence, such as the appearance of a particular comet at a particular place and time. By citing certain laws, together with suitable initial conditions, we can deduce the occurrence of the event-to-be-explained. By employing observational data collected when his comet appeared in 1682, Halley predicted its return in 1759.[1] These data, along with the laws employed in the deduction, subsequently provided an explanation of that appearance of the comet. This explanation could be described as *an argument to the effect that the event-to-be-explained was to be expected by virtue of the explanatory facts*. The key to this sort of explanation is *nomic expectability*. An event that is quite unexpected in the absence of knowledge of the explanatory facts is rendered expectable on the basis of lawful connections with other facts. Nomic expectability as thus characterized is clearly an epistemological concept. On this view, we can say that there is a relation of *logical necessity* between the laws and the initial conditions on the one hand, and the event-to-be-explained on the other—though it would be more accurate to say that the relation of logical necessity holds between the explanans-statements and the explanandum-statement.

2. Modal Conception

Under the same circumstances we can say, alternatively, that because of the lawful relations between the antecedent conditions and the event-to-be-explained there is a relation of *nomological necessity* between them. In Laplace's *Essay*, the discussion of determinism is introduced by the following remarks:

All events, even those which on account of their insignificance do not seem to follow the great laws of nature, are a result of it just as necessarily as the revolutions of the sun. In ignorance of the ties which unite such events to the entire system of the universe, they have been made to depend upon final causes or upon hazard . . . but these imaginary causes have gradually receded with the widening bounds of knowledge and disappear entirely before sound philosophy, which sees in them only the expression of our ignorance of the true causes (1951: 3).

---

[1] Actually Halley did not make a very precise prediction, for he did not take account of the perturbations in the orbit due to Jupiter and Saturn. This was done by Clairaut; see (Laplace, 1951: 6).

Nomological necessity, it might be said, derives from the laws of nature in much the same way as logical necessity rests upon the laws of logic. *In the absence of knowledge of the explanatory facts, the explanandum-event* (the appearance of the comet) was something that *might not have occurred for all we know; given the explanatory facts it had to occur.* The explanation exhibits the nomological necessity of the fact-to-be-explained, given the explanatory facts. Viewing the matter in this way, one need not maintain that an explanation is an argument showing that the explanandum-event had to occur, given the initial conditions. Although a deductive argument can be constructed (as in the epistemic account) within which a relation of logical entailment obtains, an explanation need not be regarded as such an argument, or as any kind of argument at all. In comparing the epistemic and modal conceptions, it is important to be clear on the roles of the two kinds of necessity. In the epistemic conception, the relation of *logical* necessity obtains between the entire explanans and the explanandum by virtue of the laws of deductive logic. In the modal conception, the relation of *physical* necessity holds between particular antecedent conditions and the explanandum-event by virtue of the general laws, which we are taking to be part of the explanans.[2]

3. Ontic Conception

There is still another way of looking at Laplacian explanations. If the universe is, in fact, deterministic, then nature is governed by strict laws that *constitute* natural regularities. Law-statements describe these regularities. Such regularities endow the world with patterns that can be discovered by scientific investigation, and that can be exploited for purposes of scientific explanation. To explain an event—to relate the event-to-be-explained to some antecedent conditions by means of laws—is to fit the explanandum-event into a discernible pattern. This view seems to

---

[2] The contrast being suggested is well illustrated by a controversy between Hempel and Scriven concerning the role of laws in scientific explanation. As we have seen, Hempel (1965) insists that general laws be present in the explanans. Scriven (though he is not a proponent of the modal conception) argues that a set of particular antecedent conditions may constitute an adequate explanation of a particular event; consequently, the explanans need not include reference to any general laws. A law that provides a connection between the explanans and the explanandum constitutes a 'role-justifying ground' for the explanation by showing, roughly speaking, that the explanans is explanatorily relevant to the explanandum. For Hempel, the laws of logic—which provide the relation of relevance of the explanans to the explanandum—are not part of the explanation, but can be called upon to justify the claim that a given explanans has explanatory force with respect to some explanandum. Scriven invokes similar considerations to argue that general laws of nature should remain outside of scientific explanations to be called upon, if necessary, to support the claim that a given explanation is adequate. See Scriven (1959) for details; Hempel's reply is given in his (1965: 359–64).

be present in Laplace's thought, for he remarks that comets 'seemed to oppose the order of nature' before we knew how to explain them, but that subsequent 'knowledge of the laws of the system of the world' provided understanding of them (1951: 5). Moreover, as noted previously, he speaks of 'the ties which unite such events to the entire system of the universe' (1951: 3). Because of the universal (non-statistical) character of the laws involved in Laplacian explanations, we can also say that given certain portions of the pattern of events, and given the lawful relations to which the constituents of the patterns conform, other portions of the pattern of events must have certain characteristics. Looking at explanation in this way, we might say that *to explain an event is to exhibit it as occupying its* (nomologically necessary) *place in the discernible patterns of the world.*

These three general conceptions of scientific explanation all seem to go back at least to Aristotle. We have already remarked on his identification of certain sorts of syllogisms as explantions; this conforms to the epistemic conception that regards explanations as deductive arguments.[3] He seems to be expressing the modal conception when he remarks that 'the proper object of unqualified scientific knowledge is something which cannot be other than it is' (*Posterior Analytics*, 1. 2. 71^{b}14–16). And in the same context, discussing the nature of the syllogism that yields 'scientific knowledge', he says, 'The premises must be the causes of the conclusion, better known than it, and prior to it; its causes, since we possess scientific knowledge of a thing only when we know its cause; prior, in order to be causes; antecedently known, this antecedent knowledge being not our mere understanding of the meaning, but knowledge of the fact as well' (ibid. 71^{b}29–33). These remarks suggest an ontic conception.

In the twentieth century, we still find the same three notions figuring prominently in philosophical discussions of scientific explanation. The *epistemic conception* represents the currently 'received view', which has been advocated by such influential philosophers as Braithwaite, Hempel, Nagel, and Popper. It was succinctly formulated by Hempel in the follow-

---

[3] In the *Posterior Analytics* (1928, 1. 2. 71^{b}18–24), Aristotle writes: 'By demonstration I mean a syllogism productive of scientific knowledge, a syllogism, that is, the grasp of which is *eo ipso* such knowledge. Assuming that my thesis as to the nature of scientific knowledge is correct, the premises of demonstrated knowledge must be true, primary, immediate, better known than and prior to the conclusion, which is further related to them as effect to cause. Unless these conditions are satisfied, the basic truths will not be 'appropriate' to the conclusion. Syllogism there may indeed be without these conditions, but such syllogism, not being productive of scientific knowledge, will not be demonstration.' Jeffrey (1969) offers an illuminating comparison between this Aristotelian view and Hempel's D-N account.

ing way: '[An] explanatory account may be regarded as an argument to
the effect that the event to be explained . . . was to be expected by reason
of certain explanatory facts. These may be divided into two groups: (i)
particular facts and (ii) uniformities expressed by general laws' (1962*a*:
10).

The *modal conception* has been clearly affirmed by D. H. Mellor: 'The
thesis is that we call for explanation only of what, although we know it
is so, might have been otherwise for all else of some suitable sort we
know' (1976: 234). In what does an explanation consist?

We want to know why what might not have happened nonetheless did. Causal
explanation closes the gap by deducing what happened from known earlier events
and deterministic laws. So in this respect it satisfies the demand for explanation:
what follows from what is true must also be true. Given the causal explanans,
things *could not have happened otherwise* than the explanandum says (1976: 235,
italics added).

G. H. von Wright gives concise expression to this same conception:
'What makes a deductive-nomological explanation "explain", is, one
might say, that it tells us why *E had* to be (occur), why *E* was *necessary*
once the basis [body of explanatory facts] is there and the laws are
accepted' (1971: 13, italics in original). This same view can be found
explicitly in C. S. Peirce (1932: ii. 776).

The *ontic conception* is the one for which I shall be arguing. In Salmon
(1977: 162), I offered the following characterization: 'To give scientific
explanations is to show how events . . . fit into the causal structure of the
world.' Hempel summarizes the import of his major monographic essay,
'Aspects of Scientific Explanation' (1965*a*: 488), in rather similar terms:
'The central theme of this essay has been, briefly, that all scientific
explanation involves, explicitly or by implication, a subsumption of its
subject matter under general regularities; that it seeks to provide a systematic understanding of empirical phenomena by showing that they fit
into a nomic nexus.' I find this statement by Hempel in almost complete
accord with the viewpoint I shall be advocating; my suggestion for
modification would be to substitute the words '*how* they fit into a *causal*
nexus' for '*that* they fit into a *nomic* nexus'. It seems to me that Hempel
began the 'Aspects' article with statements clearly indicating that he
embraced the epistemic conception, but he ended with a summary that
seems closer to the ontic conception. Because these three conceptions
had not been explicitly formulated and distinguished at the time of his
writing, he was, I think, unaware of any conflict. As we shall see in
subsequent chapters, there are profound differences, especially in the
context of statistical explanation.

Those philosophers who have adopted the ontic conception of scientific explanation have generally regarded the pattern into which events are to be fit as a causal pattern. This feature of the view is brought out explicitly in the quotation from Aristotle. It certainly was present in Laplace's mind when he wrote, 'Present events are connected with preceding ones by a tie based upon the evident principle that a thing cannot occur without a cause which produces it' (1951: 3), and 'We ought then to regard the present state of the universe as the effect of its anterior state and as the cause of the one which is to follow' (1951: 4). It was also explicit in my formulation quoted previously. Hempel, however, does not share this notion; for him the pattern is lawful (nomic), but the laws involved need not be causal laws (1965: 352–4). In view of well-known Humean problems associated with causality, it might *seem* desirable to try to avoid reference to causal laws in dealing with scientific explanation. Nevertheless, I shall try to show that we need not purge the causal notions; indeed, I shall argue that they are required for an adequate theory of scientific explanation. In order to implement the causal version of the ontic conception, however, it will be necessary to examine the nature of causal relations with considerable care, and to show how they can be employed unobjectionably in a theory of scientific explanation.[4]

The foregoing three ways of thinking about scientific explanation may *seem* more or less equivalent—with somewhat distinct emphases perhaps—but hardly more than different verbal formulations. This is true as long as we are talking about the kind of explanation that involves appeal to universal laws only. A striking divergence will appear, however, when we consider explanations that invoke statistical (or probabilistic) laws. In the deterministic framework of Laplace's thought, all the fundamental laws of nature are taken to be strictly universal; any appeal to probabilities is merely a reflection of human ignorance. In twentieth-century science, the situation is quite different. There is a strong presumption in contemporary physics that some of the basic laws of nature may be irreducibly statistical—that probability relations *may* constitute a fundamental feature of the physical world. There is, to be sure, some disagreement as to whether determinism is true or false—whether modern physics requires an indeterministic interpretation. I do not want to prejudge this issue. In the attempt to elaborate a philosophical theory of scientific explanation, it seems to me, we must try to construct one that will be viable in either case. Therefore, we must leave open the possibility that some scientific explanations will be unavoidably statistical.

[4] See Salmon (1984: chs. 5–7).

This means that we must pay careful attention to the nature of statistical explanation.

AN OUTLINE OF STRATEGY

Much of the contemporary literature on scientific explanation arises directly or indirectly in response to the classic 1948 Hempel–Oppenheim paper, 'Studies in the Logic of Explanation'.[5] In it the authors attempt to provide a precise explication of what has come to be known as the deductive-nomological or D-N model of scientific explanation. They did not invent this mode of scientific explanation, nor were they the first philosophers to attempt to characterize it. As mentioned previously, its roots go back at least to Aristotle, and it is strongly suggested in such works as Arnauld's *The Art of Thinking* (*Port-Royal Logic*) and Laplace's *Philosophical Essay on Probabilities*. In none of the anticipations by these or other authors, however, do we find the precision and detail of the Hempel–Oppenheim account. One might almost say that 1948 marks the division between the prehistory and the history of the philosophical study of scientific explanation. When other such influential philosophers as R. B. Braithwaite (1953), Ernest Nagel (1961), and Karl R. Popper (1935, 1959) espoused a similar account of deductive explanation, it achieved virtually the status of a 'received view'.[6]

According to the 'received view', particular facts are explained by subsuming them under general laws, while general regularities are explained by subsumption under still broader laws. If a particular fact is successfully subsumed under a lawful generalization, it is, on this view, completely explained. One can legitimately ask for an explanation of the general law that figures in the explanation, but an explanation of the general law would be a different and additional explanation, not an essential part of the original explanation of the particular fact. For example, to explain why this particular penny conducts electricity, it suffices to point out that it is composed of copper and that all copper objects conduct electricity. If we are asked to explain why copper con-

[5] See Rescher (1970) for an extensive bibliography on scientific explanation up to the date of its publication.

[6] Although Popper's *Logik der Forschung* (1935) contains an important anticipation of the D-N model, it does not provide as precise an analysis as was embodied in Hempel and Oppenheim (1948). Moreover, Popper's views on scientific explanation were not widely influential until the English translation (Popper, 1959) of his 1935 book appeared. It is for these reasons that I chose 1948, rather than 1935, as the critical point of division between the history and the prehistory of the subject.

ducts electricity, we may give a further *distinct* explanation in terms of the fact that copper is a metal with conduction electrons that are not tightly bound to individual atoms and are free to move when an electric potential is applied.

Most proponents of this subsumption theory maintain that some events can be explained statistically by subsumption under statistical laws in much the same way that other events—such as the fact that the penny just inserted behind the fuse conducts electricity—are explained by appeal to universal laws. Thus we can explain the fact that a particular window was broken by pointing out that it was struck by a flying baseball, even though not all, but only most, windows so struck will shatter.

Although I disagreed from the beginning with the proponents of the *standard* subsumption view about the nature of the relation of subsumption of particular facts under universal or statistical generalizations, I did for some time accept the notion that *suitable* subsumption under generalizations is sufficient to explain particular facts. In *Statistical Explanation and Statistical Relevance* (Salmon *et al.*, 1971), I tried to give a detailed account of what seemed to me the appropriate way to subsume facts under general laws for purposes of explanation. This effort led to the elaboration of the statistical-relevance (S-R) model of scientific explanation. As the name suggests, statistical-relevance relations play a key role in this model of scientific explanation.

Subsequent reflection has convinced me that subsumption of the foregoing sort is only part—not all—of what is involved in the explanation of particular facts. It now seems to me that explanation is a two-tiered affair. At the most basic level, it is necessary, for purposes of explanation, to subsume the event-to-be-explained under an appropriate set of statistical-relevance relations, much as was required under the S-R model. At the second level, it seems to me, the statistical relevance relations that are invoked at the first level must be explained in terms of *causal* relations. The explanation, on this view, is incomplete until the causal components of the second level have been provided. This constitutes a sharp divergence from the approach of Hempel, who explicitly rejects the demand for causal laws (1965: 352–4).

It would be advisable, I believe, to adopt an approach similar to one suggested by Wolfgang Stegmüller (1973: 345), who characterized the kind of subsumption under statistical-relevance relations provided by the S-R model as 'statistical analysis' rather than 'statistical explanation'. The latter term is reserved for the entity that comprises both the statistical-relevance level and the causal level as well. As Humphreys (1981, 1983) and Rogers (1981) persuasively argue, statistical analyses have important uses, but they fall short of providing genuine scientific understanding. To emphasize this point, I shall use the term *S-R basis* to refer to the statistical component of an explanation.[7]

STATISTICAL EXPLANATION AND ITS MODELS

The philosophical theory of scientific explanation first entered the twentieth century in 1962, for that was the year of publication of the earliest bona-fide attempt to provide a systematic account of statistical explanation in science.[8] Although the need for some sort of inductive or statistical form of explanation had been acknowledged earlier, Hempel's essay 'Deductive-Nomological *vs.* Statistical Explanation' (1962) contained the first sustained and detailed effort to provide a precise account of this mode of scientific explanation. Given the pervasiveness of statistics in virtually every branch of contemporary science, the late arrival of statistical explanation in philosophy of science is remarkable. Hempel's initial treatment of statistical explanation had various defects, some of which he attempted to rectify in his comprehensive essay 'Aspects of Scientific Explanation' (1965*a*). Nevertheless, the earlier article did show unmistakably that the construction of an adequate model for statistical explanation involves many complications and subtleties that may have been largely unanticipated. Hempel never held the view—expressed by some of the more avid devotees of the D-N model—that *all* adequate scientific explanations must conform to the deductive-nomological pattern. The 1948 Hempel–Oppenheim paper explicitly notes the need for an inductive or statistical model of scientific explanation in order to account for some types of legitimate explanation that actually occur in the various sciences (Hempel, 1965: 250–1). The task of carrying out the construction was, however, left for another occasion. Similar passing remarks about the need for inductive or statistical accounts were made by other authors as well, but the project was not undertaken in earnest until 1962—a striking delay of fourteen years after the 1948 essay.

[7] In relinquishing the thesis that the S-R model provides an adequate characterization of scientific explanation, I accept as valid most of the criticisms levelled against it by Achinstein (1983). These criticisms do not, however, undermine the utility of the S-R basis as a foundation for scientific explanations.

[8] Ilkka Niiniluoto (1981: 444) suggests that 'Peirce should be regarded as the true founder of the theory of inductive-probabilistic explanation' on account of this statement, 'The statistical syllogism may be conveniently termed the explanatory syllogism' (Peirce, 1932: ii. 716). I am inclined to disagree, for one isolated and unelaborated statement of that sort can hardly be considered even the beginnings of any geniune theory.

One can easily form the impression that philosophers had genuine feelings of ambivalence about statistical explanation. A vivid example can be found in Carnap's *Philosophical Foundations of Physics* (1966), which was based upon a seminar he offered at UCLA in 1958.[9] Early in the first chapter, he says:

The general schema involved in *all explanation* can be expressed symbolically as follows:

1. $(x)(Px \supset Qx)$
2. Pa
3. Qa

The first statement is the universal law that applies to any object x. The second statement asserts that a particular object a has the property P. These two statements taken together enable us to derive logically the third statement: object a has the property Q (1966; 7–8, italics added).

After a single intervening paragraph, he continues: 'At times, in giving an explanation, the only *known* laws that apply are statistical rather than universal. In such cases, we must be content with a statistical explanation' (1966: 8, italics added). Further down on the same page, he assures us that 'these are genuine explanations', and on the next page he points out that 'In quantum theory . . . we meet with statistical laws that may not be the result of ignorance; they may express the basic structure of the world.' I must confess to a reaction of astonishment at being told that all explanations are deductive-nomological, but that some are not, because they are statistical. This lapse was removed from the subsequent paperback edition (Carnap, 1974), which appeared under a new title.

Why did it take philosophers so long to get around to providing a serious treatment of statistical explanation? It certainly was not due to any absence of statistical explanations in science. In antiquity, Lucretius (1951: 66–8) had based his entire cosmology upon explanations involving spontaneous swerving of atoms, and some of his explanations of more restricted phenomena can readily be interpreted as statistical. He asks, for example, why it is that Roman housewives frequently become pregnant after sexual intercourse, while Roman prostitutes to a large extent avoid doing so. Conception occurs, he explains, as a result of a collision between a male seed and a female seed. During intercourse the prostitutes wiggle their hips a great deal, but wives tend to remain

---

[9] As Carnap reports in the preface, the seminar proceedings were recorded and transcribed by his wife. Martin Gardner edited—it would probably be more accurate to say 'wrote up'—the proceedings and submitted them to Carnap, who rewrote them extensively. There is little doubt that Carnap saw and approved the passages I have quoted.

passive; as everyone knows, it is much harder to hit a moving target (1951: 170).[10] In the medieval period, St Thomas Aquinas asserted:

The majority of men follow their passions, which are movements of the sensitive appetite, in which movements of heavenly bodies can cooperate: but few are wise enough to resist these passions. Consequently astrologers are able to foretell the truth in the majority of cases, especially in a general way. But not in particular cases; for nothing prevents man resisting his passions by his free will. (1947: 1: Qu. 115, a. 4, *ad* Obj. 3)

Astrological explanations are, therefore, of the statistical variety. Leibniz, who like Lucretius and Aquinas was concerned about human free will, spoke of causes that incline but do not necessitate (1951: 515; 1965: 136).

When, in the latter half of the nineteenth century, the kinetic-molecular theory of gases emerged, giving rise to classical statistical mechanics, statistical explanations became firmly entrenched in physics. In this context, it turns out, many phenomena that *for all practical purposes* appear amenable to strict D-N explanation—such as the melting of an ice cube placed in tepid water—must be admitted *strictly speaking* to be explained statistically in terms of probabilities almost indistinguishable from unity. On a smaller scale, Brownian motion involves probabilities that are, both theoretically and practically, definitely less than one. Moreover, two areas of nineteenth-century biology, Darwinian evolution and Mendelian genetics, provide explanations that are basically statistical. In addition, nineteenth-century social scientists approached such topics as suicide, crime, and intelligence by means of 'moral statistics' (Hilts: 1973).

In the present century, statistical techniques are used in virtually every branch of science, and we may well suppose that most of these disciplines, if not all, offer statistical explanations of some of the phenomena they treat. The most dramatic example is the statistical interpretation of the equations of quantum mechanics, provided by Max Born and Wolfgang Pauli in 1926–7; with the aid of this interpretation, quantum theory explains an impressive range of physical facts.[11] What is even more important is that this interpretation brings in statistical considerations at the most basic level. In nineteenth-century science, the use of probability reflected limitations of human knowledge; in quantum mechanics, it

---

[10] Lucretius writes: 'A woman makes conception more difficult by offering a mock resistance and accepting Venus with a wriggling body. She diverts the furrow from the straight course of the ploughshare and makes the seed fall wide of the plot. These tricks are employed by prostitutes for their own ends, so that they may not conceive *too frequently* and be laid up by pregnancy' (1951: 170, italics added).

[11] See Wessels (1982), for an illuminating discussion of the history of the statistical interpretation of quantum mechanics.

looks as if probability may be an ineluctable feature of the physical world. The Nobel laureate physicist Leon Cooper expresses the idea in graphic terms: 'Like a mountain range that divides a continent, feeding water to one side or the other, the probability concept is the divide that separates quantum theory from all of physics that preceded it' (1968: 492). Yet it was not until 1962 that any philosopher published a serious attempt at characterizing a statistical pattern of scientific explanation.

Inductive-Statistical Explanation

When it became respectable for empirically minded philosophers to admit that science not only describes and predicts, but also explains, it was natural enough that primary attention should have been directed to classic and beautiful examples of deductive explanation. Once the D-N model had been elaborated, either of two opposing attitudes might have been taken toward inductive or statistical explanation by those who recognized the legitimacy of explanations of this general sort. It might have been felt, on the one hand, that the construction of such a model would be a simple exercise in setting out an analogue to the D-N model or in relaxing the stringent requirements for D-N explanation in some straightforward way. It might have been felt, on the other hand, that the problems in constructing an appropriate inductive or statistical model were so formidable that one simply did not want to undertake the task. Some philosophers may unreflectingly have adopted the former attitude; the latter, it turns out, is closer to the mark.

We should have suspected as much. If D-N explanations are deductive arguments, inductive or statistical explanations are, presumably, inductive arguments. This is precisely the tack Hempel took in constructing his inductive-statistical or I-S model. In providing a D-N explanation of the fact that this penny conducts electricity, one offers an explanans consisting of two premises: the particular premiss that this penny is composed of copper, and the universal law-statement that all copper conducts electricity. The explanandum-statement follows deductively. To provide an I-S explanation of the fact that I was tired when I arrived in Melbourne for a visit in 1978, it could be pointed out that I had been travelling by air for more than twenty-four hours (including stop-overs at airports), and almost everyone who travels by air for twenty-four hours or more becomes fatigued. The explanandum gets strong inductive support from those premises; the event-to-be-explained is thus subsumed under a statistical generalization.

It has long been known that there are deep and striking disanalogies between inductive and deductive logic.[12] Deductive entailment is transitive; strong inductive support is not. Contraposition is valid for deductive entailments; it does not hold for high probabilities. These are *not* relations that hold in some approximate way if the probabilities involved are high enough; once we abandon strict logical entailment, and turn to probability or inductive support, they break down entirely. But much more crucially, as Hempel brought out clearly in his 1962 essay, the deductive principle that permits the addition of an arbitrary term to the antecedent of an entailment does not carry over at all into inductive logic. If A entails B, then $A \cdot C$ entails B, whatever C may happen to stand for. However, no matter how high the probability of B given A, there is no constraint whatever upon the probability of B given both A and C. To take an extreme case, the probability of a prime number being odd is one, but the probability that a prime number smaller than 3 is odd has the value zero. For those who feel uneasy about applying probability to cases of this arithmetical sort, we can readily supply empirical examples. A 30-year-old Australian with an advanced case of lung cancer has a low probability of surviving for five more years, even though the probability of surviving to age 35 for 30-year-old Australians in general is quite high. It is *this* basic disanalogy between deductive and inductive (or probabilistic) relations that gives rise to what Hempel called *the ambiguity of inductive-statistical explanation*—a phenomenon that, as he emphasized, has no counterpart in D-N explanation. His *requirement of maximal specificity* was designed expressly to cope with the problem of this ambiguity.

Hempel illustrates the ambiguity of I-S explanation, and the need for the requirement of maximal specificity, by means of the following example (1965: 394–6). John Jones recovers quickly from a streptococcus infection, and when we ask why, we are told that he was given penicillin, and that almost all strep infections clear up quickly after penicillin is administered. The recovery is thus rendered probable relative to these explanatory facts. There are, however, certain strains of streptococcus bacteria that are resistant to penicillin. If, in addition to the above facts, we were told that the infection is of the penicillin-resistant type, then we would have to say that the prompt recovery is rendered *improbable* relative to the available information. It would clearly be scientifically unacceptable to ignore such relevant evidence as the penicillin-resistant character of the infection; the requirement of maximal

[12] These are spelled out in detail in Salmon (1965a). See Salmon (1967: 109–11) for a discussion of the 'almost-deduction' conception of inductive inference.

specificity is designed to block statistical explanations that thus omit relevant facts. It says, in effect, that when the class to which the individual case is referred for explanatory purposes—in this instance, the class of strep infections treated by penicillin—is chosen, we must not know how to divide it into subsets in which the probability of the fact to be explained differs from its probability in the entire class. If it has been ascertained that this particular case involved the penicillin-resistant strain, then the original explanation of the rapid recovery would violate the requirement of maximal specificity, and for that reason would be judged unsatisfactory.[13]

Hempel conceived of D-N explanations as valid deductive arguments satisfying certain additional conditions. Explanations that conform to his inductive-statistical or I-S model are correct inductive arguments also satisfying certain additional restrictions. Explanations of both sorts can be characterized in terms of the following four conditions:

1. The explanation is an argument with correct (deductive or inductive) logical form,
2. At least one of the premises must be a (universal or statistical) law,
3. The premises must be true, and
4. The explanation must satisfy the requirement of maximal specificity.

This fourth condition is automatically satisfied by D-N explanations by virtue of the fact that their explanatory laws are universal generalizations. If all A are B, then obviously there is no subset of A in which the probability of B is other than one. This condition has crucial importance with respect to explanations of the I-S variety. In general, according to Hempel (1962a: 10), an explanation is an argument (satisfying these four conditions) to the effect that the event-to-be-explained was to be expected by virtue of certain explanatory facts. In the case of I-S explanations, this means that the premiss must lend high inductive probability to the conclusion—that is, the explanandum must be highly probable with respect to the explanans.

Explanations of the D-N and I-S varieties can therefore be schematized as follows (Hempel, 1965: 336, 382):

(D-N) $C_1, C_2, \ldots, C_j$ (particular explanatory conditions)

$\underline{L_1, L_2, \ldots, L_k}$ (general laws)

    E (fact-to-be-explained)

[13] The requirement of maximal specificity, as formulated by Hempel in his (1965) and revised in his (1968), does not actually do the job. Nevertheless, this was clearly its intent. (See Salmon (1984: ch. 3).)

The single line separating the premises from the conclusion signifies that the argument is deductively valid.

(I-S) $C_1, C_2, \ldots, C_j$ (particular explanatory conditions)

$$\underline{\underline{L_1, L_2, \ldots, L_k}} \;\; [r] \quad \text{(general laws, at least one statistical)}$$

E (fact-to-be-explained)

The double lines separating the premises from the conclusion signifies that the argument is inductively correct, and the number r expresses the degree of inductive probability with which the premises support the conclusion. It is presumed that r is fairly close to one.[14]

The high-probability requirement, which seems such a natural analogue of the deductive entailment relation, leads to difficulties in two ways. First, there are arguments that fulfil all the requirements imposed by the I-S model, but that patently do not constitute satisfactory scientific explanations. One can maintain, for example, that people who have colds will probably get over them within a fortnight if they take vitamin C, but the use of vitamin C may not explain the recovery, since almost all colds clear up within two weeks regardless. In arguing for the use of vitamin C in the prevention and treatment of colds, Linus Pauling (1970) does not base his claims upon the high probability of avoidance or quick recovery; instead, he urges that massive doses of vitamin C have a bearing upon the probability of avoidance or recovery—that is, the use of vitamin C is relevant to the occurrence, duration, and severity of colds. A *high* probability of recovery, given use of vitamin C, does not confer explanatory value upon the use of this drug with respect to recovery. An *enhanced* probability value does indicate that the use of vitamin C may have some explanatory force. This example, along with a host of others which, like it, fulfil all of Hempel's requirements for a correct I-S explanation, shows that fulfilling these requirements does not constitute a sufficient condition for an adequate statistical explanation.

At first blush, it might seem that the type of relevance problem illustrated by the foregoing example was peculiar to the I-S model, but Henry Kyburg (1965) showed that examples can be found which demonstrate that the D-N model is infected with precisely the same difficulty. Consider a sample of table salt that dissolves upon being placed in water. We ask why it dissolves. Suppose, Kyburg suggests, that someone has

---

[14] It should be mentioned in passing that Hempel (1965: 380–1) offers still another model of scientific explanation that he characterizes as deductive-statistical (D-S). In an explanation of this type, a statistical regularity is explained by deducing it from other statistical laws. There is no real need, however, to treat such explanations as a distinct type, for they fall under the D-N schema just given, provided we allow that at least one of the laws may be statistical. In the present context, we are concerned only with statistical explanations of non-deductive sorts.

cast a dissolving spell upon it—that is, someone wearing a funny hat waves a wand over it and says, 'I hereby cast a dissolving spell upon you.' We can then 'explain' the phenomenon by mentioning the dissolving spell—without for a moment believing that any actual magic has been accomplished—and by invoking the true universal generalization that all samples of table salt that have been hexed in this manner dissolve when placed in water. Again, an argument that satisfies all the requirements of Hempel's model patently fails to qualify as a satisfactory scientific explanation because of a failure of relevance. Given Hempel's characterizations of his D-N and I-S models of explanation, it is easy to construct any number of 'explanations' of either type that invoke some irrelevancy as a purported explanatory fact.[15] This result casts serious doubt upon the entire epistemic conception of scientific explanation, as outlined at the beginning of the chapter, in so far as it takes all explanations to be arguments of one sort or another.

The diagnosis of the difficulty can be stated very simply. Hempel's requirement of maximal specificity (RMS) guarantees that *all* known relevant facts must be included in an adequate scientific explanation, but there is no requirement to insure that *only* relevant facts will be included. The foregoing examples bear witness to the need for some requirement of this latter sort. To the best of my knowledge, the advocates of the 'received view' have not, until recently, put forth any such additional condition, nor have they come to terms with counter-examples of these types in any other way.[16] James Fetzer's *requirement of strict maximal specificity*, which rules out the use in explanations of laws that mention nomically irrelevant properties (Fetzer, 1981: 125–6), seems to do the job. In fact, in Salmon (1979: 691–4), I showed how Reichenbach's theory of nomological statements could be used to accomplish the same end.

The second problem that arises out of the high-probability requirement is illustrated by an example furnished by Michael Scriven (1959: 480). If someone contracts paresis, the straightforward explanation is that he was infected with syphilis, which had progressed through the primary, secondary, and latent stages without treatment with penicillin. Paresis is one form of tertiary syphilis, and it never occurs except in syphilitics.

---

[15] Many examples are presented and analysed in Salmon *et al.* (1971: 33–40). Nancy Cartwright (1983: 26–7) errs when she attributes to Hempel the requirement that a statistical explanation increase the probability of the explanandum; this thesis, which I first advanced in Salmon (1965), was never advocated by Hempel. Shortly thereafter (1983: 28–9), she provides a correct characterization of the relationships among my views and those of Hempel and Suppes.

[16] In Hempel's most recent discussion of statistical explanation, he appears to maintain the astonishing view that although such examples have *psychologically* misleading features, they do qualify as *logically* satisfactory explanations (1977: 107–11).

Yet far less than half of those victims of untreated latent syphilis ever develop paresis. Untreated latent syphilis is the explanation of paresis, but it does not provide any basis on which to say that the explanandum-event was to be expected by virtue of these explanatory facts. Given a victim of latent untreated syphilis, the odds are that he will *not* develop paresis. Many other examples can be found to illustrate the same point. As I understand it, mushroom poisoning may afflict only a small percentage of individuals who eat a particular type of mushroom (Smith, 1958: Introduction), but the eating of the mushroom would unhesitatingly be offered as the explanation in instances of the illness in question. The point is illustrated by remarks on certain species in a guide for mushroom hunters (Smith, 1958: 34, 185):

Helvella infula: Poisonous to some, but edible for most people. Not recommended.
Chlorophyllum molybdites: Poisonous to some but not to others. Those who are not made ill by it consider it a fine mushroom. The others suffer acutely.

These examples show that high probability does not constitute a necessary condition for legitimate statistical explanations. Taking them together with the vitamin C example, we must conclude—provisionally, at least—that a high probability of the explanandum relative to the explanans is neither necessary nor sufficient for correct statistical explanations, even if all Hempel's other conditions are fulfilled. Much more remains to be said about the high-probability requirement, for it raises a host of fundamental philosophical problems.[17]

Given the problematic status of the high-probability requirement, it was natural to attempt to construct an alternative treatment of statistical explanation that rests upon different principles. As I argued in Salmon (1965), statistical relevance, rather than high probability, seems to be the key explanatory relationship. This starting point leads to a conception of scientific explanation that differs fundamentally and radically from Hempel's I-S account. In the first place, if we are to make use of statistical relevance relations, our explanations will have to make reference to at least two probabilities, for statistical relevance involves a difference between two probabilities. More precisely, a factor C is statistically relevant to the occurrence of B under circumstances A if and only if

(1) $$P(B \mid A \cdot C) \neq P(B \mid A)$$

or

(2) $$P(B \mid A \cdot C) \neq P(B \mid A \cdot \overline{C})$$

---

[17] See Salmon (1984: ch. 4) for further discussion.

Conditions (1) and (2) are equivalent to one another, provided that C occurs with a non-vanishing probability within A; since we shall not be concerned with the relevance of factors the probabilities of which are zero, we may use either (1) or (2) as our definition of statistical relevance. We say that C is positively relevant to B if the probability of B is greater in the presence of C; it is negatively relevant if the probability of B is smaller in the presence of C. For instance, heavy cigarette smoking is positively relevant to the occurrence of lung cancer, at some later time, in a 30-year-old Australian male; it is negatively relevant to survival to the age of 70 for such a person.

In order to construct a satisfactory statistical explanation, it seems to me, we need a *prior probability* of the occurrence to be explained, as well as one or more *posterior probabilities*. A crucial feature of the explanation will be the comparison between the prior and posterior probabilities. In Hempel's case of the streptococcus infection, for instance, we might begin with the probability, in the entire class of people with streptococcus infections, of a quick recovery. We realize, however, that the administration of penicillin is statistically relevant to quick recovery, so we compare the probability of quick recovery among those who have received penicillin with the probability of quick recovery among those who have not received penicillin. Hempel warns, however, that there is another relevant factor, namely, the existence of the penicillin-resistant strain of bacteria. We must, therefore, take that factor into account as well. Our original reference class has been divided into four parts: (1) infection by non-penicillin-resistant bacteria, penicillin given; (2) infection by non-penicillin-resistant bacteria, no penicillin given; (3) infection by penicillin-resistant bacteria, penicillin given; (4) infection by penicillin-resistant bacteria, no penicillin given. Since the administration of penicillin is irrelevant to quick recovery in case of penicillin-resistant infections, the subclasses (3) and (4) of the original reference class should be merged to yield (3') infection by penicillin-resistant bacteria. If John Jones is a member of (1), we have an explanation of his quick recovery, according to the S-R approach, not because the probability is high, but, rather, because it differs significantly from the probability in the original reference class. We shall see later what must be done if John Jones happens to fall into class (3').

By way of contrast, Hempel's earlier high-probability requirement demands only that the posterior probability be sufficiently large—whatever that might mean—but makes no reference at all to any prior probability. According to Hempel's abstract model, we ask, 'Why is individual x a member of B?' The answer consists of an inductive argument having the following form:

$$P(B \mid A) = r$$

$$\frac{x \text{ is an } A}{x \text{ is a } B} \quad [r]$$

As we have seen, even if the first premiss is a statistical law, r is high, the premisses are true, and the requirement of maximal specificity has been fulfilled, our 'explanation' may be patently inadequate, due to failure of relevancy.

In Salmon (1970: 220–1), I advocated what came to be called the statistical-relevance or S-R model of scientific explanation. At that time, I thought that anything that satisfied the conditions that define that model would qualify as a legitimate scientific explanation. I no longer hold that view. It now seems to me that the statistical relationships specified in the S-R model constitute the *statistical basis* for a bona-fide scientific explanation, but that this basis must be supplemented by certain *causal factors* in order to constitute a satisfactory scientific explanation.[18] In this chapter, however, I shall confine attention to the statistical basis, as articulated in terms of the S-R model. Indeed, from here on I shall speak, not of the S-R model, but, rather, of the *S-R basis*.[19]

Adopting the S-R approach, we begin with an explanatory question in a form somewhat different from that given by Hempel. Instead of asking, for instance, 'Why did x get well within a fortnight?' we ask, 'Why did this person with a cold get well within a fortnight?' Instead of asking, 'Why is x a B?' we ask, 'Why is x, which is an A, also a B?' The answer—at least for preliminary purposes—is that x is also a C, where C is *relevant* to B within A. Thus we have a prior probability $P(B \mid A)$—in this case, the probability that a person with a cold (A) gets well within a fortnight (B). Then we let C stand for the taking of vitamin C. We are interested in the posterior probability $P(B \mid A \cdot C)$ that a person with a cold who takes vitamin C recovers within a fortnight. If the prior and posterior probabilities are equal to one another, the taking of vitamin C can play no role in explaining why this person recovered from the cold

---

[18] See Salmon (1984: chs. 5–9).

[19] I am extremely sympathetic to the thesis, expounded in Humphreys (1983), that probabilities—including those appearing in the S-R basis—are important tools in the construction of scientific explanations, but that they do not constitute any part of a scientific explanation *per se*. This thesis allows him to relax considerably the kinds of maximal specificity or homogeneity requirements that must be satisfied by statistical or probabilistic explanations. A factor that is statistically relevant may be causally irrelevant because, for example, it does not convert any contributing causes to counteracting causes or vice versa. This kind of relaxation is attractive in a theory of scientific explanation, for factors having small statistical relevance often seem otiose. Humphreys' approach does not show, however, that such relevance relations can be omitted from the S-R basis; on the contrary, the S-R basis must include such factors in order that we may ascertain whether they can be omitted from the causal explanation or not.

within the specified period of time. If the posterior probability is not equal to the prior probability, then *C* may, under certain circumstances, furnish part or all of the desired explanation. A large part of the purpose of the present book is to investigate the way in which considerations that are statistically relevant to a given occurrence have or lack explanatory import.

We cannot, of course, expect that every request for a scientific explanation will be phrased in canonical form. Someone might ask, for example, 'Why did Mary Jones get well in no more than a fortnight's time?' It might be clear from the context that she was suffering from a cold, so that the question could be reformulated as, 'Why did this person who was suffering from a cold get well within a fortnight?' In some cases, it might be necessary to seek additional clarification from the person requesting the explanation, but presumably it will be possible to discover what explanation is being called for. This point about the form of the explanation-seeking question has fundamental importance. We can easily imagine circumstances in which an altogether different explanation is sought by means of the same initial question. Perhaps Mary had exhibited symptoms strongly suggesting that she had mononucleosis; in this case, the fact that it was only an ordinary cold might constitute the explanation of her quick recovery. A given why-question, constructed in one way, may elicit an explanation, while otherwise construed, it asks for an explanation that cannot be given. 'Why did the Mayor contract paresis?' might mean, 'Why did this adult human develop paresis?' or, 'Why did this syphilitic develop paresis?' On the first construal, the question has a suitable answer, which we have already discussed. On the second construal, it has no answer—at any rate, we cannot give an answer—for we do not know of any fact in addition to syphilis that is relevant to the occurrence of paresis. Some philosophers have argued, because of these considerations, that scientific explanation has an unavoidably pragmatic aspect (e.g. van Fraassen, 1977, 1980). If this means simply that there are cases in which people ask for explanations in unclear or ambiguous terms, so that we cannot tell what explanation is being requested without further clarification, then so be it. No one would deny that we cannot be expected to supply explanations unless we know what it is we are being asked to explain. To this extent, scientific explanation surely has pragmatic or contextual components. Dealing with these considerations is, I believe, tantamount to choosing a suitable reference class with respect to which the prior probabilities are to be taken and specifying an appropriate sample space for purposes of a particular explanation. More will be said about these two items in the next section.[20]

---

[20] For an extended discussion of van Fraassen's theory, see Salmon (1984: ch. 4).

The Statistical-Relevance Approach

Let us now turn to the task of giving a detailed elaboration of the S-R basis. For purposes of initial presentation, let us construe the terms *A*, *B*, *C*, ... (with or without subscripts) as referring to classes, and let us construe our probabilities in some sense as relative frequencies. This *does not mean* that the statistical-relevance approach is tied in any crucial way to a frequency theory of probability. I am simply adopting the heuristic device of picking examples involving frequencies because they are easily grasped. Those who prefer propensities, for example, can easily make the appropriate terminological adjustments, by speaking of chance set-ups and outcomes of trials where I refer to reference classes and attributes. With this understanding in mind, let us consider the steps involved in constructing an S-R basis for a scientific explanation:

1. We begin by selecting an appropriate reference class *A* with respect to which the prior probabilities $P(B_i \mid A)$ of the B_is are to be taken.

2. We impose an *explanandum-partition* upon the initial reference class *A* in terms of an exclusive and exhaustive set of attributes $B_1, \ldots, B_m$; this defines a sample space for purposes of the explanation under consideration. (This partition was not required in earlier presentations of the S-R model.)

3. Invoking a set of statistically relevant factors $C_1, \ldots, C_s$, we partition the initial reference class *A* into a set of mutually exclusive and exhaustive cells $A \cdot C_1, \ldots, A \cdot C_s$. The properties $C_1, \ldots, C_s$ furnish the *explanans-partition*.

4. We ascertain the associated probability relations:
   prior probabilities
   $$P(B_i \mid A) = p_i$$
   $$\text{for all } i \ (1 \leqslant i \leqslant m)$$
   posterior probabilities
   $$P(B_i \mid A \cdot C_j) = p_{ij}$$
   $$\text{for all } i \text{ and } j \ (1 \leqslant i \leqslant m) \text{ and } (1 \leqslant j \leqslant s)$$

5. We require that each of the cells $A \cdot C_j$ be homogeneous with respect to the explanandum-partition $\{B_i\}$; that is, none of the cells in the partition can be further subdivided in any manner relevant to the occurrence of any B_i. (This requirement is somewhat analogous to Hempel's requirement of maximal specificity, but as we shall see, it is a much stronger condition.)

6. We ascertain the relative sizes of the cells in our explanans-partition in terms of the following marginal probabilities:

$$P(C_j \mid A) = q_j$$

(These probabilities were not included in earlier versions of the S-R model; the reasons for requiring them now will be discussed later in this chapter.)

7. We require that the explanans-partition be a maximal homogeneous partition, that is—with an important exception to be noted later—for $i \neq k$ we require that $p_{ji} \neq p_{jk}$. (This requirement assures us that our partition in terms of $C_1, \ldots, C_m$ does not introduce any irrelevant subdivision in the initial reference class A.)

8. We determine which cell $A \cdot C_j$ contains the individual x whose possession of the attribute B_i was to be explained. The probability of B_i within the cell is given in the list under 4.

Consider in a rather rough and informal manner the way in which the foregoing pattern of explanation might be applied in a concrete situation; an example of this sort was offered by James Greeno (1971: 89–90). Suppose that Albert has committed a delinquent act—say, stealing a car, a major crime—and we ask for an explanation of that fact. We ascertain from the context that he is an American teenager, and so we ask, 'Why did this American teenager commit a serious delinquent act?' The prior probabilities, which we take as our point of departure, so to speak, are simply the probabilities of the various degrees of juvenile delinquency (B_i) among American teenagers (A)—that is, $P(B_i \mid A)$. We will need a suitable explanandum-partition; Greeno suggests B_1 = no criminal convictions, B_2 = conviction for minor infractions only, B_3 = conviction for a major offence. Our sociological theories tell us that such factors as sex, religious background, marital status of parents, type of residential community, socio-economic status, and several others are relevant to delinquent behaviour. We therefore take the initial reference class of American teenagers and divide it into males and females; Jewish, Protestant, Roman Catholic, no religion; parents married, parents divorced, parents never married; urban, suburban, rural place of residence; upper, middle, lower class; and so forth. Taking such considerations into account, we arrive at a large number s of cells in our partition. We assign probabilities of the various degrees of delinquent behaviour to each of the cells in accordance with 4, and we ascertain the probability of a randomly selected American teenager belonging to each of the cells in accordance with 6. We find the cell to which Albert belongs—for example, male, from a Protestant background, parents divorced, living in a suburban area, belonging to the middle class. If we have taken into account all of the relevant factors, and if we have correctly ascertained the probabilities associated with the various cells of our partitions, then we have an S-R basis for the explanation of Albert's delinquency that conforms to the foregoing schema. If it should turn out (contrary to what I believe actually to be the case) that the probabilities of the various types of delinquency are the same for males and for females, then we would not use sex in partitioning our original reference class. By condition 5 we must employ *every* relevant factor; by condition 7 we must employ *only* relevant factors. In many concrete situations, including the present examples, we know that we have not found all relevant considerations; however, as Noretta Koertge (1975) rightly emphasized, that is an ideal for which we may aim. Our philosophical analysis is designed to capture the notion of a fully satisfactory explanation.

Nothing has been said, so far, concerning the rationale for conditions 2 and 6, which are here added to the S-R basis for the first time. We must see why these requirements are needed. Condition 2 is quite straightforward; it amounts to a requirement that the sample space for the problem at hand be specified. Both the explanans-partition and the explanandum-partition are needed to measure the information transmitted in any explanatory scheme. This is a useful measure of the explanatory value of a theory. In addition, van Fraassen's contrast class, which is the same as our explanandum-partition,[21] is needed in some cases to specify precisely what explanation is being sought. In dealing with the question, 'Why did Albert steal a car?' we used Greeno's suggested explanandum-partition. If, however, we had used different partitions (contrast classes), other explanations might have been called forth. Suppose that the contrast class included: Albert steals a car, Bill steals a car, Charlie steals a car, and so forth. Then the answer might have involved no sociology whatever; the explanation might have been that, among the members of his gang, Albert is most adept at hot-wiring. Suppose, instead, that the contrast class had included: Albert steals a car, Albert steals a diamond ring, Albert steals a bottle of whisky, and so forth. In that case, the answer might have been that he wanted to go joyriding.

The need for the marginal probabilities mentioned in 6 arises in the following way. In many cases, such as the foregoing delinquency example, the terms C_j that furnish the explanans-partition of the initial reference class are conjunctive. A given cell is determined by several distinct factors: sex *and* religious background *and* marital status of par-

---

[21] For a discussion of Greeno's information-theoretic approach, and van Fraassen's treatment of why-questions, see Salmon (1984: ch. 4).

ents *and* type of residential community *and* socio-economic status *and* ... which may be designated D_k, E_n, F_r, ... These factors will be the probabilistic contributing causes and counteracting causes that tend, respectively, to produce or prevent delinquency. In attempting to understand the phenomenon in question, it is important to know how each factor is relevant—whether positively or negatively, and how strongly—both in the population at large and in various subgroups of the population. Consider, for example, the matter of sex. It may be that within the entire class of American teenagers (A) the probability of serious delinquency (B_3) is greater among males than it is among females. If so, we would want to know by how much the probability among males exceeds the probability among females and by how much it exceeds the probability in the entire population. We also want to know whether being male is always positively relevant to serious delinquency, or whether in combination with certain other factors, it may be negatively relevant or irrelevant. Given two groups of teenagers—one consisting entirely of boys and the other entirely of girls, but alike with respect to all of the other factors—we want to know how the probabilities associated with delinquency in each of the two groups are related to one another. It might be that in each case of two cells in the explanandum-partition that differ from one another only on the basis of gender, the probability of serious delinquency in the male group is greater than it is in the female group. It might turn out, however, that sometimes the two probabilities are equal, or that in some cases the probability is higher in the female group than it is in the corresponding male group. Relationships of all of these kinds are logically possible.

It is a rather obvious fact that each of two circumstances can individually be positively relevant to a given outcome, but their conjunction can be negatively relevant or irrelevant. Each of two drugs can be positively relevant to good health, but taken together, the combination may be detrimental—for example, certain antidepressive medications taken in conjunction with various remedies for the common cold can greatly increase the chance of dangerously high blood pressure (Goodwin and Guze, 1979). A factor that is a contributing cause in some circumstances can be a counteracting cause in other cases. Problems of this sort have been discussed, sometimes under the heading of 'Simpson's paradox', by Nancy Cartwright (1983: essay 1) and Bas van Fraassen (1980: 108, 148–51). In Salmon, 1975, I have spelled out in detail the complexities that arise in connection with statistical-relevance relations. The moral is that we need to know not only how the various factors D_k, E_n, F_r, ..., are relevant to the outcome, B_i, but now the relevance of each of them is affected by the presence or absence of the others. Thus, for instance, it is possible that being female might in general be negatively relevant to delinquency, but it might be positively relevant among the very poor.

Even if all of the prior probabilities $P(B_i \mid A)$ and all of the posterior probabilities $P(B_i \mid A \cdot C_j)$ furnished under condition 4 are known, it is not possible to deduce the conditional probabilities of the B_is with respect to the individual conjuncts that make up the C_js or with respect to combinations of them. Without these conditional probabilities, we will not be in a position to ascertain all of the statistical-relevance relations that are required. We therefore need to build in a way to extract that information. This is the function of the marginal probabilities $P(C_j \mid A)$ required by condition 6. If these are known, such conditional probabilities as $P(B_i \mid A \cdot D_k)$, $P(B_i \mid A \cdot E_n)$, and $P(B_i \mid A \cdot D_k \cdot E_n)$ can be derived.[22] When 2 and 6 are added to the earlier characterization of the S-R model (Salmon *et al.*, 1971), then, I believe, we have gone as far as possible in characterizing scientific explanations at the level of statistical-relevance relations.

Several features of the new version of the S-R basis deserve explicit mention. It should be noted, in the first place, that conditions 2 and 3 demand that the entire initial reference class A be partitioned, while

---

[22] Suppose e.g. that we wish to compute $P(B_i \mid A \cdot D_k)$, where $D_k = C_{j_1} \vee \ldots \vee C_{j_q}$, the cells C_{j_r} being mutually exclusive. This can be done as follows. We are given $P(C_j \mid A)$ and $P(B_i \mid A \cdot C_j)$. By the multiplication theorem,

$$P(D_k \cdot B_i \mid A) = P(D_k \mid A) \times P(B_i \mid A \cdot D_k)$$

Assuming $P(D_k \mid A) \neq 0$, we have,

$$* \qquad P(B_i \mid A \cdot D_k) = P(D_k \cdot B_i \mid A) / P(D_k \mid A)$$

By the addition theorem

$$P(D_k \mid A) = \sum_{r=1}^{q} P(C_{j_r} \mid A)$$

$$P(D_k \cdot B_i \mid A) = \sum_{r=1}^{q} P(C_{j_r} \cdot B_i \mid A)$$

By the multiplication theorem

$$P(D_k \cdot B_i \mid A) = \sum_{r=1}^{q} P(C_{j_r} \mid A) \times P(B_i \mid A \cdot C_{j_r})$$

Substitution in (*) yields the desired relation:

$$P(B_i \mid A \cdot D_k) = \frac{\sum_{r=1}^{q} P(C_{j_r} \mid A) \times P(B_i \mid A \cdot C_{j_r})}{\sum_{r=1}^{q} P(C_{j_r} \mid A)}$$

conditions 4 and 6 require that *all* the associated probability values be given. This is one of several respects in which it differs from Hempel's I-S model. Hempel requires only that the individual mentioned in the explanandum be placed within an appropriate class, satisfying his requirement of maximal specificity, but he does not ask for information about any class in either the explanandum-partition or the explanans-partition to which that individual does not belong. Thus he might go along in requiring that Bill Smith be referred to the class of American male teenagers coming from a Protestant background, whose parents are divorced, and who is a middle-class suburban dweller, and in asking us to furnish the probability of his degree of delinquency within that class. But why, it may surely be asked, should we be concerned with the probability of delinquency in a lower-class, urban-American, female teenager from a Roman Catholic background whose parents are still married? What bearing do such facts have on Bill Smith's delinquency? The answer, I think, involves serious issues concerning scientific generality. If we ask why this American teenager becomes a delinquent, then, it seems to me, we are concerned with *all* of the factors that are relevant to the occurrence of delinquency, and with the ways in which these factors are relevant to that phenomenon (cf. Koertge, 1975). To have a satisfactory scientific answer to the question of why this A is a B_i—to achieve full scientific understanding—we need to know the factors that are relevant to the occurrence of the various B_is for *any* randomly chosen or otherwise unspecified member of A. It was mainly to make good on this desideratum that requirement 6 was added. Moreover, as Greeno and I argued in *Statistical Explanation and Statistical Relevance*, a good measure of the value of an S-R basis is the gain in information furnished by the complete partitions and the associated probabilities. This measure cannot be applied to the individual cells one at a time.

A fundamental philosophical difference between our S-R basis and Hempel's I-S model lies in the interpretation of the concept of homogeneity that appears in condition 5. Hempel's requirement of maximal specificity, which is designed to achieve a certain kind of homogeneity in the reference classes employed in I-S explanations, is *epistemically relativized*. This means, in effect, that we must not *know* of any way to make a relevant partition, but it certainly does not demand that no possibility of a relevant partition can exist unbeknown to us. As I view the S-R basis, in contrast, condition 5 demands that the cells of our explanans-partition be *objectively* homogeneous; for this model, homogeneity is not epistemically relativized. Since this issue of epistemic relativization versus objective homogeneity is discussed at length in Salmon

(1984: ch. 3), it is sufficient for now merely to call attention to this complex problem.[23]

Condition 7 has been the source of considerable criticism. One such objection rests on the fact that the initial reference class A, to which the S-R basis is referred, may not be maximal. Regarding Kyburg's hexed salt example, mentioned previously, it has been pointed out that the class of samples of table salt is not a maximal homogeneous class with respect to solubility, for there are many other chemical substances that have the same probability—namely, unity—of dissolving when placed in water. Baking soda, potassium chloride, various sugars, and many other compounds have this property. Consequently, if we take the maximality condition seriously, it has been argued, we should not ask, 'Why does this sample of table salt dissolve in water?' but, rather, 'Why does this sample of matter in the solid state dissolve when placed in water?' And indeed, one can argue, as Koertge (1975) has done persuasively, that to follow such a policy often leads to significant scientific progress. Without denying her important point, I would nevertheless suggest, for purposes of elaborating the formal schema, that we take the initial reference class A as given by the explanation-seeking why-question, and look for relevant partitions within it. A significantly different explanation, which often undeniably represents scientific progress, may result if a different why-question, embodying a broader initial reference class, is posed. If the original question is not presented in a form that unambiguously determines a reference class A, we can reasonably discuss the advantages of choosing a wider or a narrower class in the case at hand.

Another difficulty with condition 7 arises if 'accidentally'—so to speak—two different cells in the partition, $A \cdot C_i$ and $A \cdot C_j$, happen to have equal associated probabilities p_{ki} and p_{kj} for all cells B_k in the explanandum-partition. Such a circumstance might arise if the cells are

---

[23] Cartwright (1983: 27) asserts that on Hempel's account, 'what counts as a good explanation is an objective, person-independent matter', and she applauds him for holding that view. I find it difficult to reconcile her characterization with Hempel's repeated emphatic assertion (prior to 1977) of the doctrine of essential epistemic relativization of inductive-statistical explanation. In addition, she complains that my way of dealing with problems concerning the proper formulation of the explanation-seeking why-question—that is, problems concerning the choice of an appropriate initial reference class—'makes explanation a subjective matter' (ibid. 29). 'What explains what', she continues, 'depends on the laws and facts true in our world, and cannot be adjusted by shifting our interest or our focus' (ibid.). This criticism seems to me to be mistaken. Clarification of the question is often required to determine what it is that is to be explained, and this may have pragmatic dimensions. However, once the explanandum has been unambiguously specified, on my view, the identification of the appropriate explanans is fully objective. I am in complete agreement with Cartwright concerning the desirability of such objectivity; moreover, my extensive concern with objective homogeneity is based directly upon the desire to eliminate from the theory of statistical explanation such subjective features as epistemic relativization.

determined conjunctively by a number of relevant factors, and if the differences between the two cells cancel one another out. It might happen, for example, that the probabilities of the various degrees of delinquency—major offence, minor offence, no offence—for an upper-class, urban, Jewish girl would be equal to those for a middle-class, rural, Protestant boy. In this case, we might want to relax condition 7, allowing $A \cdot C_i$ and $A \cdot C_j$ to stand as separate cells, provided they differ with respect to at least two of the terms in the conjunction, so that we are faced with a fortuitous cancelling of relevant factors. If, however, $A \cdot C_i$ and $A \cdot C_j$ differed with respect to only one conjunct, they would have to be merged into a single cell. Such would be the case if, for example, among upper-class, urban-dwelling, American teenagers whose religious background is atheistic and whose parents are divorced, the probability of delinquent behaviour were the same for boys as for girls. Indeed, we have already encountered this situation in connection with Hempel's example of the streptococcus infection. If the infection is of the penicillin-resistant variety, the probability of recovery in a given period of time is the same whether penicillin is administered or not. In such cases, we want to say, there is no relevant difference between the two classes—not that relevant factors were cancelling one another out. I bring this problem up for consideration at this point, but I shall not make a consequent modification in the formal characterization of the S-R basis, for I believe that problems of this sort are best handled in the light of causal-relevance relations. Indeed, it seems advisable to postpone detailed consideration of the whole matter of regarding the cells $A \cdot C_j$ as being determined conjunctively until causation has been explicitly introduced into the discussion. Humphreys (1981, 1983) and Rogers (1981) provide useful suggestions for handling just this issue.[24]

Perhaps the most serious objection to the S-R model of scientific explanation—as it was originally presented—is based upon the principle that *mere* statistical correlations explain nothing. A rapidly falling barometric reading is a sign of an imminent storm, and it is *highly correlated* with the onset of storms, but it certainly does not *explain* the occurrence of a storm. The S-R approach does, however, have a way of dealing with examples of this sort. A factor C, which is relevant to the occurrence of B in the presence of A, may be screened off in the presence of some additional factor D; the screening-off relation is defined by equations (3) and (4), which follow. To illustrate, given a series of days (A) in some particular locale, the probability of a storm occurring (B) is

---

[24] See Salmon (1984: ch. 9).

in general quite different from the probability of a storm if there has been a recent sharp drop in the barometric reading (C). Thus C is statistically relevant to B within A. If, however, we take into consideration the further fact that there is an actual drop in atmospheric pressure (D) in the region, then it is irrelevant whether that drop is registered on a barometer. In the presence of D and A, C becomes irrelevant to B; we say that D screens off C from B—in symbols,

(3) $P(B \mid A \cdot C \cdot D) = P(B \mid A \cdot D)$

However, C does not screen off D from B, that is,

(4) $P(B \mid A \cdot C \cdot D) \neq P(B \mid A \cdot C)$

for barometers sometimes malfunction, and it is the atmospheric pressure, not the reading on the barometer *per se*, that is directly relevant to the occurrence of the storm. A factor that has been screened off is irrelevant, and according to the definition of the S-R basis (condition 7), it is not to be included in the explanation. The falling barometer does not explain the storm.

Screening off is frequent enough and important enough to deserve further illustration. A study, reported in the news media a few years ago, revealed a positive correlation between coffee drinking and heart disease, but further investigation showed that this correlation results from a correlation between coffee drinking and cigarette smoking. It turned out that cigarette smoking screened off coffee drinking from heart disease, thus rendering coffee drinking statistically (as well as causally and explanatorily) irrelevant to heart disease. Returning to a previous example for another illustration, one could reasonably suppose that there is some correlation between low socio-economic status and paresis, for there may be a higher degree of sexual promiscuity, a higher incidence of venereal disease, and less likelihood of adequate medical attention if the disease occurs. But the contraction of syphilis screens off such factors as degree of promiscuity, and the fact of syphilis going untreated screens off any tendency to fail to get adequate medical care. Thus when an individual has latent untreated syphilis, all other such circumstances as low socio-economic status are screened off from the development of paresis.

As the foregoing examples show, there are situations in which one circumstance or occurrence is correlated with another because of an indirect causal relationship. In such cases, it often happens that the more proximate causal factors screen off those that are more remote. Thus 'mere correlations' are replaced in explanatory contexts with correlations that are intuitively recognized to have explanatory force. In *Statistical*

Explanation and Statistical Relevance, where the S-R model of statistical explanation was first explicitly named and articulated, I held out some hope (but did not try to defend the thesis) that all the causal factors that play any role in scientific explanation could be explicated in terms of statistical-relevance relations—with the screening-off relation playing a crucial role. As explained in Salmon (1984: ch. 6) I no longer believe this is possible. A large part of the material in the present book is devoted to an attempt to analyse the nature of the causal relations that enter into scientific explanations, and the manner in which they function in explanatory contexts. After characterizing the S-R model, I wrote:

> One might ask on what grounds we can claim to have characterized explanation. The answer is this. When an explanation (as herein explicated) has been provided, we know exactly how to regard any *A* with respect to the property *B*. We know which ones to bet on, which to bet against, and at what odds. We know precisely what degree of expectation is rational. We know how to face uncertainty about an *A*'s being a *B* in the most reasonable, practical, and efficient way. We know every factor that is relevant to an *A* having the property *B*. We know exactly what weight should have been attached to the prediction that this *A* will be a *B*. We know all of the regularities (universal and statistical) that are relevant to our original question. What more could one ask of an explanation? (Salmon *et al.*, 1971: 78)

The answer, of course, is that we need to know something about the causal relationships as well.

In acknowledging this egregious shortcoming of the S-R model of scientific explanation, I am not abandoning it completely. The attempt, rather, is to supplement it in suitable ways. While recognizing its incompleteness, I still think it constitutes a sound basis upon which to erect a more adequate account. And at a fundamental level, I still think it provides important insights into the nature of scientific explanation.

In the introduction to *Statistical Explanation and Statistical Relevance*, I offered the following succinct comparison between Hempel's I-S model and the S-R model:

I-S model: an explanation is an *argument* that renders the explanandum *highly probable*

S-R model: an explanation is an *assembly of facts statistically relevant* to the explanandum, *regardless of the degree of probability* that results

It was Richard Jeffrey (1969) who first explicitly formulated the thesis that (at least some) statistical explanations are not arguments; it is beautifully expressed in his brief paper, 'Statistical Explanation *vs.* Statistical Inference', which was reprinted in *Statistical Explanation and Statistical Relevance*. In Salmon (1965: 145–6), I had urged that positive relevance rather than high probability is the desideratum in statistical explanation. In Salmon (1970), I expressed the view, which many philosophers found weird and counter-intuitive (e.g. L. J. Cohen, 1975), that statistical explanations may even embody *negative* relevance—that is, an explanation of an event may, in some cases, show that the event to be explained is less probable than we had initially realized. I still do not regard that thesis as absurd. In an illuminating discussion of the explanatory force of positively and negatively relevant factors, Paul Humphreys (1981) has introduced some felicitous terminology for dealing with such cases, and he has pointed to an important constraint. Consider a simple example. Smith is stricken with a heart attack, and the doctor says, '*Despite* the fact that Smith exercised regularly and had given up smoking several years ago, he contracted heart disease *because* he was seriously over-weight.' The 'because' clause mentions those factors that are positively relevant and the 'despite' clause cites those that are negatively relevant. Humphreys refers to them as *contributing causes* and *counteracting causes*, respectively. In further discussions of causal explanation,[25] we will want to say that a complete explanation of an event must make mention of the causal factors that tend to prevent its occurrence as well as those that tend to bring it about. Thus it is *not* inappropriate for the S-R basis to include factors that are negatively relevant to the explanandum-event. As Humphreys points out, however, we would hardly consider as appropriate a putative explanation that had only negative items in the 'despite' clause and no positive items in the 'because' category. 'Despite the fact that Jones never smoked, exercised regularly, was not overweight, and did not have elevated levels of triglycerides and cholesterol, he died of a heart attack' would hardly be considered an acceptable *explanation* of his fatal illness.

Before concluding this chapter on models of statistical explanation, we should take a brief look at the deductive-nomological-probabilistic (D-N-P) model of scientific explanation offered by Peter Railton (1978). By employing well-established statistical laws, such as that covering the spontaneous radioactive decay of unstable nuclei, it is possible to deduce the fact that a decay-event for a particular isotope has a certain probability of occurring within a given time interval. For an atom of carbon 14 (which is used in radiocarbon dating in archaeology, for example), the probability of a decay in 5,730 years is 1/2. The explanation of *the probability of the decay-event* conforms to Hempel's deductive-nomological pattern. Such an explanation does not, however, explain the actual

---

[25] See Salmon (1984: ch. 9).

occurrence of a decay, for, given the probability of such an event—however high or low—the event in question may not even occur. Thus the explanation does not qualify as an argument to the effect that the event-to-be-explained was to be expected with deductive certainty, given the explanans. Railton is, of course, clearly aware of the fact. He goes on to point out, nevertheless, that if we simply attach an 'addendum' to the deductive argument stating that the event-to-be-explained did, in fact, occur in the case at hand, we can claim to have a probabilistic *account*—which is not a deductive or inductive argument—of the occurrence of the event. In this respect, Railton is in rather close agreement with Jeffrey (1969) that some explanations are not arguments. He also agrees with Jeffrey in emphasizing the importance of exhibiting the physical mechanisms that lead up to the probabilistic occurrence that is to be explained. Railton's theory—like that of Jeffrey—has some deep affinities to the S-R model. In including a reference to physical mechanisms as an essential part of his D-N-P model, however, Railton goes beyond the view that statistical-relevance relations, in and of themselves, have explanatory import. His theory of scientific explanation can be more appropriately characterized as causal or mechanistic. It is closely related to the two-tiered causal-statistical account that I am attempting to elaborate as an improvement upon the S-R model.

Although, with Kyburg's help, I have offered what seem to be damaging counter-examples to the D-N model—for instance, the one about the man who explains his own avoidance of pregnancy on the basis of his regular consumption of his wife's birth control pills (Salmon *et al.*, 1971: 34)—the major emphasis has been upon statistical explanation, and that continues to be the case in what follows. Aside from the fact that contemporary science obviously provides many statistical explanations of many types of phenomena, and that any philosophical theory of statistical explanation has only lately come forth, there is a further reason for focusing upon statistical explanation. As I maintained initially, we can identify three distinct approaches to scientific explanation that do not seem to differ from one another in any important way as long as we confine our attention to contexts in which all the explanatory laws are universal generalizations. I argue in Salmon (1984: ch. 4), however, that these three general conceptions of scientific explanation can be seen to differ radically from one another when we move on to situations in which statistical explanations are in principle the best we can achieve. Close consideration of statistical explanations, with sufficient attention to their causal ingredients, provides important insight into the underlying philosophical questions relating to our scientific understanding of the world.

REFERENCES

Achinstein, Peter (1983), *The Nature of Explanation* (New York: Oxford University Press).

Aquinas, St Thomas (1947), *Summa Theologica* (New York: Benziger Brothers).

Arnauld, Antoine (1964), *The Art of Thinking* (*Port-Royal Logic*) (Indianapolis: Bobbs-Merrill).

Braithwaite, R. B. (1953), *Scientific Explanation* (Cambridge: Cambridge University Press).

Carnap, Rudolf (1966), *Philosophical Foundations of Physics*, ed. Martin Gardner (New York: Basic Books).

—— (1974), *An Introduction to the Philosophy of Science*, reprint, with corrections, of Carnap (1966) (New York: Basic Books).

Cartwright, Nancy (1983), *How the Laws of Physics Lie* (New York: Oxford University Press).

Cohen, L. J. (1975), 'Comment', in Stephan Körner (ed.), *Explanation* (Oxford: Basil Blackwell), 152–9.

Fetzer, James H. (1981), *Scientific Knowledge* (Dordrecht: D. Reidel).

Goodwin, Donald W., and Guze, Samuel B. (1979), *Psychiatric Diagnosis*, 2nd edn. (Oxford: Oxford University Press).

Greeno, James G. (1971), 'Explanation and Information', in Salmon *et al.*, (1971), 89–104.

Hempel, Carl G. (1962), 'Deductive-Nomological *vs.* Statistical Explanation', in Herbert Feigl and Grover Maxwell (eds.), *Minnesota Studies in the Philosophy of Science* (Minneapolis: University of Minnesota Press), iii. 98–169.

—— (1962a), 'Explanation in Science and in History', in Robert G. Colodny (ed.), *Frontiers in Science and Philosophy* (Pittsburgh: University of Pittsburgh Press), 7–34.

—— (1965), *Aspects of Scientific Explanation and Other Essays in the Philosophy of Science* (New York: Free Press).

—— (1965a), 'Aspects of Scientific Explanation', in Hempel (1965), 331–496.

—— (1977), 'Nachwort 1976: Neuere Ideen zu den Problemen der statistischen Erklärung', in Hempel, *Aspekte wissenschaftlicher Erklärung* (Berlin/New York: Walter de Gruyter), 98–123.

Hempel, Carl G., and Oppenheim, Paul (1948), 'Studies in the Logic of Explanation', *Philosophy of Science*, 15, 135–75; reprinted, with added Postscript, in Hempel (1965).

Hilts, Victor (1973), 'Statistics and Social Science', in Ronald N. Giere and Richard S. Westfall (eds.), *Foundations of Scientific Method: The Nineteenth Century* (Bloomington, Ind.: Indiana University Press), 206–33.

Humphreys, Paul (1981), 'Aleatory Explanations', *Synthese*, 48, 225–32.

—— (1983), 'Aleatory Explanations Expanded', in Peter Asquith and Thomas Nickles (eds.), *PSA 1982* (East Lansing, Mich.: Philosophy of Science Association).

Jeffrey, Richard C. (1969), 'Statistical Explanation *vs.* Statistical Inference', in Nicholas Rescher (ed.), *Essays in Honor of Carl G. Hempel* (Dordrecht: D. Reidel). Reprinted in Salmon *et al.* (1971), 19–28.

Koertge, Noretta (1975), 'An Exploration of Salmon's S-R Model of Explanation', *Philosophy of Science*, 42, 270–4.

Kyburg, Henry E., jun. (1965), 'Comment', *Philosophy of Science*, 32, 147–51.

Laplace, P. S. (1951), *A Philosophical Essay on Probabilities* (New York: Dover Publications).

Leibniz, G. W. (1951), 'The Theodicy: Abridgement of the Argument Reduced to Syllogistic Form', in Philip P. Wiener (ed.), *Leibniz Selections* (New York: Charles Scribner's Sons), 509–22.

—— (1965), 'A Vindication of God's Justice Reconciled with His Other Perfections and All His Actions', in Paul Schrecker (ed.), *Monadology and Other Philosophical Essays* (Indianapolis: Bobbs-Merrill), 114–47.

Lucretius (1951), *On the Nature of the Universe* (Baltimore: Penguin Books).

Mellor, D. H. (1976), 'Probable Explanation', *Australasian Journal of Philosophy*, 54, 231–41.

Nagel, Ernest (1961), *The Structure of Science* (New York: Harcourt, Brace, and World).

Niiniluoto, Ilkka (1981), 'Statistical Explanation Reconsidered', *Synthese*, 48, 437–72.

Pauling, Linus (1970), *Vitamin C and the Common Cold* (San Francisco: W. H. Freeman).

Popper, Karl R. (1935), *Logik der Forschung* (Vienna: Springer).

—— (1959), *The Logic of Scientific Discovery* (New York: Basic Books), translation of Popper (1935).

Railton, Peter (1978), 'A Deductive-Nomological Model of Probabilistic Explanation', *Philosophy of Science*, 45, 206–26.

Rescher, Nicholas (1970), *Scientific Explanation* (New York: Free Press).

Rogers, Ben (1981), 'Probabilistic Causality, Explanation, and Detection', *Synthese*, 48, 201–23.

Salmon, Wesley C. (1965), 'The Status of Prior Probabilities in Statistical Explanation', *Philosophy of Science* 32, 137–46.

—— (1965a), 'Consistency, Transitivity, and Inductive Support', *Ratio*, 7, 164–9.

—— (1970), 'Statistical Explanation', in Robert G. Colodny (ed.), *The Nature and Function of Scientific Theories* (Pittsburgh: University of Pittsburgh Press), 173–231. Reprinted in Salmon *et al.* (1971).

—— (1975), 'Confirmation and Relevance', in Grover Maxwell and Robert M. Anderson, jun. (eds.), *Minnesota Studies in the Philosophy of Science* (Minneapolis: University of Minnesota Press), vi. 3–36.

—— (1977), 'A Third Dogma of Empiricism', in Robert Butts and Jaakko Hintikka (eds.), *Basic Problems in Methodology and Linguistics* (Dordrecht: D. Reidel), 149–66.

—— (1979), *Hans Reichenbach: Logical Empiricist* (Dordrecht: D. Reidel).

Salmon, Wesley C., with contributions by Jeffrey, Richard C., and Greeno, James G. (1971), *Statistical Explanation and Statistical Relevance* (Pittsburgh: University of Pittsburgh Press).

Scriven, Michael (1959), 'Explanation and Prediction in Evolutionary Theory', *Science* 130, 477–82.

Smith, A. H. (1958), *The Mushroom Hunter's Guide* (Ann Arbor: University of Michigan Press).

Stegmüller, Wolfgang (1973), *Problems und Resultate der Wissenschaftstheorie und Analytischen Philosophie* (Berlin/New York: Springer-Verlag), iv.

van Fraassen, Bas C. (1977), 'The Pragmatics of Explanation', *American Philosophical Quarterly*, 14, 143–50.

van Fraassen, Bas C. (1980), *The Scientific Image* (Oxford: Clarendon Press).

von Wright, Georg Henrik (1971), *Explanation and Understanding* (Ithaca, NY: Cornell University Press).

Wessels, Linda (1982), 'The Origins of Born's Statistical Interpretation', in P. Asquith and R. N. Giere (eds.), *PSA 1980* (East Lansing, Mich.: Philosophy of Science Association), ii. 187–200.

IV

TOWARDS AN ARISTOTELIAN THEORY OF SCIENTIFIC EXPLANATION

B. BRODY

In this paper, I consider a variety of objections against the covering-law model of scientific explanation, show that Aristotle was already aware of them and had solutions for them, and argue that these solutions are correct. These solutions involve the notions of non-Humean causality and of essential properties. There are a great many familiar objections, both methodological and epistemological, to introducing these concepts into the methodology of science, but I show that these objections are based upon misunderstandings of these concepts.

Let us begin by considering the following explanation of why it is that sodium normally combines with chlorine in a ratio of one-to-one:[1]

(A) 1. sodium normally combines with bromine in a ratio of one-to-one
    2. everything that normally combines with bromine in a ratio of one-to-one normally combines with chlorine in a ratio of one-to-one

---

    3. therefore, sodium normally combines with chlorine in a ratio of one-to-one

This purported explanation meets all of the requirements laid down by Hempel's covering law model for scientific explanation (1965: 248–9). After all, the law to be explained is deduced from two other general laws which are true and have empirical content. Nevertheless, this purported explanation seems to have absolutely no explanatory power. And even if one were to say, as I think it would be wrong to say, that it does have at least a little explanatory power, why is it that it is not as good an explanation of the law in question (that sodium normally combines with chlorine in a ratio of one-to-one) as the explanation of that law in terms of the atomic structure of sodium and chlorine and the theory of chemical

B. Brody, 'Towards an Aristotelian Theory of Scientific Explanation', *Philosophy of Science*, 39 (1972), 20–31. Reprinted by permission of the author and the Philosophy of Science Association.
[1] I first called attention to the problems raised by this type of explanation in Brody (1967).

bonding? The covering law model, as it stands, seems to offer us no answer to that question.

A defender of the covering-law model would, presumably, offer the following reply: both of these explanations, each of which meets the requirements of the model, are explanations of the law in question, but the explanation in terms of atomic structure is to be preferred to the explanation in terms of the way that sodium combines with bromine because the former contains in its explanans more powerful laws than the latter. The laws about atomic structure and the theory of bonding are more powerful than the law about the ratio with which sodium combines with bromine because more phenomena can be explained by the former than by the latter.

I find this answer highly unsatisfactory, partially because I do not see that (A) has any explanatory power at all. But that is not the real problem. The real problem is that this answer leaves something very mysterious. I can see why, on the grounds just mentioned, one would prefer to have laws like the ones about atomic structure rather than laws like the one about the ratio with which sodium and bromine combine. But why does that make explanations in terms of the latter type of laws less preferable? Or to put the question another way, why should laws that explain more explain better?

So much for my first problem for the covering-law model, a problem with its account of the way in which we explain scientific laws. I should now like to raise another problem for it, a problem with its account of the way in which we explain particular events. Consider the following three explanations:

(B) 1. If the temperature of a gas is constant, then its pressure is inversely proportional to its volume
   2. at time t_1, the volume of the container c was v_1 and the pressure of the gas in it was p_1
   3. the temperature of the gas in c did not change from t_1 to t_2
   4. the pressure of the gas in container c at t_2 is $2p_1$

   5. the volume of x at t_2 is $\frac{1}{2}v_1$

(C) 1. if the temperature of a gas is constant, then its pressure is inversely proportional to its volume
   2. at time t_1, the volume of the container c was v_1 and the pressure of the gas in it was p_1
   3. the temperature of the gas in c did not change from t_1 to t_2
   4. the volume of the gas in container c at t_2 is $\frac{1}{2}v_1$

   5. the pressure of c at t_2 is $2p_1$

(D) 1. if the temperature of a gas is constant, then its pressure is inversely proportional to its volume
2. at time t_1, the volume of the container c was v_1 and the pressure of the gas in it was p_1
3. the temperature of the gas in c did not change from t_1 to t_2
4. by t_2, I had so compressed the container by pushing on it from all sides that its volume was $\frac{1}{2}v_1$

---

5. the pressure of c at t_2 is $2p_1$

All three of these purported explanations meet all the requirements of Hempel's model. The explanandum, in each case, is deducible from the explanans which, in each case, contains at least one true general law with empirical content. And yet, there are important differences between the three. My intuitions are that (B) is no explanation at all (thereby providing us with a clear counter-example to Hempel's model), that (C) is a poor explanation, and that (D) is a much better one. But if your intuitions are that (B) is still an explanation, even if not a very good one, that makes no difference for now. The important point is that there is a clear difference between the explanatory power of these three explanations, and the covering-law model provides us with no clue as to what it is.

These problems and counter-examples are not isolated cases. I shall give, later on in this paper after I offer my own analysis and solution of them, a recipe for constructing loads of additional problems and counter-examples. Now the existence of these troublesome cases led me to suspect that there is something fundamentally wrong with the whole covering-law model and that a new approach to the understanding of scientific explanation is required. At the same time, however, I felt that this model, which fits so many cases and seems so reasonable, just could not be junked entirely. This left me in a serious dilemma, one that I only began to see my way out of after I realized that Aristotle, in the *Posterior Analytics*, had already seen these problems and had offered a solution to them, one that contained both elements of Hempel's model and some other elements entirely foreign to it. So let me begin my presentation of my solution to these problems by looking at some aspects of Aristotle's theory of scientific explanation.

Aristotle (ibid. 1. 13), wanted to draw a distinction between knowledge of the fact (knowledge that p is so) and knowledge of the reasoned fact (knowledge why p is so) and he did so by asking us to consider the following two arguments, the former of which only provides us with knowledge of the fact while the latter of which provides us with knowledge of the reasoned fact:

(E) 1. the planets do not twinkle
     2. all objects that do not twinkle are near the earth

     3. therefore, the planets are near the earth

(F) 1. the planets are near the earth
     2. all objects that are near the earth do not twinkle

     3. therefore, the planets do not twinkle

The interesting thing about this point, for our purposes, is that while both of these arguments fit Hempel's model,[2] only one of them, as Aristotle has already seen, provides us with an explanation of its conclusion. Moreover, his account of why this is so seems just right:

> of two reciprocally predicable terms the one which is not the cause may quite easily be the better known and so become the middle term of the demonstration. . . . This syllogism, then, proves not the reasoned fact but only the fact; since they are not near because they do not twinkle. The major and middle of the proof, however, may be reversed, and then the demonstration will be of the reasoned fact . . . since its middle term is the proximate cause (78^a28–78^b3).

In other words, nearness is the cause of not twinkling, and not vice versa, so the nearness of the planets to the earth explains why they do not twinkle, but their not twinkling does not explain why they are near the earth.

It is important to note that such an account is incompatible with the logical empiricists' theory of causality as constant conjunction. After all, given the truth of the premises of (E) and (F), nearness and non-twinkling are each necessary and sufficient for each other, so, on the constant conjunction account each is equally the cause of the other.[3] And even if the constant conjunction account is supplemented in any of the usual ways, nearness and non-twinkling would still equally be the cause of each other. After all, both 'all near celestial objects twinkle' and 'all twinkling celestial objects are near' contain purely qualitative predicates, have a potentially infinite scope, are deducible from higher-order scientific generalizations and support counterfactuals. In other words, both of these generalizations are lawlike generalizations, and not mere accidental ones, so each of the events in question is, on a sophisticated Humean account, the cause of the other. So Aristotle's account presupposes the falsity of

---

[2] Leaving aside the question, irrelevant for us now, about the truth of these premises; we shall, throughout this discussion, just assume that they are true.

[3] Unless one adds the requirement that the cause must be before the effect, the normal way of drawing an asymmetry between causes and effects when each are necessary and sufficient for the other, in which case neither is the cause of the other and Aristotle's account still will not do.

the constant conjunction account of causality. But that is okay. After all, the very example that we are dealing with now, where nearness is clearly the cause of non-twinkling but not vice versa, shows us that the constant conjunction theory of causality (even in its normal more sophisticated versions) is false.

Now if we apply Aristotle's account to our example with the gas, we get a satisfactory account of what is involved there. The decrease in volume (due, itself, to my pressing on the container from all sides) is the cause of the increase in pressure, but not vice versa, so the former explains the latter but not vice versa. And Aristotle's account also explains a phenomenon called to our attention by Bromberger (1969), namely, that while we can deduce both the height of a flag-pole from the length of the shadow it casts and the position of the sun in the sky and the length of the shadow it casts from the height of the flag-pole and the position of the sun in the sky, only the latter deduction can be used in an explanation. It is easy to see why this is so; it is the sun striking at a given angle the flag-pole of the given height that causes its shadow to have the length that it does, but the sun striking the flag-pole when its shadow has the length of the shadow is surely not the cause of the height of the flag-pole.

Generalizing this point, we can add a new requirement for explanation: a deductive-nomological explanation of a particular event is a satisfactory explanation of the event when (besides meeting all Hempel's requirements) its explanans contains essentially a description of the event which is the cause of the event described in the explanandum. If they do not, then it may not be a satisfactory explanation. And similarly, a deductive-nomological explanation of a law is a satisfactory explanation of that law when (besides meeting all of Hempel's requirements) every event which is a case of the law to be explained is caused by an event which is a case of one (in each case, the same) of the laws contained essentially in the explanans.[4]

It might be thought that what we have said so far is sufficient to explain why it is that (A) is not an explanation and why it is that the explanation of sodium and chlorine's combining in a one-to-one ratio in terms of the atomic structure of sodium and chlorine is an explanation. After all, no event which is a case of sodium and chlorine combining in a one-to-one ratio is caused by any event which is a case of sodium and bromine combining in a one-to-one ratio. So, given our requirements, deduction

---

[4] We have made this condition sufficient, but not necessary, for reasons that will emerge below. It will also be seen there that Aristotle, who had a broader notion of cause, could have made it necessary as well.

(A), even though it meets all of Hempel's requirements, need not be (and indeed is not) an explanation. But every event which is a case of sodium and chlorine combining in a one-to-one ratio is caused by the sodium and chlorine in question having the atomic structure that they do (after all, if they had a different atomic structure, they would combine in a different ratio). So an explanation involving the atomic structure would meet our new requirement and would therefore be satisfactory.

The trouble with this account is that it incorrectly presupposes that it is the atomic structure of sodium and chlorine that cause them to combine in a one-to-one ratio. A whole essay would be required to show, in detail, what is wrong with this presupposition; I can, here, only briefly indicate the trouble and hope that this brief indication will be sufficient for now: a given case of sodium combining with chlorine is the same event as that sodium combining with that chlorine in a one-to-one ratio, and, like all other events, that event has only one cause.[5] It is, perhaps, that event which brings it about that the sodium and chlorine are in proximity to each other under the right conditions. That is the cause of the event in question, and not the atomic structure of the sodium and chlorine in question (which, after all, were present long before they combined). To be sure, these atomic structures help explain one aspect of the event in question, the ratio in which they combine, but that does not make them the cause of the event.[6]

To say that the atomic structure of the atoms in question is not the cause of their combining in a one-to-one ratio is *not* to say that a description of that structure is not an essential part of any causal explanation of their combining. It obviously is. But equally well, to say that a description of it is a necessary part of any causal explanation is *not* to say that it is (or is part of) the cause of their combining. There is a difference, after all, between causal explanations and causes and between parts of the former and parts of the latter. Similarly, to say that the atomic structure is not the cause of their combining is *not* to say that that event had no cause; indeed, we suggested one (the event which brought about the proximity of the atoms) and others can also be suggested (the event of the atoms acquiring certain specific electrical and quantum mechanical properties). It is only to say that the atomic structure is not the cause.

Keeping these two points in mind, we can see that all that we said before was that the perfectly satisfactory explanation, in terms of the

---

[5] To be sure, e_1 can have as causes both e_2 and e_3 (where $e_2 \neq e_3$) but only when either e_2 is the cause of e_3 or e_3 the cause of e_2. That exception is of no relevance here.

[6] It might, at least, be maintained that they are still the cause of that aspect of the event. But that is just a confusion—it is events, and not their aspects, that have causes.

atomic structure of the atoms, of their combining in a one-to-one ratio does not meet the condition just proposed because it contains no description of the event which caused the combining to take place. But since it obviously is a good explanation, some additional types of explanations must be allowed for.

So Aristotle's first suggestion, while quite helpful in solving some of our problems, does not solve all of them. There is, however, another important suggestion that he makes that will, I believe, solve the rest of them. Aristotle says:

Demonstrative knowledge must rest on necessary basic truths; for the object of scientific knowledge cannot be other than it is. Now attributes attaching essentially to their subjects attach necessarily to them. . . . It follows from this that premisses of the demonstrative syllogism must be connections essential in the sense explained: for all attributes must inhere essentially or else be accidental, and accidental attributes are not necessary to their subjects (*Posterior Analytics* 74^b 5–12).

There are many aspects of this passage that I do not want to discuss now. But one part of it seems to me to suggest a solution to our problem. It is the suggestion that a demonstration can be used as an explanation (can provide us with 'scientific knowledge') when at least one of the explanans essential to the derivation states that a certain class of objects has a certain property, and (although the explanans need not state this) that property is possessed by those objects essentially.

Let us, following that suggestion, now look at our two proposed explanations as to why sodium combines with chlorine in a ratio of one-to-one. In one of them, we are supposed to explain this in terms of the fact that sodium combines with bromine in a one-to-one ratio. In the other explanation, we are supposed to explain this in terms of the atomic structure of sodium and chlorine. Now in both of these cases, we can demonstrate from the fact in question (and certain additional facts) that sodium does combine with chlorine in a one-to-one ratio. But there is an important difference between these two proposed explanations. The atomic structure of some chunk of sodium or mass of chlorine is an essential property of that object. Something with a different atomic number would be (numerically) a different object. But the fact that it combines with bromine in a one-to-one ratio is not an essential property of the sodium chunk, although it may be true of every chunk of sodium. One can, after all, imagine situations[7] in which it would not combine in

---

[7] Even ones in which all currently believed scientific laws hold, but in which the initial conditions are quite different from the ones that now normally hold.

that ratio but in which it would still be (numerically) the same object. Therefore, one of our explanans, the one describing the atomic structure of sodium and chlorine, contains a statement that attributes to the sodium and chlorine a property which is an essential property of that sodium and chlorine (even if the statement does not say that it is an essential property), while the other of our explanans, the one describing the way in which sodium combines with bromine, does not. And it is for just this reason that the former explanans, but not the latter, explains the phenomenon in question.

Generalizing this Aristotelian point, we can set down another requirement for explanations as follows: a deductive-nomological explanation of a particular event is a satisfactory explanation of that event when (besides meeting all of Hempel's requirements) its explanans contains essentially a statement attributing to a certain class of objects a property had essentially by that class of objects (even if the statement does not say that they have it essentially) and when at least one object involved in the event described in the explanandum is a member of that class of objects. If this requirement is unfulfilled, then it may not be a satisfactory explanation. And similarly, a deductive nomological explanation of a law is a satisfactory explanation of that law when (besides meeting all of Hempel's requirements) each event which is a case of the law which is the explanandum, involves an entity which is a member of a class (in each case, the same class) such that the explanans contain a statement attributing to that class a property which each of its members have essentially (even if the statement does not say that they have it essentially).

It is important to note that such an account is incompatible with the logical empiricist conception of theoretical statements as instruments and not as statements describing the world. For after all, many of these essential attributions are going to be theoretical statements, and they can hardly be statements attributing to a class of objects an essential property if they are not really statements at all. But that is okay, for it just gives us one more reason for rejecting an account, more notable for the audacity of its proponents in proposing it than for its plausibility or for the illumination it casts.

There are two types of objection to essential explanations that we should deal with immediately. The first really has its origin in Duhem's critique of the idea that scientific theories explain the observable world. Duhem (1914: ch. 1) argued that if we view a theory as an explanation of an observable phenomenon, we would have to suppose that the theory gives us an account of the physical reality underlying what we observe. Such claims about the true nature of reality are, however, empirically unverifiable metaphysical hypotheses, which scientists should shun, and therefore we must not view a theory as an attempt to explain what we observe. Now contemporary theories of explanation, like the deductive-nomological model, avoid this problem by not requiring of an explanation that its explanans describe the true reality underlying the observed explanandum. But if we now claim that a deductive-nomological explanation is a satisfactory explanation when (among other possibilities) its explanans describe essential properties of some objects involved in the explanandum event, are we not introducing these disastrous, because empirically undecidable, issues about the true nature of the reality of these objects into science? After all, the scientist will now have to decide, presumably by non-empirical means, whether the explanans do describe the essence of the objects in question.

The trouble with this objection is that it just assumes, without any arguments, that claims about the essences of objects would have to be empirically undecidable claims, claims that could be decided only upon the basis of metaphysical assumptions. This presupposition, besides being unsupported, just seems false. After all, the claim that the essential property of sodium is its atomic number (and not its atomic weight, or its colour, or its melting point) can be defended empirically, partially by showing that for this property, unlike the others just mentioned, there are no obvious cases of sodium which do not have it, and partially by showing how all objects that have this property behave alike in many important respects while objects which do not have this property in common do not behave alike in these important respects. Now a lot more has to be said about the way in which we determine empirically the essence of a given object (or of a given type of object), and we will return to this issue below, but enough has been said, I think, to justify the claim that the idea that scientific explanation is essential explanation does not mean that scientific explanation involves empirically undecidable claims.

It should be noted, by the way, that this idea of the discovery of essences by empirical means is not new to us. It was already involved in Aristotle's theory of *epagoge* (*Posterior Analytics*, 2. 19). I do not now want to enter into the question as to exactly what Aristotle had in mind, if he did have anything exact in mind, when he was describing that process. It is sufficient to note that he, like all other true adherents of the theory of essential explanation, saw our knowledge of essences as the result of reflection upon what we have observed and not as the result of some strange sort of metaphysical knowledge.

The second objection to essential explanations has been raised by Popper. He writes:

The essentialist doctrine that I am contesting is solely the doctrine that science aims at ultimate explanation; that is to say, an explanation which (essentially, or, by its very nature) cannot be further explained, and which is in no need of any further explanation. Thus my criticism of essentialism does not aim at establishing the non-existence of essences; it merely aims at showing the obscurantist character of the role played by the idea of essences in the Galilean philosophy of science (1963: 105).

Popper's point really is very simple. If our explanans contain a statement describing essential properties (e.g. sodium has the following atomic structure . . .), then there is nothing more to be said by way of explaining these explanans themselves. After all, what could we say by way of answering the question 'Why does sodium contain the atomic structure that it does?' So the use of essential explanations leads us to unexplainable explanans, and therefore to no new insights gained in the search for explanations of these explanans, and therefore to scientific sterility. Therefore, science should reject essential explanations.

There are, I believe, two things wrong with this objection. To begin with, Popper assumes that essential explanations will involve unexplainable explanans, and this is usually only partially true. Consider, after all, our explanation of sodium combining with chlorine in a one-to-one ratio in terms of the atomic structure of sodium and chlorine. The explanans of that explanation, besides containing statements attributing to sodium and chlorine their essences (namely, their atomic number), also contain the general principles of chemical bonding, and these are not unexplainable explanans since they are not claims about essences. In general, even essential explanations leave us with some part (usually the most interesting part) of their explanans to explain, and they do not therefore lead to sterility in future enquiry. But, in addition, even if we did have an essential explanation all of the explanans of which were essential statements and therefore unexplainable explanans, what are we to do according to Popper? Should we reject the explanation? Should we keep it but believe that it is not an essential explanation? Neither of these strategies seems very plausible in those cases where we have good reasons both for supposing that the explanation is correct and for supposing that the explanans do describe the essential properties of the objects in question. It cannot, after all, be a good scientific strategy to reject what we have good reasons to accept. So even if Popper's claim about their sterility for future scientific enquiry is true for some essential explanations, I cannot see that it gives us any reasons for rejecting these explanations, or for rejecting their claim to be essential explanations, when these explanations and claims are empirically well supported.

There is, of course, a certain point to Popper's objection, a point that I gladly concede. As is shown by his example from the history of gravitational theory, people may rush to treat a property as essential, without adequate empirical evidence for that claim, and then it may turn out that they were wrong. They may even have good evidence for the claim that the property is essential and still be wrong. In either case, enquiry has been blocked where it should not have been blocked. We should certainly therefore be cautious in making claims about essential properties and should, even when we make them on the basis of good evidence, realize that they may still be wrong. But these words of caution are equally applicable to all scientific claims; the havoc wreaked by false theories that lead enquiry along mistaken paths can be as bad as the havoc wreaked by false essential claims that block enquiry. And since they are only words of caution, they should not lead us to give up either theoretical explanations in general or essential explanations in particular.

Let us see where we now stand. We have, so far, rejected Hempel's requirements for an explanation on the grounds that they are not sufficient and we have suggested two alternative Aristotelian conditions such that, for the set of explanations meeting Hempel's requirements, any explanation meeting either of these requirements is an adequate explanation.[8] Doing this is sufficient to help us deal with the problem of self-explanation, another problem that the covering-law model has had difficulty with. As Hempel already recognized in his original presentation of the covering-law model, we need some additional requirement to rule out such obvious self-explanations as

(G) $$(x)Px$$
$$\frac{Qa}{Qa}$$

and slightly less obvious self-explanations such as

(H) $$(x)(Px \cdot Qa)$$
$$\frac{Qa \vee \sim Qa}{Qa}$$

He proposed a simple solution to that problem but Eberle, Kaplan, and Montague showed that it would not do (1961). Consider, they said, the

---

[8] We have not, however, required as a necessary condition that any explanation must meet one of these two conditions. This is so, partly because of the problem of statistical explanations, but partly because of the possibility, raised by Aristotle, that there are additional types of explanations. After all, our two conditions let in explanations in terms of Aristotle's efficient and formal causes. We still have to consider, but will not in this paper, possible explanations in terms of what he would call material and final causes.

following example of a bad explanation, that meets all Hempel's requirements, of an object having a property H. Let us take any true law of the
form $(x)Fx$ (where there is no connection between an object having the
property F and having the property H). From that law it follows that
(where G is any third unrelated property)

(1) $(x)(y)[Fx \lor (Gy \supset Hy)]$

It also follows from Ha, the fact to be explained, that

(2) $(Fb \lor \sim Ga) \supset Ha$

But from these two true statements, we can derive Ha, and this derivation, a subtle form of self-explanation, meets all Hempel's requirements,
so Hempel still had not solved the problem of self-explanation. Now
there exist several syntactic solutions to this problem, solutions that are
partly *ad hoc* and partly intuitively understandable (see Kaplan (1961)
and Kim (1963)). As such, they are not entirely satisfactory. Writing
about one of them, Hempel (1965: 295) admits that, 'it would be desirable to ascertain more clearly to what extent the additional requirement
is justifiable, not on the *ad hoc* grounds that it blocks those proofs, but
in terms of the rationale of scientific explanation.' Now our theory offers
a simple, non-*ad hoc*, solution to this problem. The derivation used by
Eberle, Kaplan, and Montague does not meet either of our two conditions. Neither (1) nor (2) describe the cause of Ha. And neither (1) nor
(2) ascribe an essential property to the members of a certain class of
objects of which a is a member. Therefore, although their derivation
meets all Hempel's requirements, it need not be (and indeed is not) an
adequate explanation.

By now, the advantages of our theory are obvious. It provides intuitively satisfactory, non-*ad hoc*, solutions to problems that the covering-
law model cannot handle. And at the same time, it incorporates (by
keeping Hempel's requirements) the elements of truth in the covering-
law model. It only remains, therefore, to consider the one serious objection to this whole Aristotelian theory, an objection that we have already
touched upon when we dealt with Duhem. Given what we mean by
'causality' and 'essence', can we ever know that e_1 is the cause of e_2 or
that P is an essential property of e_1, and if so, how can we know this?

This problem can be sharpened considerably. There is no problem, in
principle, about our coming to know that events of type E_1 are constantly
conjoined with events of type E_2. All that we have to do is to observe
that this is so in enough varied cases. And if 'e_1 causes e_2' only means
that 'e_1 is an event of type E_1 and e_2 is an event of type E_2 such that

E_1 is constantly conjoined with E_2', we can easily see how we could come to know that e_1 is the cause of e_2. But if, as our Aristotelean account demands, 'e_1 causes e_2' means something more than that, can we know that it is true, and if we can, how can we know that it is true? Similarly, there is no problem, in principle, about our coming to know that objects of type O_1 have a certain property P_1 in common. All that we have to do is to observe that this is so in enough varied cases. And if 'P_1 is an essential property of o_1' only means that 'o_1 is an object of type O_1 and all objects of type O_1 have P_1', we can easily see how we could come to know that P_1 is an essential property of o_1. But if, as our Aristotelian theory demands, 'P_1 is an essential property of o_1' means something more than that, can we know that this is true, and if we can, how can we know that this is true?

There is an important difference between these questions. If we conclude that we cannot, or do not, know the truth of statements of the form 'e_1 causes e_2' or 'P_1 is an essential property of o_1' (where these statements are meant in the strong sense required by our theory), then our theory must be rejected. After all, knowledge of the truth of statements of that form is, according to our theory, a necessary condition for knowing that we have (although not for having) adequate explanations. And we obviously do know, in at least some cases, that a given explanation is adequate. So if we cannot, or do not, know statements of the above form, our theory is false. However, if we conclude that we can, and do, know the truth of statements of the above-mentioned form, but we don't know how we know their truth, then all that we have left is a research project, namely, to find out how we know their truth; what we do not have is an objection to our theory.

This is an extremely important point. I shall show, in a moment, that we do have, and *a fortiori* can have, knowledge of these statements. But, to be quite frank, I have no adequate account (only the vague indications mentioned above when talking about Duhem) of how we have this knowledge. So, on the basis of this last point, I conclude that the Aristotelian theory of explanation faces a research problem about knowledge (hence the title of this paper), but no objection about knowledge.

Now for the proof that we do, and *a fortiori* can, have knowledge of the above-mentioned type. Our examples will, I am afraid, be familiar ones. It seems obvious that we know that

1. if the temperature of a gas is constant, then an increase in its pressure is invariably accompanied by an inversely proportional decrease in its volume

2. if the temperature of a gas is constant, then a decrease in its volume is invariably accompanied by an inversely proportional increase in its pressure
3. if the temperature of a gas is constant, then an increase in its pressure does not cause an inversely proportional decrease in its volume
4. if the temperature of a gas is constant, then a decrease in its volume does cause an inversely proportional increase in its pressure

Here we have causal knowledge of the type required, since the symmetry between (1) and (2) and the asymmetry between (3) and (4) show that the causal knowledge that we have when we know (3) and (4) is not mere knowledge about constant conjunctions. Similarly, it seems obvious that we know that

1. all sodium has the property of normally combining with bromine in a one-to-one ratio
2. all sodium has the property of having the atomic number 11
3. the property of normally combining with bromine in a one-to-one ratio is not an essential property of sodium
4. the property of having the atomic number 11 is an essential property of sodium

Here we have essential knowledge of the type required, since the symmetry between (1) and (2) and the asymmetry between (3) and (4) show that the essential knowledge that we have when we know (3) and (4) is not mere knowledge about all members of a certain class having a certain property.

I conclude, therefore, that we have every good reason to accept, but none to reject, the Aristotelian theory of explanation sketched in this paper. And I also conclude that it therefore behoves us to find out how we have the type of knowledge mentioned above, the type of knowledge that lies behind our knowledge that certain explanations that we offer really are adequate explanations.

REFERENCES

Brody, B. A. (1967), 'Natural Kinds and Real Essences', *Journal of Philosophy.*
Bromberger, S. (1969), 'Why Questions', in Brody (ed.), *Introductory Readings in the Philosophy of Science* (Englewood Cliffs, NJ: Prentice Hall).
Duhem, P. (1914), *La Théorie physique, son objet et sa structure* (Paris: Chevalier et Rivière).
Eberle, R. A., Kaplan, D., and Montague, R. (1961), 'Hempel and Oppenheim on Explanation', *Philosophy of Science.*

Hempel, C. G. (1965), *Aspects of Scientific Explanation* (New York: Free Press).

Kaplan, D. (1961), 'Explanation Revisited', *Philosophy of Science.*

Kim, J. (1963), 'Discussion: On the Logical Conditions of Deductive Explanation', *Philosophy of Science.*

Popper, K. R. (1963), *Conjectures and Refutations* (London: Routledge & Kegan Paul).

ON AN ARISTOTELIAN MODEL OF SCIENTIFIC EXPLANATION[*]

TIMOTHY McCARTHY

I shall discuss two suggestions due to Baruch Brody (1972) as to how Hempel's deductive-nomological (D-N) criteria for explanation may be so augmented as to afford sufficient conditions for explanatory adequacy. These suggestions, with respect to explanation of particular events, concern (a) the inclusion of descriptions of the cause of the event to be explained in the explanans, and (b) the inclusion of sentences ascribing essential properties to objects involved in the event to be explained in the explanans. The (a)- and (b)-conditions are discussed in sections 1 and 2 respectively.

Brody's discussion is based on certain metaphysical presuppositions, that is, associated with the (a)-conditions, a non-Humean construal of causation; and associated with the (b)-conditions, an Aristotelian essentialism. The latter has been the theme of one critical study by Stemmer (1973). However, I do not want to dispute these foundations here; my concern is to show that, even if they are assumed, they cannot support the structure Brody tries to erect upon them.

1. THE CAUSAL CRITERION

The D-N model, construed as providing necessary *and sufficient* conditions of adequate explanation, asserts that a satisfactory explanation of a particular event e is a deductively valid argument to a sentence $D(e)$ describing e from a set S of sentences such that (1) all sentences of S are true and have empirical content,[1] and (2) S contains at least one law l

T. McCarthy, 'Discussion on an Aristotelian Model of Scientific Explanation', *Philosophy of Science*, 44 (1977), 159–66. Copyright © 1977 by the Philosophy of Science Association. Reprinted by permission of the author and the Philosophy of Science Association.

[*] Thanks are due to Professor Peter Achinstein for helpful comments on an earlier draft.

[1] Condition (1) does not hold for sentences expressing logical and mathematical truths.

and singular sentence s such that neither $S - \{l\}$ nor $S - \{s\}$ logically implies $D(e)$.[2]

Now these, as is well known, will not do as sufficient conditions for satisfactory explanation. Brody, following a suggestion in Aristotle, proposes to shore up Hempel's conditions with the following emendation:

(C) A deductive-nomological explanation of a particular event is a satisfactory explanation of that event when, besides meeting Hempel's requirements, its explanans essentially contains a description of the cause of the event described in the explanandum. (1972: 23)

The explanans, S, of a D-N derivation of a sentence p essentially contains a sentence q if S without q does not constitute the explanans of any D-N derivation of p. Accordingly, it suffices to refute (C) to produce a D-N derivation d of p from a set $S(d)$ of sentences such that d satisfies Hempel's conditions and such that $S(d)$ contains a sentence describing the cause of the event described by p without which $S(d)$ does not logically imply p, and such that d is devoid of explanatory import. It will be seen that such derivations exist in abundant supply. However, before passing to counter-examples, I want briefly to develop the rationale for (C).

Brody asks us to consider the following pair of arguments from Aristotle:

(1) The planets do not twinkle
All objects which do not twinkle are near the earth
∴ The planets are near the earth

(2) All objects near the earth do not twinkle
The planets are near the earth
∴ The planets do not twinkle

Both (1) and (2) are acceptable D-N explanations.[3] However, Brody says of (1) that it is either no explanation at all, or a very poor one. Brody thinks (2) is significantly better. I tend to agree. But if this is accepted, we see that the D-N model provides no clue as to where the relevant difference between them may lie. This is where Aristotle comes in. Aristotle characterizes the difference between (1) and (2) as follows:

Of two reciprocally predicable terms the one that is not the cause may quite easily be the better known and so become the middle term of the demonstration . . . this syllogism [(1)] then proves not the reasoned fact but only the fact; since they are

---

[2] 'Law' here is to be understood in Hempel's way as a true universal conditional involving only purely qualitative predicates, and supporting counterfactuals.

[3] Except for the singular reference in the sentences purporting to function as laws. However, it is easily seen that these cases may be restructured so that this is avoided.

not near because they do not twinkle. The major and middle of the proof, however, may be reversed, and then the demonstration will be of the reasoned fact . . . since its middle is the proximate cause [as in (2)] (1901: 78^{a}28–^{b}3).

The gist of Aristotle's remarks is aptly given by Brody: most basically, 'nearness is the cause of not twinkling, and not vice versa, so the nearness of the planets to the earth explains why they do not twinkle, but their twinkling does not explain why they are near the earth' (1972: 23).

I take it that the point here is that to explain an event e, it is necessary, but not sufficient, to show that a sentence describing e is the *terminus* of some demonstration (D-N derivation); such a demonstration will, however, suffice to explain e if, besides showing that e must occur (given the explanans), it shows also, in a causal sense, why e occurs. Brody concludes that it is sufficient to explain e to provide a D-N derivation d of a sentence describing e which depends on a description of e's cause in an ineliminable way; the derivation will thus involve essentially reference to the cause of e, and so should constitute a satisfactory explanation of e. One might suppose that the idea is to mirror the *causal* dependence of e on its cause by the *deductive* dependence in d of a description of e upon a description of e's cause. That is an interesting idea; immediately, however, we may begin to suspect a gap in the argument. The basic worry may be put in this way: why should it follow, merely because a D-N derivation of a sentence describing e ineliminably involves, *in some way or other*, a description of e's cause that this description functions in the derivation to show (causally) why e occurs? No obvious reason exists why a D-N derivation of a sentence describing e could not depend on a description of e's cause in some way quite unrelated to the causal dependence of e on that cause.

In the following derivation we have, I believe, just such a case. Let e be any event, let '$D(e)$' represent any sentence describing e, and '$C(e)$' any sentence describing e's cause. Let '$(x)(Ax \supset Bx)$' represent any law irrelevant to the occurrence of e. Finally, let o be any object such that Ao. Then the following derivation satisfies (C):

(3) $(x)(Ax \supset Bx)$
                        $C(e) \, \& \, Ao$
                        $\dfrac{-Bo \vee -C(e) \vee D(e)}{D(e)}$

It is clear that (3) satisfies Hempel's conditions. It also satisfies Brody's causal condition in (C), since it makes ineliminable use of a description of the cause of e. However, I believe that it is clear that instances of (3) are typically not explanations at all, or very unsatisfactory ones. The reason is precisely that the logical dependence of '$D(e)$' on '$C(e)$' has nothing at all to do with the causal dependence of e on the event described by '$C(e)$', because the law mediating the deductive relation between '$C(e)$' and '$D(e)$' is causally irrelevant to the occurrence of e. Thus (C) is refuted.[4]

However, the historian of the D-N model will have noticed that (3) assumes a well-known, truth-functional form which is easily ruled out on the basis of an additional syntactical requirement proposed by Kim (1963). Let the singular sentences in the explanans of a D-N derivation of a sentence p be put into a complete conjunctive normal form in those atomic sentences which occur essentially in the given singular sentences. The requirement is that none of the conjuncts be logical consequences of p. Now, I will not pause to consider Hempel's question (1965: 295) concerning the extent to which this additional condition may be justified on non-*ad hoc* grounds; indeed, Brody apparently offers (C) *in replacement of* Kim's condition in the light of the difficulty of providing such a justification (1972: 29). In any case, Kim and others have pointed out that with respect to the unsupplemented D-N model, the new condition is trivializable. It may be seen as well that (C) *with Kim's further condition in force* succumbs to the same fate. For, let '$C(o)$' represent a sentence describing the cause of o's turning black (say, from having been put in black paint). Consider now the following derivation:

(4) All crows are black
$(x)(y)(x$ turns the colour of y & y is black $. \supset . x$ turns black)
$C(o)$ & Henry is a crow
$- C(o) \vee o$ turns the colour of Henry

---

o turns black

Again, it is clear that (4) satisfies Hempel's conditions. Moreover, by the use of '$C(o)$' it satisfies (C). It may also be verified that none of the conjuncts of the complete conjunctive normal form of (4)'s singular premisses is logically implied by its explanandum; thus Kim's condition is satisfied. Surely, however, (4) has no more explanatory import than (3).

There is another style of counter-example to the pure D-N model which can be adapted to Brody's causal model. Imagine a D-N derivation of a sentence 'Ao' which results from citing the facts that a certain instrument, M, has predicted that Ao, and that M satisfies a certain condition 'W' such that it is a law, in Hempel's sense, that for any x, y, if y is an

---

[4] Notice also that no assumptions have been made here concerning the form of descriptions either of e or of the relevant cause of e. Thus, no restrictions on such descriptions will suffice to insulate (C) from counter-instances of this kind.

instrument such that $W(y)$ and y predicts that Ax, then Ax. Such a derivation would typically have little or no explanatory force. It is easy to see, however, that this sort of trivialization besets (C) as well. For, consider a case in which the *cause* of the event predicted causes the prediction of that event; e.g., in which the cause of o's being A causes the instrument M to forecast Ao. Specifically, imagine that o's being F causes o's being A, and that it is a law that, for any x, anything of type T standing in relation R to x predicts that Ax if Fx, and further that if anything of type T standing in relation R to x predicts that Ax, then Ax. Call these 'L_1' and 'L_2,' respectively. We now obtain the following derivation, which satisfies (C):

(5)
$$L_1, L_2$$
$$Fo$$
$$\underline{T(M) \,\&\, R(o, M)}$$
$$Ao$$

This parallels counter-instances to Hempel's model of the 'infallible predictor' type, except that the derivation reflects the fact that the cause of the event predicted causes the prediction of that event. It is equally either no explanation at all, or a clearly unsatisfactory one.

To summarize: Brody's causal condition (C) was explained as sanctioning as an adequate explanation any D-N derivation of a sentence describing a particular event which makes use of a description of the cause of that event in an ineliminable way. We were then led to suspect a disparity between this criterion and the apparent rationale for it; and our suspicions were confirmed by various counter-examples which showed that (C) shares with Hempel's conditions a variety of trivializing consequences.

2. THE ESSENTIAL PROPERTIES CRITERION

Brody recognizes that his causal criterion is not satisfied by all satisfactory explanations. Consider, for example, the following D-N derivation:

(6) u has atomic number n_1
v has atomic number n_2
$(x)(y)(x$ has atomic number n_1 & y has atomic number $n_2 . \supset . x$ combines with y in a ratio of $1 : 1)$

$\underline{}$

u combines with v in a ratio of $1 : 1$[5]

---

[5] 'u' and 'v' are schematic for mass terms (e.g. 'sodium'), so that the explanandum in (6) is schematic for a chemical law.

It may be tempting to suppose that we have in (6) a derivation satisfying Brody's causal condition for the explanation of *laws*, namely:

(C$_L$) A deductive-nomological explanation of a law is a satisfactory explanation of that law when (besides meeting Hempel's requirements) every event which is a case of the law to be explained is caused by an event which is a case of one (in each case, the same) of the laws contained essentially in the explanans (1972: 23–4)

For, it might seem, every event which is a case of u and v combining in a 1 : 1 ratio is caused by the samples of u and v in question having the atomic structures that they do. But this, Brody asserts, would be incorrect. The reason, he says (rightly, I think), is that one cannot assert the *structures* of u and v to be causes at all. It would take us too far afield to examine this point in any detail; for the purposes at hand, it may simply be granted. Brody makes it in order to set the stage for a second supplementation of the D-N conditions, one he takes to provide a second set of (sufficient) conditions for explanatory adequacy. And here, again, Aristotle is taken to have known all along; in (1901: 74^{b}5–12) he writes:

Demonstrative knowledge must rest on necessary basic truths; for the object of knowledge cannot be other than it is. Now attributes attaching essentially to their subjects attach necessarily to them . . . it follows that premises of a demonstrative syllogism must be connections essential in the sense explained; for all attributes must inhere essentially or else be accidental, and accidental attributes are not essential to their subjects.

Brody takes this passage to suggest the following criterion for the adequate explanation of particular events:

(E) A D-N derivation of a sentence describing a particular event e is a satisfactory explanation of e if, besides meeting Hempel's conditions, its explanans essentially contains a sentence ascribing a property to an object or objects involved in e which is had essentially by that object or those objects

There is a related criterion for adequate explanation of laws:

(E$_L$) A D-N derivation of a law is a satisfactory explanation of that law when, besides meeting Hempel's conditions, each event which is a case of the law which is the explanandum involves an entity which is a member of a class (in each case, the same class) such that the explanans contains a statement attributing to the members of that class a property which each of them has essentially

Now a predicate 'F' expresses an essential property of u if necessarily F holds of u. We see, then, that (6) counts as a satisfactory explanation on the present criteria if enduring structural traits such as atomic number count as *essential properties*, i.e. properties which are essential properties of any object they apply to at all. This is an assumption which Brody in fact makes; again, for present purposes, it may simply be granted. I shall criticize only the criterion (E) having to do with the explanation of particular events; that related difficulties arise for (E_L) should be apparent.

It suffices to refute (E) to construct a D-N derivation of a sentence describing a particular event which makes deductively ineliminable use of a sentence ascribing an essential property to one or more objects involved in that event which does not constitute a satisfactory explanation. To begin with, it is easily seen that truth-functional difficulties arise for (E) precisely as for (C). In fact, derivation (3) is effective in the present setting if we suppose that o is an object involved in e, and that 'Ao' ascribes an essential property to o. If e is taken to be an event whose occurrence is explanatorily unrelated to o's having the essential property in question, a counter-example to (E) is generated. Slightly more subtle, but effective against (E) with Kim's further condition in force, is the following:

(7) All grass is green
   $(x)(y)(x$ has the atomic number of chlorine & y is grass $. \supset . x$ has the colour of $y)$
   $(x)(y)(x$ is the colour of y & y is green $. \supset . x$ is green$)$
   u has the atomic number of chlorine, v is grass

   ———————————————————————————————————

   u is green

It may be verified that (7) satisfies Hempel's conditions. Further, it depends on a sentence ascribing an essential property to the object u involved in the explanandum; thus, it satisfies (E) as well. However (7), if an explanation at all, is seen to lack explanatory force (at least) for the reason that, while there may be an explanatory connection between u's having the atomic number of chlorine and its being the colour it is, that connection is not spelled out in the derivation, which depends essentially on certain considerations having no explanatory relevance to the explanandum. A different sort of case is provided by:

(8) $(x)(x$ has atomic number 63 $. \supset . x$ is a good thermal conductor$)$
   $(x)(x$ is a good thermal conductor $. \supset . x$ is a good electrical conductor$)$
   u has atomic number 63

   ———————————————————————————————————

   u is a good electrical conductor

This derivation satisfies (E), yet lacks explanatory import inasmuch as it shows no explanatory connection between the essential property ascribed to u and the property ascribed to u in the explanandum (for it is in no sense—either causally or essentially—because an object is a good thermal conductor that it is a good electrical conductor; both dispositions result from an underlying microstructure).

There may be other styles of counter-example to the unsupplemented D-N conditions which can be adapted to Brody's causal and essential models. But the further adaptation may not be worth the effort. Enough has been said, I believe, to show that Brody's supplementations of the D-N conditions succumb to the same sorts of considerations as the D-N model itself. All three isolate classes of sentences (laws, causal statements, ascriptions of essential properties) which figure prominently in many satisfactory explanations, and all three attempt to guarantee the explanatory adequacy of a derivation by forcing it to depend upon sentences from one or more of these classes. And in no case is it made clear why the desired guarantee should be forthcoming; why it should be expected that the logical dependence of an argument upon sentences of these kinds will ensure its explanatory adequacy.

REFERENCES

Aristotle (1901), *Posterior Analytics*, tr. E. S. Bouchier (Oxford: Blackwell).
Brody, B. A. (1972), 'Towards an Aristotelian Theory of Scientific Explanation', *Philosophy of Science*, 39, 20–31.
Hempel, C. G. (1965), *Aspects of Scientific Explanation* (New York: Free Press).
Kim, J. (1963), 'On the Logical Conditions of Deductive Explanation', *Philosophy of Science*, 30, 286–91.
Stemmer, N. (1973), 'Brody's Defense of Essentialism', *Philosophy of Science*, 40, 393–6.

V

CAN THERE BE A MODEL OF EXPLANATION?*

PETER ACHINSTEIN

1. INTRODUCTION

Since 1948, when Hempel and Oppenheim[1] published their pioneering article, various models of explanation have appeared. But each has had its counter-examples, and observers of the philosophical scene may wonder whether models of the kind sought are really possible. Are their proponents engaged in a fruitless task of inquiry?

Hempel and other modellists are particularly concerned with explanations that answer questions of the form

(1) Why is it the case that p?[2]

The sentence replacing 'p' in (1) Hempel calls the *explanandum*. It describes the phenomenon, or event, or fact, to be explained. The answer to an explanation-seeking why-question of form (1) Hempel calls the *explanans*. It is a sentence, or set of sentences, that provides the explanation. We can speak of the explanans as explaining the explanandum. And we can say that an explanans *potentially* explains an explanandum when, if the sentences of the explanans were true, the explanans would correctly explain the explanandum.

Thus, if the explanation-seeking why-question is

Q: Why is it the case that this metal expanded?

then the explanandum is

P. Achinstein, 'Can There Be a Model of Explanation?', in *Theory and Decision*, 13 (1981), 201–27. Copyright © 1981 by D. Reidel Publishing Co., Dordrecht. Reprinted by permission of Kluwer Academic Publishers.

* Material in this paper is from a book I am writing on the nature of explanation. I am indebted to the National Science Foundation for support of research. [Since published as *The Nature of Explanation* (Oxford: Oxford University Press, 1983).]

[1] Carl G. Hempel and Paul Oppenheim, 'Studies in the Logic of Explanation', reprinted in Hempel, *Aspects of Scientific Explanation* (New York, 1965), 245–90.
[2] See Hempel, *Aspects*, 334.

(2) This metal expanded

If, in reply to Q, an explainer claims that

(3) This metal was heated; and all metals that are heated expand

then (3) is the explanans for the explanandum (2). And (3) potentially explains (2) if, given the truth of (3), (3) would correctly explain (2).

A *model* of explanation is a set of necessary and sufficient conditions that determine whether the explanans correctly explains the explanandum (where the explanation-seeking question is of form (1)). It can also be described as a set of conditions that determine whether the explanans potentially explains the explanandum. If the conditions are satisfied by a given explanans and explanandum, then the former correctly explains the latter, provided that the former is true.

In what follows, my concern will be with models as sets of sufficient (rather than necessary) conditions for correct explanations; and as providing such conditions for explanations of particular events or facts rather than of general laws. Most of the counter-examples in the literature have been raised against models construed in this way. I shall argue that one important reason for the failure of these models is that their proponents want to impose requirements which, in effect, destroy the efficacy of their models.

2. TWO REQUIREMENTS FOR A MODEL

The first is that no singular sentence or conjunction of such sentences in the explanans can entail the explanandum.[3] I will call this the No-Entailment-By-Singular-Sentence requirement, or NES for short. What is the justification for it?

There are, I suggest, three reasons that modellists I have in mind support it. First, it precludes certain 'self-explanations' and 'partial self-explanations'. Suppose we want to explain why a particular metal expanded. Assume that the explanandum in this case is

---

[3] [There is a more adequate technical formulation in *The Nature of Explanation*, p. 159 of the 1985 paperback edn.] The notion of a singular sentence which Hempel and others use in characterizing their models is often employed in a more or less intuitive way. And when there is an attempt to make this notion more precise difficulties emerge, as Hempel himself is aware (see his *Aspects*, 356). Hempel suggests that the concept can be adequately defined for a formalized language containing quantificational notation; but there are problems even here (see my *Law and Explanation* (Oxford, 1971), 36–7). However, the kinds of cases I will be concerned with are quite simple and would, I think, be classified by modellists both as singular sentences and as ones to be excluded from the explanation in question.

(1) This metal expanded

The NES requirement precludes (1) itself from being or, being part of, an explanans for (1). It also precludes from an explanans for (1) sentences such as 'This metal was heated and expanded', and 'This metal expanded, and all metals that are heated expand', which would be regarded as partial self-explanations of (1).

Secondly, modellists emphasize the importance of general laws in an explanans. Such laws provide an essential link between the singular sentences of the explanans and the singular sentence that constitutes the explanandum. Intuitively, to explain a particular event involves relating it to other particular events via a law; if the singular sentences of the explanans themselves entail the explanandum laws become unnecessary, on this view.

Thirdly, the NES requirement in effect removes from an explanans certain sentences which involve explanatory connectives such as 'explains', 'because', 'on account of', 'due to', 'reason', and 'causes'. Let me call sentences in which such terms connect phrases or other sentences *explanation-sentences*. Here are some examples:

(2) This metal's being heated explains why it expanded
This metal expanded because it was heated
This metal expanded on account of its being heated
The reason this metal expanded is that it was heated
This metal's expanding is due to its being heated
The fact that this metal was heated caused it to expand

NES precludes any of the explanation-sentences in (2) from being, or being part of, an explanans for (1), since each of them is a singular sentence that entails (1).[4] Without a condition such as NES one could simply require, e.g., that an explanans for an explanandum p be a singular explanation-sentence of the form 'q explains (why) p' or 'p because q'. Any such explanans, if it were true, would correctly explain the explanandum p. Of course, any of the the six sentences in (2), if true, could be cited in correctly explaining why this metal expanded. Modellists need not deny this. Their claim is that the sentences in (2) do not correctly explain (1) in the right sort of way. They would exclude sentences in (2)

---

[4] To classify them as singular, of course, is not to deny that they have certain implicitly general features which further analysis might separate from the singular ones. Some might claim that NES should be applied only to completely singular sentences—ones with no generality present at all (cf. Hempel). The sentences in (2) must first be reduced. For example, in the case of the last the reduction might be: 'this metal was heated; this metal expanded; and whenever a metal is heated it expands'. If the latter is now taken to be the reduced form of the explanans, then the reduced explanans contains a singular sentence—'this metal expanded'—which entails the explanandum (1).

from an explanans for (1) because they think that an adequate explanans for (1) must reconstruct the sentences in (2) so that the explanatory connectives in the latter are, in effect, analysed in non-explanatory terms. One of the purposes of a model of explanation is to define terms such as 'explains', 'because', 'reason', and 'cause', and not to allow them to be used as primitives within an explanans. By providing some analysis of explanation modellists want to show why it is that this metal's being heated explains why it expanded. There is little enlightenment in saying that this is so because (2) is true.

NES does not exclude all singular explanation-sentences (or all explanatory connectives) from an explanans.[5] But by precluding those that entail the explanandum it does eliminate ones that, from the viewpoint of the modellists, most seriously reduce the possibility of philosophical enlightenment from the resulting explanation. (Whether such modellists would advocate a broader requirement eschewing all explanation-sentences from an explanans is a possibility I shall not discuss.)

NES also excludes certain sentences from an explanans that do not explicitly invoke explanatory connectives but are importantly like those that do which are excluded. Suppose we want to explain why the motion of this particle was accelerated. Our explanandum is

(3) The motion of this particle was accelerated

Consider the explanans

(4) An electrical force accelerated the motion of this particle

Although (4) itself contains no explicit explanatory connective such as 'explains', 'because', or 'causes', it nevertheless carries a causal implication concerning the event to be explained. It is roughly equivalent to the following explanation-sentences which do have such connectives:

(5) An electrical force caused the motion of this particle to be accelerated. The motion of this particle was accelerated because of the presence of an electrical force

NES precludes (4) as well as (5) from an explanans for (3), since (4) is a singular sentence that entails (3). Those who support NES would emphasize that (4), no less than (5), invokes an essentially explanatory connection between an explanans-event and the explanandum-event which it is the task of a model of explanation to explicate.

The second requirement, which I shall call the *a priori* requirement, is that the only *empirical* consideration in determining whether the expla-

---

[5] e.g., it permits the following: Explanandum: 'An event of type C occurred.' Explanans: 'An event of type A caused one of type B; whenever an event of type B occurs so does one of type C.'

nans correctly explains the explanandum is the truth of the explanans; all other considerations are *a priori*. Accordingly, whether an explanans potentially explains an explanandum is a matter that can be settled by *a priori* means (e.g. by appeal to the meanings of words, and to deductive relationships between sentences). A model must thus impose conditions on potential explanations the satisfaction of which can be determined non-empirically. A condition such as this would therefore be precluded:

> An explanans potentially explains an explanandum only if there is a (true) universal or statistical law relating the explanans and explanandum

Whether there is such a law is not an *a priori* matter.

The idea is that a model of explanation should require that sufficient information be incorporated into the explanans that it becomes an *a priori* question whether the explanans, if true, would correctly explain the explanandum. There is an analogy between this and what various logicians and philosophers say about the concepts of *proof* and *evidence*.

Often a scientist will claim that a proposition q can be proved from a proposition p, or that e is evidence that h is true, even though the scientist is tacitly making additional empirical background assumptions which have a bearing on the validity of the proof or on the truth of the evidence claim. If all of these assumptions are made explicit as additional premisses in the proof, or as additional conjuncts to the evidence, then whether such and such is a proof, or is evidence for a hypothesis, is settleable *a priori*. Similarly, often a scientist who claims that a certain explanans correctly explains an explanandum will be making relevant empirical background assumptions not incorporated into the explanans; if the latter are made explicit and added to the explanans, it becomes an *a priori* question whether the explanans, if true, would correctly explain the explanandum.

There is an additional similarity alleged between these concepts. A scientist would not regard a proof as correct—i.e. as proving what it purports to prove—unless its premisses are true. Nor would he regard e as evidence that h (or e as confirming or supporting h) unless e is true. (That John has those spots is not evidence that he has measles unless he does have those spots.)[6] And whether the premisses of the proof, or the evidence report, is true is, in the empirical sciences at least, not an *a priori* question. Nevertheless, deductive logicians, as well as inductive logicians in the Carnapian tradition, believe that they can isolate an *a priori* aspect of proof and evidence such that the only empirical consideration in determining whether a proof or a statement of the form 'e is

---

[6] See my 'Concepts of Evidence', *Mind* 87 (1978), 22–45.

evidence that *h'* is correct is the truth of the premises of the proof or of the *e*-statement in the evidence claim; all other considerations are *a priori*. What I have been calling the *a priori* requirement makes the corresponding claim about the concept of explanation: the only empirical consideration in determining whether an explanation is correct is the truth of the explanans.

3. MODELS PURPORTING TO SATISFY THESE REQUIREMENTS

3.1. The Basic D-N Model

Consider this model as providing a set of sufficient conditions for explanations of particular events, facts, etc. The explanation-seeking question is of the form 'Why is it the case that *p*?', where *p*, the explanandum, is a sentence describing the event to be explained. The explanans is a set containing sentences of two sorts. One sort purports to describe particular conditions that obtained prior to, or at the same time as, the event to be explained. The other are lawlike sentences (sentences that if true are laws). The model requires that the explanans entail the explanandum and that the explanans be true.

No singular sentence (or conjunction of such sentences) that entails the explanandum will appear in the explanans.[7] Any such sentence which is an explanation-sentence that someone might utter in explaining something will itself be analysed as a D-N explanation, i.e. as a deductive argument in which no premises that are singular sentences entail the conclusion. For example, if someone utters the singular explanation-sentence

(1) This metal's being heated explains why it expanded (caused it to expand, etc.)

in explaining why the metal expanded, a D-N theorist will restructure (1) as an argument such as this

(2) This metal was heated
    Any metal that is heated expands
       Therefore,
    This metal expanded

And he will identify the premises of (2) as the explanans of the explanation and the conclusion as the explanandum. The premise in (2) that is a singular sentence does not entail the conclusion.

---

[7] This is required in Hempel's informal and formal characterizations of his model. See his *Aspects*, 248, 273, 277.

The *a priori* requirement also seems to be satisfied by this model. The only empirical consideration in determining whether the explanans correctly explains the explanandum is the truth of the explanans; all other considerations are *a priori*. What are these other considerations? They are whether the explanans (but not the conjunction of singular sentences in it) deductively entails the explanandum and whether it contains at least some sentences that are lawlike. The former is not an empirical question, nor is the latter, as construed by Hempel, since whether a sentence is lawlike depends only on its syntactical form and the semantical interpretation of its terms.[8]

The D-N model as a set of sufficient conditions for particular events is very broad, and one might seek to add further restrictions. Three more limited versions will be noted.

3.2. The D-N Dispositional Model[9]

Here the explanandum is a sentence with a form such as

(3) X manifested P when conditions of type C obtained

and the explanans contains sentences of the form

(4) X has F, and conditions of type C obtained

(5) Anything with F manifests P when conditions of type C obtain

For an explanans consisting of (4) and (5) to provide a correct D-N dispositional explanation of (3) the model requires that F be a disposition-term, that (5) be lawlike, that (4) and (5) entail (3), and that (4) and (5) be true. The singular sentence in the explanans is not to entail the explanandum. And the satisfaction of all the conditions of the model, save for the truth of the explanans, is determined *a priori*. The only condition in addition to those of the basic D-N model is that F be a disposition-term, something settleable syntactically and/or semantically.

3.3. The D-N Motivational Model[10]

Here the explanandum is a sentence saying that some agent acted in a certain way. The explanans contains a singular sentence attributing a desire (motive, end) to that agent, a singular sentence attributing the

---

[8] See Hempel's discussion of lawlikeness, *Aspects*, 271–2, 292, 340. I am here giving a simplified account of the D-N model; the same considerations apply to the more complete account given by Hempel and Oppenheim in Section 7 of their classic paper.

[9] See Hempel, *Aspects*, 462.   [10] Ibid. 254.

belief to that agent that performing the act described in the explanandum is, in the circumstances, a (the best, the only) way to satisfy that desire, and a lawlike sentence relating desires, beliefs, and actions of the kind in question. For example, the explanandum might be a sentence of the form

(6) Agent X performed act A

and the explanans might contain sentences of the form

(7) X desired G

(8) X believed that doing A is, in the circumstances, a (the best, the only) way to obtain G

(9) Whenever an agent desires something G and believes that the performance of a certain act is, in the circumstances, a (etc.) way to obtain G he performs that act[11]

For an explanans consisting of (7)–(9) to provide a correct D-N *motivational* explanation of (6) the model requires that (9) be lawlike, that (7)–(9) entail (6), and that (7)–(9) be true. In this model, like the others, the singular sentences in the explanans are not to entail the explanandum, and the satisfaction of all the requirements of the model, save for the truth of the explanans, is settleable *a priori*.

3.4. Woodward's Functional Interdependence Model

This proposes adding to the basic D-N conditions the following additional necessary condition:

(10) *Condition of functional interdependence*: the law occurring in the explanans for the explanandum p must be stated in terms of variables or parameters, variations in the values of which will permit the derivation of other explananda which are appropriately different from p[12]

Suppose that the explanation-seeking question is 'Why is it the case that this pendulum has a period of 2.03 seconds?', for which the explanandum is

(11) This pendulum has a period of 2.03 seconds

Consider the following D-N argument:

(12) This pendulum is a simple pendulum

---

[11] I do not for a moment believe that (9) is true. But this is not the problem I want to deal with here.

[12] James Woodward, 'Scientific Explanation', *British Journal for the Philosophy of Science*, 30 (1979), 41–67, 46.

The length of this pendulum is 100 cm

The period T of a simple pendulum is related to the length L by the formula $T = 2\pi(L/g)^{1/2}$, where $g = 980$ cm/sec^2,

Therefore,

This pendulum has a period of 2.03 seconds

The third premiss in (12) is a law satisfying Woodward's condition (10). Its variables are the period T and the length L. And variations in the values of these variables will permit the derivation of explananda which Woodward regards as appropriately different from (11). For example, if we change the explanandum (11) to

(13) This pendulum has a period of 3.14 seconds

the law in (12) allows the derivation of (13) if the value of L is changed to 245 cm. Woodward is impressed by the fact that explanations, particularly in science, permit a variety of possible phenomena to be explained. He writes:

The laws in examples [of this sort] formulate a systematic relation between . . . variables. They show us how a range of different changes in certain of these variables will be linked to changes in others of these variables. In consequence, these generalizations are such that when the variables in them assume one set of values (when we make certain assumptions about boundary and initial conditions) the explananda in the . . . explanations are derivable, and when the variables in them assume other sets of values, a range of other explananda is derivable (ibid.).

The satisfaction of condition (10) is settleable *a priori*. If this condition is the only one to be added to those of the basic D-N model, then the resulting model satisfies the *a priori* requirement and NES.[13]

4. VIOLATION OF THE *A PRIORI* REQUIREMENT

Despite the claims of these models the *a priori* requirement is not really satisfied (or else we will have to call certain explanations correct which are clearly not so). In order to show this I shall make use of some of the many counter-examples that have been employed against the D-N model. In these examples, the explanans is true, and the other D-N conditions are satisfied. Yet the explanandum-event did not occur because of the explanans-event, but for some other reason; and this can only be known empirically.

[13] Woodward does not claim that (10) plus the basic D-N model provides a set of sufficient conditions for correct explanations—a point to which I will return later. But for the moment I want to treat his model as if it did purport to provide such a set.

Consider this example:

(a) Jones ate a pound of arsenic at time t
    Anyone who eats a pound of arsenic dies within 24 hours
    Therefore,
    Jones died within 24 hours of t

Assume that the premisses of (a) are true. Then it is supposed to be settleable *a priori* whether these premisses correctly explain the conclusion. According to the D-N model all we need to determine is whether the second premiss is lawlike (let us assume that it is), and whether the conjunction of premisses (but not the first premiss alone) entails the conclusion (it does). Since these D-N conditions are satisfied, the explanans should correctly explain the explanandum; and assuming the truth of the explanans, this matter is settleable on *a priori* grounds alone, no matter what other empirical propositions are true. However, the matter is not settleable *a priori*, since Jones could have died within 24 hours of t for some unrelated reason. For example, he might have died in a car accident not brought on by his arsenic feast, which, given the information in the explanans, could only be determined empirically. Suppose he did die from being hit by a car. Then the explanans in (a) does not correctly explain the explanandum, even though all the conditions of the D-N model are satisfied. Assuming the truth of the premisses in (a) it is not settleable *a priori* whether these premisses correctly explain the explanandum.

A similar problem besets all the more specialized versions of the D-N model cited above. Thus consider these D-N arguments:

(b) Disposition example:
    That bar is magnetic, and a small piece of iron was placed near it
    Any magnetic bar is such that when a small piece of iron is placed
        near it the iron moves towards the bar
        Therefore,
    This small piece of iron moved towards the bar

Suppose, however, that a much more powerful contact force had been exerted on the small piece of iron, and that it moved towards the bar because of this force, not because the bar is magnetic. (Assume that the magnetic force is negligible by comparison with the mechanical force.)

(c) Motivational example:
    Smith desired to buy eggs and he believed that going to the store
    is the only way to buy eggs

> Whenever, etc. (law relating beliefs and desires to actions)
>   Therefore,
> Smith went to the store

But suppose Smith went to the store because he wanted to see his girl-friend who works in the store, not because he wanted to buy eggs.

> (d) Functional interdependence example:
>   This pendulum is a simple pendulum
>   The period of this pendulum is 2.03 seconds
>   The period T of a simple pendulum is related to the length L by the formula $T = 2\pi(L/g)^{1/2}$, where $g = 980 \, \text{cm/sec}^2$
>   Therefore,
>   This pendulum has a length of 100 cm.

A pendulum has the period it has because of its length, but not vice versa. (This type of counter-example is like the others in so far as it invokes an explanans-fact that is inoperative with respect to the explanandum-fact; but the case is also different because there is no intervening cause here, although there is in the others.)

With each of the D-N models considered, whether a particular example satisfies the *requirements of the model* (with the exception of the truth-requirement for the explanans) is settleable *a priori*. Yet in all of these cases, given the truth of the explanans, whether the latter correctly explains the explanandum is not settleable *a priori*. Thus in example (c), even if Smith desired to buy eggs, and he believed that going to the store is the only way to do so, and the lawlike sentence relating beliefs and desires to actions is true, it does not follow that

> (1) Smith went to the store *because* he desired to buy eggs and believed that going to the store is the only way to buy eggs

The explanans of (c) correctly explains the explanandum only if (1) is true. Yet given the truth of the explanans of (c) it is not settleable *a priori*, but only empirically, whether (1) is true.

We could, of course, see to it that the matter is settleable *a priori* by changing the motivational model so as to incorporate (1) into the explanans of (c). Assuming that this enlarged explanans is true, whether the latter correctly explains the explanandum is *a priori*—indeed, trivially so. But now, of course, the NES requirement is violated, since (1) is a singular sentence that entails the explanandum of (c).

In example (d), even if the explanans is true and Woodward's condition of functional interdependence is satisfied, it does not follow that

(2) This pendulum has a length of 100 cm *because* it is a simple pendulum with a period of 2.03 seconds and the law of the simple pendulum holds[14]

The explanans of (*d*) correctly explains the explanandum only if (2) is true. Yet assuming the truth of the explanans of (*d*), the truth-value of (2) is not settleable *a priori*, but only empirically. Assuming that there is a lawlike connection between the period and length of a simple pendulum, whether the pendulum has the period it does because of its length, or whether it has the length it does because of its period, or whether neither of these is true, is not knowable *a priori*.[15]

More generally, in the explanans in each of these models some factors are cited, together with a lawlike sentence relating these factors to the type of event to be explained. But given that the factors were present and that the lawlike sentence is true, there is no *a priori* guarantee that the event in question occurred because of those factors. Whether it did is an empirical question whose answer even the truth of the lawlike sentence does not completely determine. And if we include in the explanans a sentence to the effect that the event in question did occur because of those factors we violate the NES requirement.

It might be replied that we should tighten the conditions on the lawlike sentence in the explanans by requiring not simply that it relate the factors cited in the explanans to the type of event in the explanandum, but that it do so in an explicitly explanatory way. Thus in (a) we might require the lawlike sentence not to be simply 'Anyone who eats a pound of arsenic dies within 24 hours' but

(3) Anyone who eats a pound of arsenic dies within 24 hours because he has done so

It is settleable *a priori* whether an explanans consisting of (3) together with 'Jones ate a pound of arsenic' is such that, if true, it would correctly explain the explanandum in (*a*).

This solution, however, would not be an attractive one for D-N theorists, since (3) is just a generalized explanation-sentence containing an explanatory connective that such theorists are trying to analyse by means of their model. Moreover, tightening the lawlike sentence in this way will

---

[14] We might be willing to say that this pendulum *must* have a length of 100 cm because it is a simple pendulum with a period of 2.03 seconds, etc. But here the explanandum is different.

[15] Woodward recognizes the pendulum example (d) as a genuine counter-example to the D-N model even when supplemented by his functional interdependence condition (p. 55). He believes that a causal condition, which he does not formulate, will need to be added to the D-N model in addition to his condition. This type of proposal will be examined in section 6 when Brody's causal model is discussed.

produce many false explanations, since such tightened sentences will often be false even though their looser counterparts are true. For example, (3), construed as lawlike, is false since people who eat a pound of arsenic can die from unrelated causes. And if we weaken (3)—still keeping the explanatory clause—by writing

(4) Anyone who eats a pound of arsenic *can* die within 24 hours because he has done so

we obtain a sentence that is true but not powerful enough for the job. It is not settleable *a priori* whether an explanans consisting of (4) together with 'Jones ate a pound of arsenic at t,' if true, would correctly explain why Jones died within $t + 24$, since he could have died for a different reason even though this explanans is true.

Another possible way of tightening the conditions on the lawlike sentence in the explanans is to require that it relate spatio-temporally contiguous events. (This would mean that the explanans would have to describe an event—or chain of events—that is spatio-temporally contiguous with the explanandum-event.) Jaegwon Kim has discussed laws of this sort, and he provides schemas for them which are roughly equivalent to the following:

(5) $(x)(t)(t')$ (x has P at t, and $\mathrm{loc}(x, t)$ is spatially contiguous with $\mathrm{loc}(x, t')$, and t is temporally contiguous with $t' \rightarrow x$ has Q at t').

(6) $(x)(y)(t)(t')$ (x has P at t, and $\mathrm{loc}(x, t)$ is spatially contiguous with $\mathrm{loc}(y, t')$, and t is temporally contiguous with $t' \rightarrow y$ has Q at t').[16]

'$\mathrm{loc}(x, t)$' means the location of x at time t. Kim does not specify a precise meaning for the arrow in (5) and (6), except that it is to convey the idea of 'causal or nomological implication'.[17] Under the present proposal, the arsenic explanation (a) would be precluded, since the only law invoked in (a) is not of forms (5) or (6). It does not express a relationship between types of events that are spatio-temporally contiguous. And, indeed, the explanans-event in (a) is not spatio-temporally contiguous with the explanandum-event.

This solution, like the previous one, may succeed in excluding intervening cause counter-examples such as (a). But it would not, I think, be welcomed by D-N theorists. If the arrow in (5) and (6) is to be construed causally as meaning (something like) 'causes it to be the case that', then, as with (3), the laws in D-N explanations will be generalized explanation-sentences containing an explanatory connective that D-N theorists

[16] Kim, 'Causation, Nomic Subsumption, and the Concept of an Event', *Journal of Philosophy*, 70 (1973), 217–36. This is a modification of Kim, whose formulations need a slight repair.
[17] Ibid. 229 n. 19.

seek to define by means of their model. Furthermore, requiring laws of forms (5) or (6) in an explanans will disallow explanations that D-N theorists, and many others, find perfectly acceptable. For example, it will not permit an explanation of a particle's acceleration due to the gravitational or electrical force of another body acting over a spatial distance. It will not permit explaining why a certain amount of a chemical compound was formed by appeal simply to (macro-) laws governing chemical reactions—where the formation of that amount of the compound takes time and is not temporally contiguous with the mixing of the reactants. Nor will the present proposal suffice to preclude all of the previous counter-examples. In particular, the pendulum example (d)—in which the pendulum's length is explained by reference to its period—is not disallowed. Assuming that the arrow in (5) and (6) represents nomological but not causal implication, we can express the following 'law':

$(x)(t)(t')$ (x is a simple pendulum with a period T at time t, and $\text{loc}(x, t)$ is spatially contiguous with $\text{loc}(x, t')$, and t is temporally contiguous with $t' \to x$ has a length L at t' which is related to T by the formula $T = 2\pi(L/g)^{1/2}$)

This, being of form (5), can be used in the explanans in (d), which, when suitably modified, will permit an explanation of the pendulum's length by reference to its period. For these reasons the present proposal does not seem promising.

Our observations regarding the various D-N models can be generalized. Assume that the explanans satisfies the NES requirement. In the explanans we can describe an event of a type always associated with an event of the sort described in the explanandum. We can include a law saying that such events are invariably and necessarily associated. The truth of the explanans event-description and of the law is no guarantee that the explanandum-event occurred because of the explanans-event. It could have occurred because some event unrelated to the one in the explanans was operative whereas the explanans-event was not. And this cannot be known *a priori* from the explanans. We can make it *a priori* by including in the explanans an appropriate singular sentence that entails the explanandum (e.g. an explanation-sentence that says in effect that the explanandum is true because the explanans is, or that the explanans-event caused the explanandum-event). But then the NES requirement would be violated. Or, we can make it *a priori* by using a generalized explanation-sentence. But since this is contrary to the philosophical spirit of such models and will, in any case, tend to produce false explanations, it will not be considered a viable solution. We can also make it *a priori* by requiring laws of forms (5) or (6) and construing the arrow causally. But this too does not satisfy the intent of such models, and, in addition, will not permit wanted explanations. On the other hand, if the arrow is understood nomologically but not causally, then whether the explanans, if true, correctly explains the explanandum is, in general, not knowable *a priori*.

To avoid the kind of problem in question we can say that it is an empirical, not an *a priori*, question whether an explanans describing events and containing laws relating these types of events to the explanandum-event correctly explains the latter. Or we can include in the explanans some singular sentence—either an explanation-sentence or something like it (e.g. (4) in Section 2)—that entails the explanandum. In the first case the *a priori* requirement is violated, in the second, NES. For this reason, I suggest, D-N models which attempt to provide sufficient conditions for correct explanations in such a way as to satisfy both these requirements will not be successful.

5. EMPIRICAL MODELS

The models I shall now mention satisfy NES but overtly violate the *a priori* requirement. Their proponents seem to recognize that if the former requirement is to be satisfied it is not an *a priori* but an empirical question whether the explanans if true would correctly explain the explanandum. However, I shall argue, the empirical considerations they introduce are not of the right sort to avoid the problem discussed above.

5.1. Salmon's Statistical-Relevance (S-R) model[18]

This is a model for the explanation of particular events. Salmon construes such an explanation as answering a question of the form

(1) Why is X, which is a member of class A, a member of class B?

Although Salmon does not do so, I shall say that the explanandum in such a case has the form

(2) X, which is a member of class A, is a member of class B

which is presupposed by (1). The explanans consists of a set of empirical probability laws relating classes A and B, together with a class inclusion sentence for X, as follows:

---

[18] Wesley C. Salmon, *Statistical Explanation and Statistical Relevance* (Pittsburgh, 1971).

(3)
$$p(B, A \cdot C_1) = p_1$$
$$p(B, A \cdot C_2) = p_2$$
$$\cdot$$
$$\cdot$$
$$\cdot$$
$$p(B, A \cdot C_n) = p_n$$
$$X \in C_k \quad (1 \leqslant k \leqslant n)$$

Salmon imposes two conditions on the explanans. One is that the probability values $p_1, \ldots, p_n$ all be different. (I shall not go into the reason for this.) The other is

(4) *The homogeneity condition*:

    $A \cdot C_1, \ A \cdot C_2, \ldots, A \cdot C_n$ is a partition of A, and each $A \cdot C_i$ is homogeneous with respect to B

$A \cdot C_1, \ A \cdot C_2, \ldots, A \cdot C_n$ is a *partition* of A if and only if these sets comprise a set of mutually exclusive and exhaustive subsets of A. To say that a set A is *homogeneous* with respect to B is to say that there is no way, even in principle, to effect a partition of A that is statistically relevant to B without already knowing which members of A are also members of B. (C is statistically relevant to B within A if and only if $p(B, A \cdot C) \neq p(B, A)$.) Intuitively, if A is homogeneous with respect to B then A is a random class with respect to B. Unlike the D-N models, Salmon's model does not require that the explanans show that the event in the explanandum was to be expected, but only with what probability it was to be expected.

I shall assume that for Salmon if the explanans (3) and the explanandum (2) are true, then the explanans correctly explains the explanandum, provided that Salmon's two conditions are satisfied.[19] Let us consider a simple example in which the probabilities have the values 0 and 1. There is a wire connected in a circuit to a live battery and a working bulb, and we want to explain why the bulb is lit, or more precisely, what, if anything, the wire does which contributes to the lighting of the bulb. Putting this in Salmon's form (1), the explanatory question becomes 'Why is this wire, which is a member of the class of things connected in a circuit to a live battery and working bulb, a member of the class of circuits containing a bulb that is lit?' The explanandum is

---

[19] Salmon in his book does not explicitly say this, but this seems to be the most reasonable interpretation of his position. See e.g. his remarks on pp. 79–80 in which he is distinguishing homogeneity from epistemic homogeneity, and in which he compares his model with Hempel's. Salmon might, of course, say that he is supplying conditions only for the concept of a well-confirmed or justified explanation. But I am construing his model in a stronger sense, and in private conversation he assures me that this is the correct interpretation.

(5) This wire, which is a member of the class of things connected in a circuit to a live battery and a working bulb, is a member of the class of circuits containing a bulb that is lit

Letting

A = the class of things connected in a circuit to a live battery and a working bulb

B = the class of circuits containing a bulb that is lit

C_1 = the class of things that conduct electricity

C_2 = the class of things that don't conduct electricity

X = this wire

we can construct the following explanans for (5):

(6)
$$p(B, A \cdot C_1) = 1$$
$$p(B, A \cdot C_2) = 0$$
$$X \in A \cdot C_1$$

Salmon's two conditions are satisfied since the probability values are different, and since $A \cdot C_1$ and $A \cdot C_2$ is a partition of A and each $A \cdot C_i$ is homogeneous with respect to B. Roughly, (6) explains why the bulb is lit by pointing out that the probability that the bulb in the circuit will be lit, given that the wire conducts electricity, is 1; that the probability that it will be lit, given that the wire does not conduct electricity, is 0; and that the wire does in fact conduct electricity.

Salmon's model satisfies the NES requirement since the only singular sentence in the explanans will not entail the explanandum. (Otherwise at least one of the probability laws in the explanans would be *a priori*, not empirical.) However, the homogeneity condition prevents his model from satisfying the *a priori* requirement.[20] Whether $A \cdot C_1, A \cdot C_2, \ldots, A \cdot C_n$ is a partition of A, i.e. whether these classes have any members in common and every member of A belongs to one of them, is not in general an *a priori* question (though it happens to be in the above example). Nor is the question of whether each $A \cdot C_i$ is homogeneous with respect to B. For example, it cannot be decided *a priori* whether there is some subclass of the class of electrical conductors such that the probability of the bulb being lit is different in this subclass from what it is in the class as a whole; this is an empirical issue. Accordingly, whether the explanans (6),

---

[20] It is possible to construe Salmon's model as requiring the satisfaction of the homogeneity condition to be stated in the explanans itself. If so the model would purport to satisfy the *a priori* requirement. However, in what follows I shall continue to assume that only sentences of the type in (3) comprise the explanans, and thus that the model is an empirical one. On either construction the same difficulty will emerge.

if true, would correctly explain the explanandum (5) is not settleable *a priori*.

Does the inclusion of the empirical homogeneity condition avoid the kind of problem earlier discussed plaguing the D-N models? Unfortunately not. This can be seen if we change our circuit example a bit. Let *A* and *B* be the same classes as before. We now introduce

C_3 = the class of things that conduct heat

C_4 = the class of things that do not conduct heat

I shall make the simplifying assumption that it is a law that something conducts heat if and only if it conducts electricity. Now consider this explanans for (5):

(7) $$p(B, A \cdot C_3) = 1$$
$$p(B, A \cdot C_4) = 0$$
$$X \in A \cdot C_3$$

Although Salmon's two conditions are satisfied, (7) ought not to be regarded as a correct explanation of (5), even if (7) is true. Intuitively, if we took (7) to correctly explain (5) we would be saying that the bulb is lit because the wire conducts heat (where the probability that it is lit, given that the wire does (not) conduct heat, is 1 (0)). But this seems incorrect. The bulb is lit because the wire conducts electricity not heat, though to be sure it does both, and that it does one if and only if it does the other is a law of nature. Admittedly, by our assumption, the class $A \cdot C_3$ = the class $A \cdot C_1$. But it is not a class which explains for Salmon, but a sentence indicating that an item is a member of a class. If the class is described in one way the explanation may be correct, while not if described in another way. In sentences of the form

(8) *X*'s being an $A \cdot C_i$ (a member of the class $A \cdot C_i$)—together with such and such probability laws—correctly explains why *X*, which is a member of *A*, is a member of *B*

the '$A \cdot C$' position is referentially opaque. A sentence obtained from (8) by substituting an expression referring to the same class as 'the class $A \cdot C_i$' will not always have the same truth-value.

The kind of example here used against Salmon's model[21] is similar to those raised earlier, in the following respect. In the explanans a certain fact about the wire is cited, namely, that it conducts heat, which (under the conditions of the set-up) is nomologically associated, albeit indirectly,

---

[21] A variety of such examples, as well as other trenchant criticism, can be found in John B. Meixner, 'Homogeneity and Explanatory Depth', *Philosophy of Science* 46 (1979), 366–81.

with the fact to be explained, namely, the bulb's being lit. However, it is not the explanans-fact that is the operative one in this case but the fact that the wire conducts electricity. By invoking the homogeneity condition Salmon in effect recognizes that the question of the explanatory operativeness of the explanans is not an *a priori* matter. The problem is that his homogeneity condition is not sufficient to guarantee that the explanans-fact is operative with respect to the explanandum-fact.

5.2. Brody's Essential Property Model

Brody construes this model as providing a set of sufficient conditions for explanations of particular events. These conditions are those of the basic D-N model together with the following:

Essential property condition: 'The explanans contains essentially a statement attributing to a certain class of objects a property had essentially by that class of objects (even if the statement does not say that they have it essentially) and ... at least one object involved in the event described in the explanandum is a member of that class of objects.'[22]

For example, since Brody thinks that atomic numbers are essential properties of the elements he would regard the following explanation as correct, provided its premises are true, since the D-N conditions plus his essential property condition are satisfied:

(9) This substance is copper
    Copper has the atomic number 29
    Anything with the atomic number 29 conducts electricity
        Therefore,
    This substance conducts electricity

Brody proposes the essential property condition in order to preclude certain counter-examples to the basic D-N model. Moreover, he regards the satisfaction of this condition as an empirical matter.[23] Brody's model, in cases in which the explanans does not say explicitly that the property in question is essential, does not satisfy the *a priori* condition.[24] To know whether (9) is a correct explanation if its premises are true is not an *a*

---

[22] B. A. Brody, 'Towards an Aristotelian Theory of Scientific Explanation', *Philosophy of Science*, 39 (1972), 20–31, 26. There is a corresponding model for the explanation of laws, which I will not discuss.

[23] Ibid. 27.

[24] If the explanans explicitly says that the property is essential, then the model purports to satisfy the *a priori* condition. But I want here to consider a model that explicitly violates this condition. (In either case the model turns out to be unsatisfactory.)

priori matter, since we must know whether having the atomic number 29 is an essential property of copper; and this knowledge is empirical, according to Brody. On the other hand, the NES requirement is satisfied since the singular premisses in an explanans will not entail the explanandum.

One might object to Brody's model on grounds of obscurity in the notion of an essential property (which, by the way, he seems to distinguish from mere properties a thing has necessarily). However, the problem I want to raise is not this, and so I shall suppose the model is reasonably clear; indeed I shall stick to atomic number, which is the sort of property Brody claims to be essential to the element which has it.

Consider now the following argument which satisfies Brody's essential property condition plus the other requirements of the D-N model:

(10) Jones ate a pound of the substance in that jar
The substance in that jar is arsenic
Arsenic has the atomic number 33
Anyone who eats a pound of substance whose atomic number is 33 dies within 24 hours
Therefore,
Jones died within 24 hours of eating a pound of the substance in that jar

Suppose, however, that Jones died in an unrelated car accident, and not because he ate arsenic. Although Brody's essential property condition is satisfied, as are the conditions of the D-N model, the explanans in (10) does not, even though true, correctly explain the explanandum. Brody may in effect recognize that it is not an *a priori* but an empirical question whether a D-N explanans if true correctly explains its explanandum. Nevertheless, the empirical requirement which his model invokes—the essential property condition—is not of the right sort to avoid the kind of problem plaguing this and previous models. Like Salmon's homogeneity condition, Brody's essential property condition is not sufficient to guarantee that the explanans-fact is explanatorily operative with respect to the explanandum-fact.

Both Salmon and Brody seem to recognize that if the NES requirement is to be satisfied it is an empirical, not an *a priori*, question whether an explanans if true correctly explains an explanandum. Yet the empirical considerations their models deploy are not sufficient to ensure that if satisfied an explanation will be correct if its explanans is true. Can problems of the sort generated by these models be avoided in any way other than by abandoning NES?

6. TWO CAUSAL MODELS

The final two models I shall discuss seem to offer a solution. They satisfy NES, violate the *a priori* requirement, and yet avoid the previous problems. However, whether they provide accounts that would be welcome to most of those seeking models of explanation is quite dubious.

6.1. Brody's Causal Model

As in the case of his essential property model, Brody regards this as providing a set of sufficient conditions for explanations of particular events. These conditions are those of the basic D-N model together with the

Causal condition: The explanans 'contains essentially a description of the event which is the cause of the event described in the explanandum'.[25]

To see how this is supposed to work let us reconsider

(1) Jones ate a pound of arsenic at time *t*
    Anyone who eats a pound of arsenic dies within 24 hours
        Therefore,
    Jones died within 24 hours of *t*

The problem we noted with the basic D-N model occurs, e.g. if both premisses of (1) are true but Jones died within 24 hours of *t* for some unrelated reason. Brody's causal condition saves the day since in such a case the event described in the first premiss of the explanans was not the cause of the event described in the explanandum. Hence, on this model we cannot conclude that the explanans if true correctly explains the explanandum.

The present model, like those in Section 5, violates the *a priori* requirement. Whether the explanans if true correctly explains the explanandum is not an *a priori* question, since the causal condition must be satisfied; and whether the explanans-event caused the explanandum-event is, in general, an empirical matter. It is not completely clear whether Brody wants to exclude from the explanans itself singular causal sentences that entail the explanandum, but I shall consider that version of his model which makes this exclusion. Like the models in Section 5, this model, I shall assume, is to satisfy the NES requirement.

[25] Brody, 'Towards', 23.

Woodward is another modellist who proposes the need for a causal condition to supplement the basic D-N model and his own functional interdependence condition:

These examples suggest that a fully acceptable model of scientific explanation will need to embody some characteristically causal notions (e.g., some notion of causal priority), or some more generalized analogue of these (e.g., some notion of explanatory priority)[26]

However, unlike Brody, he leaves open the question of how such a causal condition should be formulated.[27]

6.2. The Causal–Motivational Model[28]

Here, as in the D-N motivational model (Section 3), the explanandum is a sentence saying that some agent acted in a certain way. The explanans contains a sentence attributing a desire (motive, etc.) to that agent, and a sentence attributing the belief to that agent that performing the act described in the explanandum is, in the circumstances, a (the best, the only) way to satisfy that desire. Thus the explanandum might be a sentence of the form

Agent X performed act A

and the explanans might contain sentences of the form

X desired G
X believed that doing A is, in the circumstances, a (the best, the only) way to obtain G

Unlike the D-N motivational model, however, a law in the explanans relating beliefs and desires to actions is not required. What is required is the satisfaction of a

Causal condition: X's desire and his belief (described in the explanans) caused X to perform act A

The counter-example cited earlier against the D-N motivational model is now avoided, since in that example it was not the agent's belief and desire mentioned in the explanans, but some other belief and desire, that

[26] Woodward, 'Scientific Explanation', 53.
[27] In his most recent writings, Salmon too proposes adding a causal condition to his S-R model. Unlike Brody and Woodward, he attempts to define the concept of causation he utilizes, although, by his own admission, the definition is not entirely adequate. See his 'Why Ask "Why?"?', Proceedings and Addresses of the American Philosophical Association, 51 (1977), 683–705.
[28] See C. J. Ducasse, 'Explanation, Mechanism, and Teleology', Journal of Philosophy, 22 (1925), 150–5; Donald Davidson, 'Actions, Reasons, and Causes', Journal of Philosophy, 60 (1963), 685–700; Alvin I. Goldman, A Theory of Human Action (Princeton, NJ, 1976), 78.

caused him to act. As in the case of Brody's causal model, the *a priori* requirement is not satisfied, but NES is.

I shall not here try to defend or criticize these two models.[29] I shall assume for the sake of the argument that each avoids, or can be modified so as to avoid, the kind of problem I have been concerned with. However, each does so by violating NES in spirit, whereas the earlier models satisfy this requirement both in spirit and in letter. In order to apply the present causal models one must determine the truth of sentences such as these:

(2) Jones's eating a pound of arsenic at time *t* caused him to die within 24 hours of *t*

(3) Smith's desire to buy eggs and his belief that going to the store is the only way to do so caused him to go to the store

But these are singular explanation-sentences that entail the explanandum. To be sure, neither model requires such sentences to be in the explanans. Still in each model to determine whether the explanans if true correctly explains the explanandum one has to determine the truth of such sentences. I am not criticizing the models on these grounds. But I believe that many of those who seek models of explanation will want to do so. They will say that in order to know whether the explanans in such a model correctly explains the explanandum one has to determine, independently of the truth of the explanans, the truth of sentences of a sort these modellists want to exclude from the explanans itself. Moreover, they will point out that there is not much difference between determining the truth of (2) and (3), on the one hand, and that of

Jones's eating a pound of arsenic at *t* explains why he died within 24 hours of *t* (or Jones died within 24 hours of *t* because he ate a pound of arsenic at *t*)

Smith's desire to buy eggs and his belief that going to the store is the only way to do so explains why he went to the store (or, Smith went to the store because he had this desire and belief)

on the other. Models which impose the above causal conditions, they are likely to say, provide insufficient philosophical clarification for the concept of explanation, even though the explanation-sentence expressing the causal relationship is not itself a part of the explanans.[30] If a central aim

---

[29] A cogent attack on Brody's model can be found in Timothy McCarthy, 'On an Aristotelian Model of Scientific Explanation', *Philosophy of Science*, 44 (1977), 159–66. See Davidson, 'Psychology as Philosophy', in J. Glover (ed.), *Philosophy of Mind* (Oxford, 1976), 103–4, for criticism of the causal-motivational model (which Davidson himself once supported).

[30] Indeed, some formulations of the causal-motivational model (e.g., Ducasse's and Goldman's) seem to allow as an explicit part of the explanans a singular explanation-sentence that entails the explanandum. In such cases the NES requirement is violated in letter as well as in spirit.

of modellists is to define terms such as 'explain', 'because', and 'causes', this excludes their employment as primitive notions in the explanans, or, as in the present case, in the conditions of the model.

Regardless of whether we view such criticism as important, the present models are of interest because to avoid the sorts of problems raised in Section 4 these models, unlike their D-N ancestors, require establishing the truth of empirical sentences to determine whether the explanans if true correctly explains the explanandum. In this respect they are like Salmon's S-R model and Brody's essential property model. However, unlike the latter, the empirical sentences whose truth they require establishing are themselves singular sentences that entail the explanandum.

7. CONCLUSIONS

Our discussion suggests the following conclusions: (1) If the explanans is not to contain singular sentences that entail the explanandum, then it will be an empirical not an *a priori* question whether the explanans if true correctly explains the explanandum. (2) This empirical question will involve determining the truth-values of certain singular sentences (either explanation-sentences or something akin to them) that do entail the explanandum; otherwise factors will be citable in the explanans which are not explanatorily operative with respect to the explanandum-event.

Can there be a model of explanation? Specifically, can there be a set of sufficient conditions which are such that if they are satisfied by the explanans and explanandum the former correctly explains the latter? Our discussion suggests that there can be no such model if, like D-N theorists, we insist that it satisfy both the *a priori* and the NES requirements. Moreover, it also suggests that we will not be successful in discovering a model in which (*a*) the NES requirement is satisfied, and in which (*b*) it is not an explicit condition of the model that some singular sentence be true that entails the explanandum.

It does not follow from this that an explanans which appeals to causal factors, laws, dispositions, desires and beliefs, statistically relevant factors, or essential properties cannot correctly explain an explanandum. However, models of explanation of the sort I am considering do not simply list kinds of factors that can be explanatory. Their proponents want to supply sufficient conditions for correct explanations. If it is demanded that these conditions satisfy the NES and *a priori* requirements, or NES plus (*b*) above, then I am suggesting that such models will not be forthcoming.

VI

PROBABILITY, EXPLANATION, AND INFORMATION*

PETER RAILTON

INTRODUCTION

Elsewhere I have argued that probabilistic explanation, properly so called, is the explanation of things that happen by chance: the outcomes of irreducibly probabilistic processes.[1] Probabilistic explanation proceeds, I claimed, by producing a law-based demonstration that the explanandum phenomenon had a particular probability of obtaining and noting that, by chance, it did obtain. This account will be presented in more detail below, but the bulk of this paper will be given over to a discussion of two large problems confronting such a view: (1) many seemingly acceptable explanations of chance phenomena do not make explicit use of laws; and (2) many seemingly acceptable explanations that at least purport to be probabilistic concern phenomena known or assumed to be deterministic. Problem (1) is but an instance of a very general difficulty facing law-based accounts of explanation, and the approach to solving it suggested below may readily be extended to non-probabilistic explanation. Problem (2) raises a difficulty peculiar to probabilistic explanation, but I hope to show that an appropriate solution to problem (1) offers the key to a plausible solution to problem (2).

Discussions of both problems run the risk of degenerating into merely verbal quibbles. However, I believe that the issues concerning the appli-

Peter Railton, 'Probability, Explanation, and Information', *Synthese*, 48 (1981), 233–56. Copyright © 1981 by D. Reidel Publishing Co., Dordrecht. Reprinted by permission of Kluwer Academic Publishers.

* I am very grateful to David Lewis, Lawrence Sklar, and Timothy McCarthy for helpful comments on some of the material contained in this paper. Some of the research for this paper was supported by a dissertation-writing fellowship from the Whiting Foundation at Princeton University and by a Junior Fellowship in the Society of Fellows at the University of Michigan, Ann Arbor; their support is acknowledged with thanks.

[1] In 'A Deductive-Nomological Model of Probabilistic Explanation', *Philosophy of Science*, 45 (1978), 206–26.

cation of the expressions 'explanation' and 'probabilistic explanation' which will be discussed here are genuine, and that we may gain understanding of explanatory practice by seeing how one might resolve problems (1) and (2). To revive an old way of putting it, there is a worthwhile analogy between 'explanation' and 'proof'.[2] Various things have come to be called 'proofs'—deductive arguments, experiments, testimony, and testimonials, to name a few—but when well-informed we do not allow these various things to play the same roles in our lives. When we do logic, science, jury duty, or shopping, we attend to the relevant differences among so-called 'proofs'. Similarly, various things have come to be called 'explanations'—deductive arguments, inductive arguments, diagnoses, explications of meaning, excuses, and apologies, to name a few— but when well-informed we do not allow these various things to play the same roles in our lives. When we do science, engineering, politics, and car repair we attend to the relevant differences among so-called 'explanations'. 'Probability', too, is applied to various things—degrees of belief, degrees of confirmation, relative frequencies, chancy dispositions, and the outcomes of combinatorics, among others. These differences, too, must be attended to when we look for guidance in our lives. I hope it will be clear in what follows that what one says about explanation and probabilistic explanation, and thus what one says about 'explanation' and 'probabilistic explanation', may make a difference that *is* a difference to actual practice.

1. THE DEDUCTIVE-NOMOLOGICAL-PROBABILISTIC ACCOUNT OF PROBABILISTIC EXPLANATION

The essence of probabilistic explanation, according to the deductive-nomological-probabilistic (D-N-P) account, is a deductive demonstration to the effect that an empirical theory assigned a particular probability to an explanandum phenomenon. If the relevant parts of the theory are true, the factual premises involved accurate, and the deduction valid, then the D-N-P explanation is true.[3] If the premises are only partly true, or only approximately true, then the explanation may be only partly or approximately true. (Whatever partial and approximate truth are, this is not the place to analyse them.)

I take it that quantum mechanics, under the dominant interpretation, gives us reason to think that there are irreducibly probabilistic processes

---

[2] This analogy has been most fully developed by C. G. Hempel. See his 'Aspects of Scientific Explanation', in *Aspects of Scientific Explanation and Other Essays* (New York: Free Press, 1965).

[3] Here I follow Hempel's use of 'true' for explanations. See 'Aspects', 338.

in nature, and that they are governed by probabilistic and non-probabilistic laws. Thus, the occurrence of a given alpha-decay, or of particular rates of alpha-decay, is widely held to be the result of a physically random process: spontaneous nuclear disintegration. If the dominant interpretation of quantum mechanics is right, there are no 'hidden variables' characterizing unknown initial conditions of nuclei that suffice, in conjunction with deterministic laws, to account for the occurrences of alpha-decay: two radio-nuclei may be in the same physical state at a time t_0, and may be subjected to the same environment during the time interval from t_0 to $t_0 + \tau$ and yet one may decay during τ and the other not. Although this process is physically random in the sense that there is no 'sufficient reason' for the one atom to decay but not the other, alpha-decay does obey physical laws: probabilistic laws such as those concerning barrier penetration and decay rate; deterministic laws such as those concerning the conservation of mass-energy and charge. Using certain simplifying assumptions, and substituting the appropriate values for decay energy, atomic number, and atomic weight in the Schrödinger wave-equation, it is possible to derive the decay constants of radio-elements from quantum mechanics. Moreover, it has become conventional wisdom that such decay constants are not mere statistical averages drawn from large samples, but rather physically-irreducible, single-case probabilities—probabilities 'per unit time for one nucleus'.[4] The psi-function of the wave-equation, then, is viewed as giving a complete description of the state of individual systems, and this state-description may at best determine only a probability distribution over future states.

For an example of a D-N-P explanation of a particular chance fact, let us suppose that a Uranium 238 nucleus u has alpha-decayed during a time interval of length τ which began at t_0. Current physical theory assigns a decay constant, λ_{238}, to U^{238} nuclei, which allows us to state the probability that an individual U^{238} nucleus will decay during an interval of length t in the absence of external radiation: $(1 - \exp(-\lambda_{238} \cdot t))$. If we assume that u was not excited by external radiation at t_0 or after, we may make the following deductive-nomological inference:

(1a) All U^{238} nuclei have probability $(1 - \exp(-\lambda_{238} \cdot \tau))$ to emit an alpha-particle during any interval of length τ, unless subjected to environmental radiation

4 R. D. Evans, *The Atomic Nucleus* (New York: McGraw-Hill, 1965).

(1*b*) *u* was a U^{238} nucleus at t_0 and was not subject to environmental radiation during the interval $t_0 - (t_0 + \tau)$

(1*c*) *u* had probability $(1 - \exp(-\lambda_{238} \cdot \tau))$ to alpha-decay during $t_0 - (t_0 + \tau)$

Inference (1) yields a D-N explanation of the fact that *u* had a particular probability to alpha-decay at the time in question.

According to the D-N-P account, if (1) were prefaced with a theoretical derivation of the law premiss (1*a*) from underlying quantum physics, and were followed by a parenthetic addendum to the effect that *u* did indeed alpha-decay during the interval $t_0 - (t + \tau)$, the result would be a full probabilistic explanation of *u*'s alpha-decay. Schematically, a D-N-P explanation of the fact that *e* is *G* at some time t_0 has the form (2) as follows:

(2*a*) A theoretical derivation of a probabilistic law of the form (2*b*)

(2*b*) $\forall x \, \forall t \, [F_{x,t} \rightarrow \text{Probability } (G)_{x,t} = r]$

(2*c*) F_{e,t_0}

(2*d*) Probability $(G)_{e,t_0} = r$

(2*e*) (G_{e,t_0})[5]

Explanation (2) is a *purported* probabilistic explanation in D-N-P form. If its premisses were true and its logic impeccable, it would be a *true* one as well.

It is obvious from the universal form of law (2*a*), and from the use of universal instantiation and *modus ponens* in deriving the conclusion (2*d*), that the probability involved is a single-case probability. The parenthetic addendum (2*e*) *is* the explanandum, and yet it in no way follows from the other premisses. Is this paradoxical? No, it had better not be derivable in this way, for we are supposing that G_{e,t_0} occurred by chance, and thus that its occurrence cannot be derived from initial conditions by any empirical law. Presumably, we knew that G_{e,t_0} before any explanation was offered, and so step (2*e*) brings no news. In that sense, it is dispensable. Indeed, it is dispensable in another sense as well: those holding the view that explanations are always *arguments*—and should contain nothing inessential or incidental to such arguments—will want to see the addendum gone from the D-N-P schema. I have no real quarrel with either reason for dropping it, except that it will be of use in stringing together D-N-P explanations or amalgamating them with larger explana-

---

[5] The connective '$\rightarrow$' is to be read in whatever way it should be read for laws of nature in general, e.g., as a 'strong conditional'. The parentheses enclosing the addendum (2*e*) are meant to indicate that it is not a logical consequence of the steps above it. More will be said about the nature of the theoretical derivation (2*a*) in what follows.

tions; but more on this later. Moreover, I would like to urge that we move away from the view that explanations are purely arguments and pursue instead the idea that explanations are *accounts*—accounts in which arguments play a central role, but do not tell the whole story. Again, we will return to this below. For now I leave the parenthetic addendum in place, a token of my disaffection with the 'pure argument' view.

It is more important to observe that the D-N-P account makes nothing of the *degree* of probability of the explanandum according to the explanans: in schema (2), r may take on any value in the unit interval without prejudice to explanation, as long, of course, as the probability-attribution is accurate. Thus, in contrast to Hempel's inductive-statistical model of probabilistic explanation, the D-N-P account does not require *high* probability either to establish the inductive relevance of the explanans to the explanandum (since deductive relevance takes its place), or to establish the nomic expectability of the explanandum (since nomic expectability 'with practical certainty' is replaced by nomic expectability with whatever probability the explanandum had).[6] This has the considerable advantage that we are no longer forced to say that less-than-highly-probable facts are inexplicable in principle. Some things that happen are vastly improbable, such as the decay at a given moment of an atom of U^{238} (mean life: 6.5×10^9 years), or are merely probable, such as the creation of a levorotatory form when an amino acid is synthesized (probability under standard conditions: .5); and when merely probable or vastly improbable things do occur, the appropriate explanation is that they had a particular probability of occurring, which chanced to be realized. Probabilistic explanation thus is not a second-string substitute for deterministic explanation, showing that the explanandum phenomenon *almost* had to occur. On the contrary, probabilistic explanation is a form of explanation in its own right, charged with the distinctive task of dealing with phenomena that came about by chance. As such, there is no special explanatory virtue in probabilistic explanations that show their explanandum to have been highly probable, unless the explananda in question *were* highly probable. If they were not, then any explanations purporting to establish their high probability would not be explanatorily virtuous, but explanatorily inaccurate. Probabilistic explanations conferring high probability may have other sorts of virtues, however: other things equal, they receive greater inductive support from the evidence that the explanandum phenomenon occurred; other things equal, they may support more definite policies for the future; and so on. But we should not

---

[6] See Hempel, 'Aspects', sect. 3; and Railton, sect. 5.

confuse all the virtues of explanations with explanatory virtues. Without being a better explanation, one explanation may be more easily confirmed, more fully confirmed, more readily applied, or more easily translated into modern Greek than another.

A satisfactory D-N-P explanation need not even pick out factors that *raise* the probability of the explanandum over what it would have been in their absence.[7] Consider the following example. Suppose that we are applying a herbicide to a patch of healthy milkweed, and suppose that a dose of this herbicide alters the biochemical state of milkweeds from a normal, healthy state S, in which plants have probability .9999 of surviving 24 hours, to a state S', in which there is but probability .05 of lasting that long. When we return to the milkweed patch 24 hours after spraying and find to our consternation that a particular plant which received a full dose of herbicide is still standing, how are we to explain this? Presumably, we should point out that the plant was in state S' after the spraying, that in this state it had probability .05 of surviving 24 hours, and that, by chance, it did. The spraying, then, is part of the explanation of survival even though it *lowered* the probability of survival for this plant from what it would otherwise have been. We may not wish to speak of the spraying as a *probabilistic cause* of survival, since we may want to reserve the expression 'probabilistic cause' for factors that *do* raise the probability of an event in the circumstances. Thus, for those plants failing to survive, we could speak of the spraying as a probabilistic cause of their deaths, while for the plant that survived, it would be strained at best to speak of the spraying as a probabilistic cause of its survival. This suggests that probabilistic explanation is not a mere subspecies of causal explanation.[8]

It might be objected that if probabilistic explanations do not establish high probability for their explananda or do not point to factors that raise the probability of their explananda, then how do they show why one outcome—alpha-decay, survival, or whatever—was realized rather than its opposite? The short answer is that they do not show this and should not try. If there were a reason why one probabilistic outcome of a chance process was realized rather than another, we would not be dealing with a chance process.

---

[7] Just how this sort of counterfactual is to be read is too large an issue to enter into in this context. Instead, in the example that follows I will assume what I take to be a natural reading of the relevant counterfactual.

[8] I have borrowed this example of a plant surviving herbicide spraying from Nancy Cartwright (in conversation), who used it to make a different point. I am, of course, assuming that the effect of the herbicide is genuinely probabilistic—perhaps owing its indeterminism to the chance factors (such as electron location) that influence chemical bonding.

Finally, because the D-N-P account is deductive in form and based upon lawful, physical, single-case probabilities, it can be shown not to suffer from the problems of explanatory ambiguity and epistemic relativity that have bedevilled the Hempelian inductive-statistical account.[9] If the very same fact were to be assigned two different lawful, physical, single-case probabilities by our theory, then contradiction, not ambiguity, would have to be dealt with, and we should proceed accordingly. As for epistemic relativity, incomplete knowledge may make something *look like* a genuinely probabilistic law or a genuine probabilistic explanation, but it cannot make things *be* other than they are. This is strictly parallel with non-probabilistic explanation: in both cases there are problems about distinguishing laws from mere factual regularities; in both cases there are problems about distinguishing the true from the false. This is not to say that there are no special epistemological problems in the probabilistic case, but only to say that these problems *are* epistemological and do not make successful explanation independent of either lawfulness or truth.

With this brief sketch of the D-N-P account in hand, let us turn to two important problems it faces.

2. PROBABILISTIC EXPLANATIONS WITHOUT LAWS

A difficulty that arises at once for the D-N-P account is that many of the probabilistic explanations offered and accepted in scientific discourse fall far short of providing all of the elements of schema (2). We have encountered an example already in the explanation given of a particular milkweed plant's survival:

(3) the plant was in state S' after the spraying ... in this state it had probability .05 of surviving 24 hours, and ... by chance, it did

Explanation (3) contains no theoretical derivation, no explicit law, and yet seems highly acceptable. Other less-than-full-fledged probabilistic explanations encountered in scientific contexts are bare indeed: 'Why did this muon decay?' 'Because it was unstable'; 'Why is the Geiger counter clicking as we approach that rock?' 'Because the rock contains uranium'; 'Why is this one lobster blue?' 'It is a random mutation, very rare'; and so on. Proffered probabilistic explanations may fail to include all of the elements of (2) for a variety of reasons: in certain contexts, a more

---

[9] See, for example, Hempel, 'Aspects', sect. 3, and 'Maximal Specificity and Lawlikeness in Probabilistic Explanation', *Philosophy of Science*, 35 (1968), 116–33.

elaborate explanation may be out of place—the audience may be too well-versed, not well-versed enough, not interested enough, or short on time; a more elaborate explanation may not be available even if it were appropriate—the relevant laws and facts may not be known, or may be known only qualitatively; the person offering the explanation may simply not know enough; and so on. Whatever the reason, does the D-N-P account force us to say that these less-than-full-fledged specimens are not explanations?

In many cases, that would be an intolerably strict position to take. But where should one draw the line between explanation and non-explanation? The answer lies in not drawing lines, at least at this point, and in recognizing instead a continuum of explanatoriness. The extreme ends of this continuum may be characterized as follows. At one end we find what I will call an *ideal D-N-P text*, comprising all that schema (2) involves. At the other end we find statements completely devoid of what I will call 'explanatory information'.[10] What is explanatory information? Consider an ideal D-N-P text for the explanation of a fact *p*. Now consider any statement *S* that, were we ignorant of this text but conversant with the language and concepts employed in it and in *S*, would enable us to answer questions about this text in such a way as to eliminate some degree of uncertainty about what is contained in it. To the extent that *S* enables us to give accurate answers to such questions, i.e., to the extent that it enables us to reconstruct this text or otherwise illuminates the features of this text, we may say that *S provides explanatory information concerning why p*. It is hardly novel to speak of sentences conveying information about complete texts in this way: presumably we employ such a notion whenever we talk of a piece of writing or an utterance as a summary, paraphrase, gloss, condensation, or partial description of an actual text, such as a novel, a speech, or a news report. Unfortunately, I know of no satisfactory account of this familiar and highly general notion, and so cannot appeal to one here. Nor can I begin to provide an account of my own making within *this* text. To make matters worse, I cannot even pretend to be using the well-defined notion of information employed in Wiener–Shannon information theory. However, before I say anything more about the sort of analysis that is needed for the concept of information employed here, let us look at a few examples to see of what use that concept might be.

---

[10] I owe the expression 'explanatory information' to David Lewis, who, in the course of advising my dissertation on explanation, supplied me with this notion as a way of expressing the idea of illuminating ideal explanatory texts. I do not know whether he would agree with the way I am using this notion here.

To be told that a Geiger counter is clicking because it is near a uranium-bearing rock is to be given explanatory information—assuming that the claim is true as far as it goes—since it would be part of the relevant ideal D-N-P text for the clicking that the rock in question contains uranium. Of course, the ideal D-N-P text would go on to tell us about the radioactivity of uranium, about the process of alpha-decay, about the mechanics of Geiger counters, and so on. Obviously, the amount of explanatory information provided by the brief and informal explanation given above is much less than the amount needed to reconstruct a sizeable chunk of the entire ideal D-N-P text, sufficing only to illuminate one central feature of the whole text. But to illuminate one central feature is to convey a not insignificant amount of information, and thus to be off the zero point of our continuum. Somewhat nearer that point, but still above it, is the explanation 'Why did this muon decay?' 'Because it was unstable'. This explanation accurately points to a dispositional property of the muon responsible for the decay, distinguishing this decay from disintegration due to external forces such as bombardment by other particles; and as we know from the shortcomings of extensionalist accounts of dispositional properties, it is one thing to say that the muon decayed, and something else—something more—to say that it decayed owing to a disposition to do so. The disposition in this case is probabilistic, and so the relevant ideal explanatory text would be D-N-P in form. The brief statement about instability thus answers certain questions about the form as well as the content of the relevant ideal text. The explanation of a lobster's blueness in terms of a 'random mutation, very rare' also tells us that the relevant explanatory text is D-N-P in form, and further indicates the particular probabilistic process involved—genetic mutation—and provides a qualitative characterization of the probability of blueness. Thus it falls higher on the explanatory continuum than the Geiger counter clicking or muon decay explanations. A bit higher on the continuum is what (3) says about the explanation of a particular milkweed plant's survival. (3), too, tells us that the relevant ideal text is D-N-P in form, and mentions an important factor in the process culminating in the explanandum—the spraying—but it further supplies a precise, quantitative evaluation of the probability involved. Much higher on the continuum are the explanatory remarks made earlier about the alpha-decay of U^{238} nucleus u, for they contained a probabilistic law explicitly and roughed out a theoretical derivation of this law. Higher still are the well-developed explanatory accounts one finds in physics or chemistry textbooks and monographs. These are not usually explanations of particular facts, but our characterization of explanatory information is indifferent as to whether p is a particular or general fact. Moreover, outside the context of standard teaching and reference works, scientists are often concerned with particular-fact explanation—e.g. in the course of experimentation—and, as the D-N-P account recognizes, elaborate theoretical derivations have a natural place in fully developed explanations of particular facts.

Perhaps this is the point at which to say something more about what the theoretical derivations figuring in schema (2) might look like. The place to look for guidance is plainly scientific explanatory practice itself. If one inspects the best-developed explanations in physics or chemistry textbooks and monographs, one will observe that these accounts typically include not only derivations of lower-level laws and generalizations from higher-level theory and facts, but also attempts to *elucidate the mechanisms* at work. Thus an account of alpha-decay ordinarily does more than solve the wave-equation for given radionuclei and their alpha-particles; it also provides a model of the nucleus as a potential well, shows how alpha-decay is an example of the general phenomenon of potential-barrier penetration ('tunnelling'), discusses decay products and sequences, and so on. Some simplifying assumptions are invariably made, along with an expression of hope that as we learn more about the nucleus and the forces involved we will be able to give a more realistic physical model. It seems to me implausible to follow the old empiricist line and treat all these remarks on mechanisms, models, and so on as mere *marginalia*, incidental to the 'real explanation', the law-based inference to the explanandum. I do not have anything very definite to say about what would count as 'elucidating the mechanisms at work'—probabilistic or otherwise—but it seems clear enough that an account of scientific explanation seeking fidelity to scientific explanatory practice should recognize that part of scientific ideals of explanation and understanding is a description of the mechanisms at work, where this includes, but is not merely, an invocation of the relevant laws. Theories broadly conceived, complete with fundamental notions about how nature works—corpuscularianism, action-at-a-distance theory, ether theory, atomic theory, elementary particle theory, the hoped-for unified field theory, etc.—not laws alone, are the touchstone in explanation. Of course, there *are* marginal comments to be found accompanying the explanations offered in scientific textbooks and monographs: one reads of helpful ways of visualizing, conceptualizing, applying mathematical devices, etc., which the reader is warned not to take too seriously. There is no hard-and-fast line between simplified idealizations of actual physical processes—such as the model of a gas as a collection of perfectly elastic spheres or the 'liquid-drop' model of the atomic nucleus—and more genuine marginalia or heuristics—such as the treatment of electron orbits as harmonic vibrations of strings, the analogy between electrostatics and hydrostatics, the use of topological images in describing systems with multiple equilibria, and so on. Moreover, one theory's attempt at a realistic model may serve as no more than a first-order approximation for later developments of that theory—e.g. atomic theory in this century. Genuine marginalia and heuristics are usually flagged as such, but in a developing theory there may be indefiniteness about whether a given model should be taken in a realistic and explanatory way or as an unrealistic 'picture' with instrumental value only. No analysis of explanation should be more definite on such questions than the best available theories themselves; so while we should ask of scientists whether a particular model ought to be part of ideal explanatory texts, we should be prepared to accept indefiniteness and even disagreement in the answers we receive.

At a particular time we may have nothing like the knowledge necessary to produce an ideal D-N-P text for a given chance phenomenon. Not knowing fully what the relevant ideal text looks like, we evidently will be unable in many cases to say how much explanatory information a given proffered explanation carries. But again, it is not the job of an analysis of explanation to settle questions beyond the reach of existing empirical science. Instead, it is appropriate and adequate that an analysis of explanation define what *would be* explanatory information in such cases. On the analysis given here, a proffered explanation supplies explanatory information (whether we recognize it as such or not) to the extent that it does in fact (whether we know it or not) correctly answer questions about the relevant ideal text. Whether in a given context we *regard* a proffered explanation as embodying explanatory information, in light of the interpretation we impose on it and our epistemic condition generally, is a matter for the pragmatics of explanation.

The general form of the ideal D-N-P text is meant to represent the ideal striven for in actual explanatory practice, i.e., it comprises the things that a research programme seeks to discover in developing the capacity to produce better explanations of chance phenomena. Thus, the ideal D-N-P text reflects not only an ideal of explanation, but of *scientific understanding*: we may say that we understand why a given chance phenomenon occurred to the extent that we are able, at least in principle, to produce the relevant ideal D-N-P text or texts. However, the concept of scientific understanding—and its links to ideals of explanation and to fundamental conceptions of nature—deserves more serious treatment than it can be given here. Needless to say, even if we did possess the ability to fill out arbitrarily extensive bits of ideal explanatory texts, and in this sense thoroughly understood the phenomena in question, we would not always find it appropriate to provide even a moderate portion of the relevant ideal texts in response to particular why-questions. On the contrary, we would tailor the explanatory information provided in a given context to the needs of that context; if we had the capacity to supply abitrarily large amounts of explanatory information, there would be no need to flaunt it.

As mentioned earlier, there is no ready-made analysis of information capable of doing the work here asked of the concept of information. The sort of information discussed in *information theory* is *syntactic*, a measure of the statistical rarity of signals from an observed source, and the sort discussed here is *semantic*, it is information *about* something (an ideal text).[11] However, there is an important analogy that makes the same term applicable: in both cases the amount of information carried by a 'message' is proportional to the degree to which it reduces uncertainly; thus a statistically common signal from a stable source embodies less information than a statistically rare one, and a fuller proffered explanation leaves less to be known about the relevant ideal text than a slighter one. Hence, information is a kind of *selection power* over possibilities. But here the analogy ends. Semantic equivalents, such as 'son' and 'male offspring', *as signals* may require different amounts of information to be transmitted; and there is no guarantee whatsoever that a message embodying huge amounts of syntactic information (and so requiring a great deal of channel capacity to transmit) will embody much if any semantic information—witness certain television test patterns (and perhaps certain programmes as well). It is rather elusive of me to speak of explanatory information without having an adequate theory of semantic information, but none of us has an adequate theory of semantic information and the notion still receives very heavy use. I trust that as I join the multitude it is intuitively (deceptively?) clear what this notion involves.

Perhaps a few things might be said by way of clarification. If a self-consistent sentence S implies a sentence S^*, then S contains at least as much semantic information as S^*. Thus, for example, 'p' is at least as informative as '$p \lor q$', since 'p' reduces the range of possibilities at least as much as '$p \lor q$'; and '$p \& q$' is at least as informative as 'p' or 'q' alone, since the conjunction reduces the range of possibilities at least as much as either individual conjunct. If S is a tautology, then S contains

---

[11] For a discussion of the differences between 'information about' and Wiener–Shannon information, see Colin Cherry, *On Human Communication*, second edn. (Cambridge, Mass.: MIT Press, 1966), ch. 6.

no semantic information (at least, if we take logic for granted), since S eliminates no (logical) possibilities. If S contains semantic information $I(S)$, and S^* contains semantic information $I(S^*)$, and if the conjunction of S and S^* is self-consistent, then the amount of semantic information contained in the conjunct will equal $I(S)$ plus $I(S^*)$ minus any overlap. And so on. In these properties of semantic information it may be seen that, as with syntactic information, this quantity bears something of an inverse relation to probability. Propositions with logical probability one, tautologies, convey zero semantic information; conjunction raises semantic information, but for independent events it lowers probability; disjunction lowers semantic information, but for independent events it raises probability; and so on. But a thicket of problems confronts any attempt to press an analysis of semantic information, or its formal parallels with syntactic information, much further. First, if we are going to speak of semantic information as reducing uncertainty, we must have a probability measure of the appropriate kind. The Carnap–Bar-Hillel approach to semantic information uses Carnapian logical probability, but this brings with it grave problems.[12] Second, to talk of the semantic content of a sentence is to assume some *semantic interpretation* of it, and this raises a number of issues: What sort of semantics should we use? Must semantic interpretation be a function of context? What constraints must be met in order for a semantic interpretation to exist? And so on. For our purposes I propose to ignore the general questions about semantic interpretation and speak only of uncertainty as a measure of the extent to which, in what I will call a 'standard context', one is unable to reconstruct particular or general features of the relevant ideal explanatory text. Clearly, any general analysis of semantic information would have to provide a measure of the extent to which a given sentence allows us, in a *variety* of contexts, to 'recover' or 'reconstruct' an aspect of the *world* itself. It is less problematic, though by no means easy, to treat of semantic information in a particular *kind* of context and with respect to a definite *text*. The sort of *standard context* I have in mind is the following: one understands all of the concepts and terms used in the relevant ideal text and in the proffered explanation, but is ignorant of the overall structure and the details of that ideal text. Hereinafter, when I speak of a proffered

---

[12] See Y. Bar-Hillel and R. Carnap, 'Semantic Information', *British Journal for the Philosophy of Science*, 4 (1953), 147–57. It should be noted that a proffered explanation may communicate some explanatory information in virtue of the *organization* of its components as well as their content proper. Thus a tautology by itself lacks semantic information, but it may play a role in the structure of a proffered explanation that informatively reflects the structure of the relevant ideal text. There is no obvious way of incorporating *organization* into existing accounts of semantic information.

explanation conveying explanatory information, I will mean that it would do so in the appropriate standard context.[13]

The account of explanatory information just outlined is liable to trivialization. For example, suppose that an alpha-decay occurs, and suppose that someone proffers the following 'explanation': 'the relevant ideal text contains more than 10^2 words in English'. For all that has been said here, this remark, if true (as it doubtless is), would count as conveying explanatory information and hence as some sort of an explanation. The first line of defence against this kind of degenerate case would be to say that the information must be about the *content* and not merely the *form* of the relevant ideal text. But this is not satisfactory, for it certainly is explanatory to be told, of the alpha-decay, that the relevant ideal text is probabilistic rather than deterministic in form. In order to avoid further complication, I will simply tolerate this kind of degenerate case, relegating it to the very low end of the continuum of explanatoriness. And rightfully enough: it certainly does not afford much illumination of the relevant ideal explanatory text.

In sum, proffered explanations of chance phenomena that fail to live up to the standards of the D-N-P schema may still count as explanations, even as quite good explanations, in virtue of finding another way of communicating explanatory information. In context, such explanations may be more than adequate.

What has been said here about the potential explanatoriness of statements lacking strict D-N-P form can be extended in obvious ways to non-probabilistic explanations and may be used to replace the justly criticized idea that the D-N schema provides necessary conditions for successful explanations. I would argue that the D-N schema instead provides the skeletal form for ideal explanatory texts of non-probabilistic phenomena, where these ideal texts in turn afford a yardstick against which to measure the explanatoriness of proffered explanations in

---

[13] One particularly difficult problem for a theory of semantic information, especially one that focuses on information as selection power, will be to capture some form of the *de re/de dicto* distinction. For example, we may have two proffered explanations of a given phenomenon. One correctly but sketchily describes the relevant ideal explanatory text; the other is a disjunction of a very full description of the relevant ideal explanatory text and a feeble description of some quite irrelevant ideal explanatory text. Both explanations would then be true, and we can imagine that (by some intuitive measure) both manage overall to exclude an equally large number of irrelevant possibilities—the specificity of the first disjunct of the latter making up for the failure to eliminate the second disjunct. It may seem that the former explanation is more explanatory because it is more *about* the relevant ideal text than the latter, since the former allows us to identify at least some part of the relevant ideal text, while the latter does not. Obviously this is part of a general problem about *de re* and *de dicto*; depending upon one's approach to this problem, one may come up with different accounts of semantic information. I am grateful to Timothy McCarthy for drawing my attention to this issue.

precisely the same way that ideal D-N-P texts afford a yardstick for proffered explanations of chance phenomena. Thus, proffered explanations of non-probabilistic phenomena may take various forms and yet still be successful in virtue of communicating information about the relevant ideal text. For example, an ideal text for the explanation of the outcome of a causal process would look something like this: an inter-connected series of law-based accounts of all the nodes and links in the causal network culminating in the explanandum, complete with a fully detailed description of the causal mechanisms involved and theoretical derivations of all the covering laws involved. This full-blown causal account would extend, via various relations of reduction and supervenience, to all levels of analysis, i.e., the ideal text would be closed under relations of causal dependence, reduction, and supervenience. It would be the whole story concerning why the explanandum occurred, relative to a correct theory of the lawful dependencies of the world. Such an ideal causal D-N text would be infinite if time were without beginning or infinitely divisible, and plainly there is no question of ever setting such an ideal text down on paper. (Indeed, if time is continuous, an ideal causal text might have to be non-denumerably infinite—and thus 'ideal' in a very strong sense.) But it is clear that a whole range of less-than-ideal proffered explanations could more or less successfully convey information about such an ideal text and so be more or less successful explanations, even if not in D-N form.

Is it preposterous to suggest that any such ideal could exist for scientific explanation and understanding? Has anyone ever attempted or even wanted to construct an ideal causal or probabilistic text? It is not preposterous if we recognize that the actual ideal is not to *produce* such texts, but to have the ability (in principle) to produce arbitrary parts of them. It is thus irrelevant whether individual scientists ever set out to fill in ideal texts as wholes, since within the division of labour among scientists it is possible to find someone (or, more precisely, some group) interested in developing the ability to fill in virtually any particular aspect of ideal texts—macro or micro, fundamental or 'phenomenological', stretching over experimental or historical or geological or cosmological time. A chemist may be uninterested in how the reagents he handles came into being; a cosmologist may be interested in just that; a geologist may be interested in how those substances came to be distributed over the surface of the earth; an evolutionary biologist may be interested in how chemists (and the rest of us) came into being; an anthropologist or historian may be interested in how man and material came into contact with one another. To the extent that there are links and nodes, at whatever level of analysis, which we could not even in principle fill in, we may say that we do not completely understand the phenomenon under study.

The full explanatory history of a given phenomenon may have both causal and probabilistic links; for example, the click of a Geiger counter placed near a rock is due to geological causes as well as spontaneous alpha-decay. There is no problem in saying that an ideal explanatory text for a given fact contains both D-N and D-N-P subparts, so long as we do not try to treat the very same links as both probabilistic and non-probabilistic.

As we noticed earlier, in cases of genuine probabilistic explanation there are certain why-questions that simply do not have answers—questions as to why one probability rather than another was realized in a given case. We now are in a position to say what this comes to: such why-questions are requests for information that simply is not available—no part of even the ideal explanatory text contains a sufficient reason why one probability was realized rather than another. That is, this request for further explanation is refused, not because we do not know enough, but because there is simply nothing more to be known.[14] Nor is this the only sort of case in which a why-question may be refused. In any theory, probabilistic or otherwise, there will be certain matters of ultimate brute fact or of absolutely fundamental law. Purported ideal explanatory texts based upon a theory may simply lack the material needed to provide answers to certain requests for further explanatory information. For example, current physics offers us no reason why there is no negative gravity, despite the fact that there is both negative and positive charge (perhaps this is an initial condition of our universe), nor does it offer a reason why mass-energy is conserved rather than not (this seems to be a fundamental law). At most basic level, it may be difficult to distinguish fundamental laws from initial conditions of the entire universe, but that is not the point at issue here. The point is that a theory may legitimately spurn certain requests for further explanation in both probabilistic and non-probabilistic cases, and we can now say what such spurning consists in: the *absence* of certain things from purported ideal explanatory texts based upon that theory.

Where the orthodox covering-law account of explanation propounded by Hempel and others was right has been in claiming that explanatory practice in the sciences is in a central way *law-seeking* or *nomothetic*. Where it went wrong was in interpreting this fact as grounds for saying that any successful explanation must succeed either in virtue of explicitly

---

[14] B. C. van Fraassen is among those who have argued that there is need for an analysis of the notion that certain requests for information may be refused.

invoking covering laws or by implicitly asserting the existence of such laws. It is difficult to dispute the claim that scientific explanatory practice—whether engaged in causal, probabilistic, reductive, or functional explanation—*aims* ultimately (though not exclusively) at uncovering laws. This aim is reflected in the account offered here in the structure of ideal explanatory texts: their backbone is a series of law-based deductions. But it is equally difficult to dispute the claim that many proffered explanations succeed in doing some genuine explaining *without* either using laws explicitly or (somehow) tacitly asserting their existence. This fact is reflected here in the analysis offered of explanatoriness, which is treated as a matter of providing accurate information about the relevant ideal explanatory text, where this information may concern features of that text other than laws. For definiteness, let us call the analysis of explanation offered here *the nomothetic account of scientific explanation*.

3. THE USE OF STATISTICS AND PROBABILITIES IN THE EXPLANATION OF NON-PROBABILISTIC PHENOMENA

Now that the nomothetic account of probabilistic and causal explanation has been introduced, we are in a position to deal with the second of the two problems posed at the outset. Thus far it has been assumed that the proffered probabilistic explanations we have considered are purported explanations of indeterministic phenomena, so that the relevant ideal texts (or the relevant portions of more comprehensive ideal texts) are D-N-P in form. But what if a seemingly probabilistic explanation is offered of a phenomenon brought about in a deterministic way? Must we dismiss all such explanations as worthless? Would not this make many important statistical explanations in science—e.g. in classical statistical mechanics—worthless? But surely we cannot say *that*.

Fortunately, there is no need to make such a sweeping dismissal of statistical explanations of non-random phenomena. For while statistical explanations of phenomena not due to chance cannot be explanatory in virtue of providing information about a relevant ideal D-N-P text, since there is *no* such text, they may be explanatory in virtue of providing information about a relevant ideal D-N text, since there *is* just such a thing. To see this, let us look briefly and informally at two important cases of statistical explanation of non-random phenomena: classical statistical mechanics and 'explanation by correlation'.

The foundations of statistical mechanics evoke a wealth of difficult issues that cannot be considered here. Moreover, classical statistical mechanics has been replaced by quantum statistical mechanics, so it is no longer accurate to say that thermodynamic phenomena (for example) do not involve real indeterminacy at all. To simplify matters, I will stick to the classical case, which will enable me to discuss a view that has been and continues to be widely held: if classical statistical mechanics *were* right and no indeterminacy were involved in thermodynamic phenomena, the statistical explanations of the classical theory would be genuinely explanatory. I have in mind familiar explanations of such facts as the tendency of closed systems to move towards and stay near equilibrium, the functional correlation of macroscopic thermodynamical variables for gases at equilibrium (e.g. the ideal gas law, $PV = kT$), etc., in terms of the most common micro-states in an ensemble created by permutations of molecules over a constrained phase space (Boltzmann) or in terms of average values over all possible micro-states in a constrained phase space (Gibbs). For any given gas, its particular state S at a time t will be determined solely by its molecular constitution, its initial condition, the deterministic laws of classical dynamics operating upon this initial condition, and the boundary conditions to which it has been subject. Therefore, the ideal explanatory text for its being in state S at time t will not be probabilistic, but will be a complete causal history of the time-evolution of that gas. Now if the ideal explanatory text is purely deterministic, how could one explain the fact that the gas is in state S at time t by reference to 'most probable' macro-states or 'phase averages'? After all, these 'probabilities' and 'averages' are the product of sheer combinatorics under certain limiting assumptions, with no guarantee that the gas in question met these limiting assumptions and with no regard to the fact that the gas had definite initial conditions, not a probability distribution over initial conditions. Briefly, the answer is that such appeals to combinatorics serve to illuminate a significant feature of the causal process underlying the behaviour of classical thermodynamic systems, thereby serving partially to illuminate the relevant ideal explanatory texts in particular cases. For example, various proofs in ergodic theory and related results show that if a gas is in an initial condition that obeys a relatively few constraints, it will, over infinite time, spend most of its time at or near equilibrium. This illuminates a *modal* feature of the causal processes involved and therefore a modal feature of the relevant ideal explanatory texts: this sort of causal process is such that its macroscopic outcomes are remarkably insensitive (in the limit) to wide variations in initial microstates. The stability of an outcome of a causal process in spite of significant variation in initial conditions can be informative about an ideal causal explanatory text in the same way that it is informative to learn, regarding a given causal explanation of the First World War, that a world war would have come about (according to this explanation) even if no bomb had exploded in Sarajevo. This sort of robustness or resilience of a process is important to grasp in coming to know explanations based upon it. Traditional worries about the applicability of ergodic proofs and related results to systems finite in time and degrees of freedom do not go away, on this view. But for those who have held that these worries can be dealt with, we now have a way of capturing the intuition that ergodic theory and its kin are somehow explanatory: they shed light on a modal feature of the causal processes underlying thermodynamic behaviour, thus providing information about the relevant ideal causal texts. Since we almost never know the actual initial condition of a gas in any detail, this sort of modal information about the causal process involved in the time-evolution of classical gases is especially valuable.

Statistical generalizations about macro- and micro-states of thermodynamic systems—e.g., 'closed systems are seldom found in states with measure zero (in the natural measure)'—also may play a role in explaining thermodynamic behaviour, such as the prevalence of equilibrium. However, the role they play is not that of illuminating lawful features of the processes underlying such behaviour, but rather that of providing information about the factual premises of the relevant ideal explanatory texts. Thus, it should be no mystery that statistical generalizations believed *not* to reflect underlying probabilistic laws may still be explanatory, once we recognize that they function in explanation not as ersatz laws but as summaries of information about initial conditions and boundary conditions.

It would be excessively fastidious to insist that classical statistical mechanics offers no probabilistic explanations only if this were tantamount to insisting that it could offer *no* explanations, i.e., could provide no explanatory information. We have seen that classical statistical-mechanical arguments *can* provide explanatory information concerning purely non-probabilistic processes. Thus classical statistical mechanics need hardly be dismissed as non-explanatory, although one must be careful in stating what it does and does not explain. Most crucially, it does not explain why the initial conditions should be such that we observe the macroscopic regularities in thermodynamic behaviour that we do. This shows up on the present account in the fact that these initial conditions can be traced backwards through ideal causal texts only to still earlier conditions, not to laws. Thus the prevalence of equilibrium and other such features of macroscopic behaviour must, on the classical theory, ultimately be attributed to brute fact and to the operation of deterministic laws on brute fact. But *given* a certain range of initial conditions, classical statistical mechanics does provide explanations of macroscopic behaviour. The charge of excessive fastidiousness seems doubly unearned in light of the fact that classical statistical mechanics itself tells us that the relevant ideal explanatory texts are deterministic.

The second sort of statistical explanation that is frequently offered of phenomena known or believed not to be indeterministic falls under the general (often pejorative) heading 'explanation by correlation'. To this species belong the many efforts to use the direct results of multivariate analysis and other statistical techniques as explanations of the behaviour of (often complex) causal systems. It is unnecessary in this connection to delve into the details of explanations of this kind, for the point to be made is highly general: the nomothetic account enables us to see how statements of statistical correlations may function not only as evidence for causal connections, but also as sources of information about aspects of ideal causal explanatory texts. Thus, a statement that there is a 'statistically significant' correlation between exposure to high levels of cotton fibre in the air and incidence of brown lung disease may be put forward to explain why textile workers in cotton mills experience abnormally high rates of brown lung. Here the explanation may involve an irreducibly probabilistic process, but it need not. Let us suppose for the sake of illustration that it does not; still, the statement of correlation may convey information about the relevant causal ideal text for explaining the frequency of brown lung among cotton mill workers, since it points to a substance, cotton fibre, the presence of which in the air breathed is an important element of this text. In this light, claims about the 'strength' of a correlation, or about the amount of variation in one variable that can be 'statistically explained' in terms of the variation of another, may be seen as ways of providing information about the extent and independence of the roles of various factors in relevant ideal explanatory texts.

It would be a naïve view to equate the use of statistical correlations with probabilistic explanation, and we should not accuse scientists making heavy use of correlations—often because of limitations in the available evidence, limitations on possible experimentation, and the sheer complexity of the systems studied—of offering probabilistic explanations of phenomena they suspect are deterministic. 'Explanation by correlation', of course, does not always provide genuinely explanatory information—correlations may be accidental, epiphenomenal, mediated by other variables, misleading about the direction of causation, or otherwise explanatorily defective. But even so, some such correlations may provide a limited amount of information about the relevant ideal explanatory texts. Thus, failure to detect an intervening variable may weaken an explanation greatly, but that explanation may still pick out factors with some causal role even if it misrepresents the directness and import-ance of their contribution. Clearly, 'explanation by correlation' needs further and deeper treatment than it can be given here. But I hope that it is equally clear how the nomothetic account enables us to say that state-ments of statistical correlation may function as (partial) explanations in virtue of providing (partial) information about relevant ideal explanatory texts, even when these are not probabilistic.

Finally, while it is important to recognize that statistical generaliza-tions of various sorts may provide explanatory information concerning non-probabilistic phenomena, it is important as well to recognize the difference it makes to explanatory practice whether or not a given phe-nomenon is viewed as being fundamentally probabilistic. For example, quantum mechanics, with its underlying probabilism, initially confronted a resistance from scientists quite unlike that meeting classical statistical mechanics, with its underlying determinism. Furthermore, there are dif-ferences in the research programmes supported by these two theories: in the latter case we find sustained work in physics and mathematics to account for statistical features of systems in terms of 'hidden variables', in the former case we find—after considerable debate—widespread ac-ceptance of the non-existence of 'hidden variables' and work in physics and mathematics to demonstrate the 'completeness' of the quantum the-ory. These differences in the history of theory-acceptance and theory-de-velopment are the sort of thing that a satisfactory account of scientific explanation ought to illuminate, but an account of probabilistic explana-tion that fails to demarcate 'statistical' explanations of deterministic processes (as in classical statistical mechanics) from probabilistic expla-nations of indeterministic processes (as in quantum mechanics) tends to obscure rather than clarify these differences. Unlike many of its compe-titors, the nomothetic account of probabilistic explanation—based as it is on the D-N-P model—makes precisely this demarcation, and in so doing helps us to understand these features of scientific explanatory practice. In this way, an account of explanation may itself do some explanatory work by accounting for a difference in practice in terms of a difference in theory. At first blush, one might think that whenever statistics or probabilities are involved in explanatory practice one is dealing with a form of probabilistic explanation. However, this illusion is quickly shed once one recognizes the variety of ways in which statis-tics and probabilities figure in explanatory activities. Perhaps the com-monest use of statistics and probabilities in connection with explanation is *epistemic*: they are used in the process of assembling and assessing evidence for causal and non-causal explanations alike. Somewhat less common, but still important, are such uses as those discussed above: statistics and probabilities are used in providing explanatory information about causal and non-causal processes and their initial conditions. In *some* cases of the last sort we have genuine probabilistic explanation, specifically, in those cases where information is provided about a physically indeterministic process.

Some have protested that accepting the existence of fundamentally probabilistic phenomena amounts to a resignation of intellectual responsibility, to an abandonment of curiosity. Plainly it would be irresponsible to accept without reservation the hypothesis that a process is probabilistic on the basis of scanty examination and weak evidence, but it would be irresponsible to accept *any* hypothesis without reservation in such circumstances. And surely no one can argue that acceptance of the irreducible probabilism of quantum mechanics is based upon scanty examination or weak evidence. Moreover, as we noted earlier, the explanations offered by a number of historically significant scientific theories must come to an end somewhere. Acceptance of a lawful probabilistic relationship as fundamental on the basis of overwhelming evidence is no greater bar to the growth of science than is acceptance of a lawful deterministic relationship as fundamental on the basis of equally good evidence. Certainly the history of quantum mechanics provides stronger support for this claim than any *a priori* argument possibly could: the deepening and widening of scientific knowledge under quantum theory gives the lie to those who prophesied that explanation by chance would be the enemy of enlarged inquiry and understanding.

REFERENCES

Bar-Hillel, Y. and Carnap, R., 'Semantic Information', *British Journal for the Philosophy of Science*, 4 (1953), 147–57.

Cherry, Colin, *On Human Communication*, second edn. (Cambridge, Mass.: MIT Press, 1966).

Evans, R. D., *The Atomic Nucleus* (New York: McGraw-Hill, 1965).

Hempel, C. G., 'Aspects of Scientific Explanation', in *Aspects of Scientific Explanation and Other Essays* (New York: Free Press, 1965).

—— 'Maximal Specificity and Lawlikeness in Probabilistic Explanation', *Philosophy of Science*, 35 (1968), 116–33.

Railton, P., 'A Deductive-Nomological Model of Probabilistic Explanation', *Philosophy of Science*, 45 (1978), 206–26.

VII

CAUSAL EXPLANATION[*]

DAVID LEWIS

1. CAUSAL HISTORIES

Any particular event that we might wish to explain stands at the end of a long and complicated causal history. We might imagine a world where causal histories are short and simple; but in the world as we know it, the only question is whether they are infinite or merely enormous.

An explanandum event has its causes. These act jointly. We have the icy road, the bald tire, the drunk driver, the blind corner, the approaching car, and more. Together, these cause the crash. Jointly they suffice to make the crash inevitable, or at least highly probable, or at least much more probable than it would otherwise have been. And the crash depends on each. Without any one it would not have happened, or at least it would have been very much less probable than it was.

But these are by no means all the causes of the crash. For one thing, each of these causes in turn has its causes; and those too are causes of the crash. So in turn are their causes, and so, perhaps, *ad infinitum*. The crash is the culmination of countless distinct, converging causal chains. Roughly speaking, a causal history has the structure of a tree. But not quite: the chains may diverge as well as converge. The roots in childhood of our driver's reckless disposition, for example, are part of the causal chains via his drunkenness, and also are part of other chains via his bald tire.

Further, causal chains are dense. (Not necessarily, perhaps—time might be discrete—but in the world as we mostly believe it to be.) A causal chain may go back as far as it can go and still not be complete, since it may leave out intermediate links. The blind corner and the oncoming car were not immediate causes of the crash. They caused a swerve; that and the bald tire and icy road caused a skid; that and the

David Lewis, *Philosophical Papers*, ii (Oxford and New York: Oxford University Press, 1986), 214–40. Copyright © 1986 David Lewis. Reprinted by permission of the author.

[*] This paper is descended, distantly, from my Hägerstrom Lectures in Uppsala in 1977, and more directly from my Howison Lectures in Berkeley in 1979.

driver's drunkenness caused him to apply the brake, which only made matters worse. . . . And still we have mentioned only a few of the most salient stages in the last second of the causal history of the crash. The causal process was in fact a continuous one.

Finally, several causes may be lumped together into one big cause. Or one cause may be divisible into parts. Some of these parts may themselves be causes of the explanandum event, or of parts of it. (Indeed, some parts of the explanandum event itself may be causes of others.) The baldness of the tire consists of the baldness of the inner half plus the baldness of the outer half; the driver's drunkenness consists of many different disabilities, of which several may have contributed in different ways to the crash. There is no one right way—though there may be more or less natural ways—of carving up a causal history.

The multiplicity of causes and the complexity of causal histories are obscured when we speak, as we sometimes do, of *the* cause of something. That suggests that there is only one. But in fact it is commonplace to speak of "the X" when we know that there are many X's, and even many X's in our domain of discourse, as witness McCawley's sentence "the dog got in a fight with another dog." If someone says that the bald tire was the cause of the crash, another says that the driver's drunkenness was the cause, and still another says that the cause was the bad upbringing which made him so reckless, I do not think any of them disagree with me when I say that the causal history includes all three. They disagree only about which part of the causal history is most salient for the purposes of some particular inquiry. They may be looking for the most remarkable part, the most remediable or blameworthy part, the least obvious of the discoverable parts, ... Some parts will be salient in some contexts, others in others. Some will not be at all salient in any likely context, but they belong to the causal history all the same: the availability of petrol, the birth of the driver's paternal grandmother, the building of the fatal road, the position and velocity of the car a split second before the impact.[1]

(It is sometimes thought that only an aggregate of conditions inclusive enough to be sufficient all by itself—Mill's "whole cause"—deserves to

[1] On definite descriptions that do not imply uniqueness, see "Scorekeeping in a Language Game," in my *Philosophical Papers*, i (1983), 233–49 (originally published in *Journal of Philosophical Logic*, 8 (1979), 339–59); and James McCawley, "Presupposition and Discourse Structure," in David Dinneen and Choon-kyu Oh (eds.), *Syntax and Semantics*, 11 (New York: Academic Press, 1979), 371–88. On causal selection, see Morton G. White, *Foundations of Historical Knowledge* (New York: Harper & Row, 1965), ch. 4. Peter Unger, in " The Uniqueness of Causation," *American Philosophical Quarterly*, 14 (1977), 177–88, has noted that not only "the cause of " but also the verb "caused" may be used selectively. There is something odd—inconsistent, he thinks—in saying with emphasis that each of two distinct things caused something. Even "a cause of " may carry some hint of selectivity. It would be strange, though I think not false, to say in any ordinary context that the availability of petrol was a cause of the crash.

be called "the cause." But even on this eccentric usage, we still have many deserving candidates for the title. For if we have a whole cause at one time, then also we have other whole causes at later times, and perhaps at earlier times as well.)

A causal history is a relational structure. Its *relata* are *events*: local matters of particular fact, of the sorts that may cause or be caused. I have in mind events in the most ordinary sense of the word: flashes, battles, conversations, impacts, strolls, deaths, touchdowns, falls, kisses, ... But also I mean to include events in a broader sense: a moving object's continuing to move, the retention of a trace, the presence of copper in a sample.[2]

These events may stand in various relations, for instance spatiotemporal relations and relations of part to whole. But it is their causal relations that make a causal history. In particular, I am concerned with relations of causal dependence. An event depends on others, which depend in turn on yet others, . . . ; and the events to which an event is thus linked, either directly or stepwise, I take to be its causes. Given the full structure of causal dependence, all other causal relations are given. Further, I take causal dependence itself to be counterfactual dependence, of a suitably non-backtracking sort, between distinct events: in Hume's words, "if the first . . . had not been, the second never had existed."[3] But this paper is not meant to rely on my views about the analysis of causation. Whatever causation may be, there are still causal histories, and what I shall say about causal explanation should still apply.[4]

I include relations of probabilistic causal dependence. Those who know of the strong scientific case for saying that our world is an indeterministic one, and that most events therein are to some extent matters of chance, never seriously renounce the commonsensical view that there is plenty of causation in the world. (They may preach the "downfall of causality" in their philosophical moments. But whatever that may mean, evidently it does not imply any shortage of causation.) For instance, they would never dream of agreeing with those ignorant tribes who disbelieve that pregnancies are caused by events of sexual intercourse. The causation they believe in must be probabilistic. And if, as seems likely, our

---

[2] See Lewis, "Events," *Philosophical Papers*, ii. 241–69.

[3] *An Enquiry Concerning Human Understanding*, ed. L. A. Selby-Bigge (Oxford: Oxford University Press, 1963), sect. VII, part II, p. 76.

[4] One author who connects explanation and causation in much the same way as I do, but builds on a very different account of causation, is Wesley C. Salmon. See his "Theoretical Explanation," in Stephan Körner (ed.), *Explanation* (New Haven, Conn.: Yale University Press, 1975); "A Third Dogma of Empiricism," in R. Butts and J. Hintikka (eds.), *Basic Problems in Methodology and Linguistics* (Dordrecht: Reidel, 1977); and "Why Ask 'Why?'?," *Proceedings of the American Philosophical Association*, 51 (1978), 683–705.

world is indeed thoroughly indeterministic and chancy, its causal histories must be largely or entirely structures of probabilistic causal dependence. I take such dependence to obtain when the objective chances of some events depend counterfactually upon other events: if the cause had not been, the effect would have been very much less probable than it actually was.[5] But again, what is said in this paper should be compatible with any analysis of probabilistic causation.

The causal history of a particular event includes that event itself, and all events which are part of it. Further, it is closed under causal dependence: anything on which an event in the history depends is itself an event in the history. (A causal history need not be closed under the converse relation. Normally plenty of omitted events will depend on included ones.) Finally, a causal history includes no more than it must to meet these conditions.

2. EXPLANATION AS INFORMATION

Here is my main thesis: *to explain an event is to provide some information about its causal history.*

In an act of explaining, someone who is in possession of some information about the causal history of some event—*explanatory information*, I shall call it—tries to convey it to someone else. Normally, to someone who is thought not to possess it already, but there are exceptions: examination answers and the like. Afterward, if the recipient understands and believes what he is told, he too will possess the information. The why-question concerning a particular event is a request for explanatory information, and hence a request that an act of explaining be performed.

In one sense of the word, an explanation of an event is such an act of explaining. To quote Sylvain Bromberger, "an explanation may be something about which it makes sense to ask: How long did it take? Was it interrupted at any point? Who gave it? When? Where? What were the exact words used? For whose benefit was it given?"[6] But it is not clear whether just any act of explaining counts as an explanation. Some acts of explaining are unsatisfactory; for instance the explanatory information provided might be incorrect, or there might not be enough of it, or it might be stale news. If so, do we say that the performance was no explanation at all? Or that it was an unsatisfactory explanation? The

---

[5] See Lewis, Postscript B to "Causation," *Philosophical Papers*, ii. 175–84.

[6] "An Approach to Explanation," in R. J. Butler (ed.), *Analytical Philosophy*, second series (Oxford: Blackwell, 1965).

answer, I think, is that we will gladly say either—thereby making life hard for those who want to settle, once and for all, the necessary and sufficient conditions for something to count as an explanation. Fortunately that is a project we needn't undertake.

Bromberger goes on to say that an explanation "may be something about which none of [the previous] questions makes sense, but about which it makes sense to ask: Does anyone know it? Who thought of it first? Is it very complicated?" An explanation in this second sense of the word is not an act of explaining. It is a chunk of explanatory information—information that may once, or often, or never, have been conveyed in an act of explaining. (It might even be information that never could be conveyed, for it might have no finite expression in any language we could ever use.) It is a proposition about the causal history of the explanandum event. Again it is unclear—and again we needn't make it clear—what to say about an unsatisfactory chunk of explanatory information, say one that is incorrect or one that is too small to suit us. We may call it a bad explanation, or no explanation at all.

Among the true propositions about the causal history of an event, one is maximal in strength. It is the whole truth on the subject—the biggest chunk of explanatory information that is free of error. We might call this the *whole* explanation of the explanandum event, or simply *the* explanation. (But "the explanation" might also denote that one out of many explanations, in either sense, that is most salient in a certain context.) It is, of course, very unlikely that so much explanatory information ever could be known, or conveyed to anyone in some tremendous act of explaining!

One who explains may provide not another, but rather himself, with explanatory information. He may think up some hypothesis about the causal history of the explanandum event, which hypothesis he then accepts. Thus Holmes has explained the clues (correctly or not, as the case may be) when he has solved the crime to his satisfaction, even if he keeps his solution to himself. His achievement in this case probably could not be called "an explanation"; though the chunk of explanatory information he has provided himself might be so called, especially if it is a satisfactory one.

Not only a person, but other sorts of things as well, may explain. A theory or a hypothesis, or more generally any collection of premises, may provide explanatory information (correct or incorrect) by implying it. That is so whether or not anyone draws the inference, whether or not anyone accepts or even thinks of the theory in question, and whether or not the theory is true. Thus we may wonder whether our theories explain more than we will ever realize, or whether other undreamt-of theories explain more than the theories we accept.

Explanatory information comes in many shapes and sizes. Most simply, an explainer might give information about the causal history of the explanandum by saying that a certain particular event is included therein. That is, he might specify one of the causes of the explanandum. Or he might specify several. And if so, they might comprise all or part of a cross-section of the causal history: several events, more or less simultaneous and causally independent of one another, that jointly cause the explanandum. Alternatively, he might trace a causal chain. He might specify a sequence of events in the history, ending with the explanandum, each of which is among the causes of the next. Or he might trace a more complicated, branching structure that is likewise embedded in the complete history.

An explainer well might be unable to specify fully any particular event in the history, but might be in a position to make existential statements. He might say, for instance, that the history includes an event of such-and-such kind. Or he might say that the history includes several events of such-and-such kinds, related to one another in such-and-such ways. In other words, he might make an existential statement to the effect that the history includes a pattern of events of a certain sort. (Such a pattern might be regarded, at least in some cases, as one complex and scattered event with smaller events as parts.) He might say that the causal history has a certain sort of cross-section, for instance, or that it includes a certain sort of causal chain.

If someone says that the causal history includes a pattern of events having such-and-such description, there are various sorts of description that he might give. A detailed structural specification might be given, listing the kinds and relations of the events that comprise the pattern. But that is not the only case. The explainer might instead say that the pattern that occupies a certain place in the causal history is some biological, as opposed to merely chemical, process. Or he might say that it has some global structural feature: it is a case of underdamped negative feedback, a dialectical triad, or a resonance phenomenon. (And he might have reason to say this even if he has no idea, for instance, what sort of thing it is that plays the role of a damper in the system in question). Or he might say that it is a process analogous to some other, familiar process. (So in this special case, at least, there is something to the idea that we may explain by analogizing the unfamiliar to the familiar. At this point I am indebted to David Velleman.) Or he might say that the causal process, whatever it may be, is of a sort that tends in general to produce a certain kind of effect. I say "we have lungs because they keep us alive"; my point being that lungs were produced by that process, whatever it may be, that can and does produce all manner of life-sustaining organs.

(In conveying that point by those words, of course I am relying on the shared presupposition that such a process exists. In explaining, as in other communication, literal meaning and background work together.) And I might say this much, whether or not I have definite opinions about what sort of process it is that produces life-sustaining organs. My statement is neutral between evolution, creation, vital forces, or what have you; it is also neutral between opinionation and agnosticism.

In short: information about what the causal history includes may range from the very specific to the very abstract. But we are still not done. There is also negative information: information about what the causal history does *not* include. "Why was the CIA man there when His Excellency dropped dead?—Just coincidence, believe it or not." Here the information given is negative, to the effect that a certain sort of pattern of events—namely, a plot—does not figure in the causal history. (At least, not in that fairly recent part where one might have been suspected. Various ancient plots doubtless figure in the causal histories of all current events, this one included.)

A final example. The patient takes opium and straightway falls asleep; the doctor explains that opium has a dormitive virtue. Doubtless the doctor's statement was not as informative as we might have wished, but observe that it is not altogether devoid of explanatory information. The test is that it suffices to rule out at least some hypotheses about the causal history of the explanandum. It rules out this one: the opium merchants know that opium is an inert substance, yet they wish to market it as a soporific. So they keep close watch; and whenever they see a patient take opium, they sneak in and administer a genuine soporific. The doctor has implied that this hypothesis, at least, is false; whatever the truth may be, at least it somehow involves distinctive intrinsic properties of the opium.

Of course I do not say that all explanatory information is of equal worth; or that all of it equally deserves the honorific name "explanation." My point is simply that we should be aware of the variety of explanatory information. We should not suppose that the only possible way to give some information about how an event was caused is to name one or more of its causes.

3. NON-CAUSAL EXPLANATION?

It seems quite safe to say that the provision of information about causal histories figures very prominently in the explaining of particular events. What is not so clear is that it is the whole story. Besides the causal explanation that I am discussing, is there also any such thing as non-causal explanation of particular events? My main thesis says there is not. I shall consider three apparent cases of it, one discussed by Hempel and two suggested to me by Peter Railton.[7]

First case. We have a block of glass of varying refractive index. A beam of light enters at point A and leaves at point B. In between, it passes through point C. Why? Because C falls on the path from A to B that takes light the least time to traverse; and according to Fermat's principle of least time, that is the path that any light going from A to B must follow. That seems non-causal. The light does not get to C because it looks ahead, calculates the path of least time to its destination B, and steers accordingly! The refractive index in parts of the glass that the light has not yet reached has nothing to do with causing it to get to C, but that is part of what makes it so that C is on the path of least time from A to B.

I reply that it is by no means clear that the light's passing through C has been explained. But if it has, that is because this explanation combines with information that its recipient already possesses to imply something about the causal history of the explanandum. Any likely recipient of an explanation that mentions Fermat's principle must already know a good deal about the propagation of light. He probably knows that the bending of the beam at any point depends causally on the local variation of refractive index around that point. He probably knows, or at least can guess, that Fermat's principle is somehow provable from some law describing that dependence together with some law relating refractive index to speed of light. Then he knows this: (1) the pattern of variation of the refractive index along some path from A to C is part of the causal history of the light's passing through C, and (2) the pattern is such that it, together with a pattern of variation elsewhere that is not part of the causal history, makes the path from A to C be part of a path of least time from A to B. To know this much is not to know just what the pattern that enters into the causal history looks like, but it is to know something—something relational—about that pattern. So the explanation does indeed provide a peculiar kind of information about the causal history of the explanandum, on condition that the recipient is able to supply the extra premises needed.

Second case. A star has been collapsing, but the collapse stops. Why? Because it's gone as far as it can go. Any more collapsed state would

[7] Carl G. Hempel, *Aspects of Scientific Explanation and other Essays in the Philosophy of Science* (New York: Free Press, 1965), 353; Peter Railton, *Explaining Explanation* (Ph.D. dissertation, Princeton University, 1979). I am much indebted to Railton throughout this paper, both where he and I agree and where we do not. For his own views on explanation, see also his "A Deductive-Nomological Model of Probabilistic Explanation," *Philosophy of Science*, 45 (1978), 206–26; and "Probability, Explanation, and Information," *Synthese*, 48 (1981), 233–56.

violate the Pauli Exclusion Principle. It's not that anything caused it to stop—there was no countervailing pressure, or anything like that. There was nothing to keep it out of a more collapsed state. Rather, there just was no such state for it to get into. The state-space of physical possibilities gave out. (If ordinary space had boundaries, a similar example could be given in which ordinary space gives out and something stops at the edge.)

I reply that information about the causal history of the stopping has indeed been provided, but it was information of an unexpectedly negative sort. It was the information that the stopping has no causes at all, except for all the causes of the collapse which was a precondition of the stopping. Negative information is still information. If you request information about arctic penguins, the best information I can give you is that there aren't any.

Third case. Walt is immune to smallpox. Why? Because he possesses antibodies capable of killing off any smallpox virus that might come along. But his possession of antibodies doesn't *cause* his immunity. It *is* his immunity. Immunity is a disposition, to have a disposition is to have something or other that occupies a certain causal role, and in Walt's case what occupies the role is his possession of antibodies.

I reply that it's as if we'd said it this way: Walt has some property that protects him from smallpox. Why? Because he possesses antibodies, and possession of antibodies is a property that protects him from smallpox. Schematically: Why is it that something is F? Because A is F. An existential quantification is explained by providing an instance. I agree that something has been explained, and not by providing information about its causal history. But I don't agree that any particular event has been non-causally explained. The case is outside the scope of my thesis. That which protects Walt—namely, his possession of antibodies—is indeed a particular event. It is an element of causal histories; it causes and is caused. But that was not the explanandum. We could no more explain that just by saying that Walt possesses antibodies than we could explain an event just by saying that it took place. What we did explain was something else: the fact that something or other protects Walt. The obtaining of this existential fact is not an event. It cannot be caused. Rather, events that would provide it with a truth-making instance can be caused. We explain the existential fact by identifying the truth-making instance, by providing information about the causal history thereof, or both.[8]

---

[8] For further discussion of explanation of facts involving the existence of patterns of events, see my "Events."

What more we say about the case depends on our theory of dispositions.[9] I take for granted that a disposition requires a causal basis: one has the disposition iff one has a property that occupies a certain causal role. (I would be inclined to require that this be an intrinsic property, but that is controversial.) Shall we then identify the disposition with its basis? That would make the disposition a cause of its manifestations, since the basis is. But the identification might vary from case to case. (It surely would, if we count the unactualized cases.) For there might be different bases in different cases. Walt might be disposed to remain healthy if exposed to virus on the basis of his possession of antibodies, but Milt might be so disposed on the basis of his possession of dormant antibody-makers. Then if the disposition is the basis, immunity is different properties in the cases of Walt and Milt. Or better: "immunity" denotes different properties in the two cases, and there is no property of immunity *simpliciter* that Walt and Milt share.

That is disagreeably odd. But Walt and Milt do at least share something: the existential property of having some basis or other. This is the property such that, necessarily, it belongs to an individual X iff X has some property that occupies the appropriate role in X's case. So perhaps we should distinguish the disposition from its various bases, and identify it rather with the existential property. That way, "immunity" could indeed name a property shared by Walt and Milt. But this alternative has a disagreeable oddity of its own. The existential property, unlike the various bases, is too disjunctive and too extrinsic to occupy any causal role. There is no event that is essentially a having of the existential property; *a fortiori*, no such event ever causes anything. (Compare the absurd double-counting of causes that would ensue if we said, for instance, that when a match struck in the evening lights, one of the causes of the lighting is an event that essentially involves the property of being struck in the evening or twirled in the morning. I say there is no such event.) So if the disposition is the existential property, then it is causally impotent. On this theory, we are mistaken whenever we ascribe effects to dispositions.

[9] See the discussions of dispositions and their bases in D. M. Armstrong, *A Materialist Theory of the Mind* (London: Routledge & Kegan Paul, 1968), 85–8; Armstrong, *Belief, Truth and Knowledge* (Cambridge: Cambridge University Press, 1973), 11–16; Elizabeth W. Prior, Robert Pargetter, and Frank Jackson, "Three Theses about Dispositions," *American Philosophical Quarterly*, 19 (1982), 251–7; and Elizabeth W. Prior, *Dispositions* (Aberdeen: Aberdeen University Press, 1985). See also my "Events," sect. VIII. Parallel issues arise for functionalist theories of mind. See my "An Argument for the Identity Theory" (originally published in *Journal of Philosophy*, 63 (1966), 17–25) and "Mad Pain and Martian Pain" (originally published in Ned Block (ed.), *Readings in Philosophy of Psychology*, i (Harvard University Press, 1980), 216–22), both reprinted in *Philosophical Papers*, i; and Jackson, Pargetter, and Prior, "Functionalism and Type–Type Identity Theories," *Philosophical Studies*, 42 (1982), 209–25.

Fortunately we needn't decide between the two theories. Though they differ on the analysis of disposition-names like "immunity," they agree about what entities there are. There is one genuine event—Walt's possession of antibodies. There is a truth about Walt to the effect that he has the existential property. But there is no second event that is essentially a having of the existential property, but is not essentially a having of it in any particular way. Whatever "Walt's immunity" may denote, it does not denote such an event. And since there is no such event at all, there is no such event to be non-causally explained.

4. GENERAL EXPLANATION

My main thesis concerns the explanation of particular events. As it stands, it says nothing about what it is to explain general kinds of events. However, it has a natural extension. All the events of a given kind have their causal histories, and these histories may to some extent be alike. Especially, the final parts of the histories may be much the same from one case to the next, however much the earlier parts may differ. Then information may be provided about what is common to all the parallel causal histories—call it *general explanatory information* about events of the given kind. To explain a kind of event is to provide some general explanatory information about events of that kind.

Thus explaining why struck matches light in general is not so very different from explaining why some particular struck match lit. In general, and in the particular case, the causal history involves friction, small hot spots, liberation of oxygen from a compound that decomposes when hot, local combustion of a heated inflammable substance facilitated by this extra oxygen, further heat produced by this combustion, and so on.

There are intermediate degrees of generality. If we are not prepared to say that every event of such-and-such kind, without exception, has a causal history with so-and-so features, we need not therefore abjure generality altogether and stick to explaining events one at a time. We may generalize modestly, without laying claim to universality, and say just that quite often an event of such-and-such kind has a causal history with so-and-so features. Or we may get a bit more ambitious and say that it is so in most cases, or at least in most cases that are likely to arise under the circumstances that prevail hereabouts. Such modest generality may be especially characteristic of history and the social sciences; but it appears also in the physical sciences of complex systems, such as meteorology and geology. We may be short of known laws to the effect that storms with feature X always do Y, or always have a certain definite probability of doing Y. Presumably there are such laws, but they are too complicated to discover either directly or by derivation from first principles. But we do have a great deal of general knowledge of the sorts of causal processes that commonly go on in storms.

The pursuit of general explanations may be very much more widespread in science than the pursuit of general laws. And not necessarily because we doubt that there are general laws to pursue. Even if the scientific community unanimously believed in the existence of powerful general laws that govern all the causal processes of nature, and whether or not those laws were yet known, meteorologists and geologists and physiologists and historians and engineers and laymen would still want general knowledge about the sorts of causal processes that go on in the systems they study.

5. EXPLAINING WELL AND BADLY

An act of explaining may be more or less satisfactory, in several different ways. It will be instructive to list them. It will *not* be instructive to fuss about whether an unsatisfactory act of explaining, or an unsatisfactory chunk of explanatory information, deserves to be so-called, and I shall leave all such questions unsettled.

1. An act of explaining may be unsatisfactory because the explanatory information provided is unsatisfactory. In particular, it might be misinformation: it might be a false proposition about the causal history of the explanandum. This defect admits of degree. False is false, but a false proposition may or may not be close to the truth.[10] If it has a natural division into conjuncts, more or fewer of them may be true. If it has some especially salient consequences, more or fewer of those may be true. The world as it is may be more or less similar to the world as it would be if the falsehood were true.

2. The explanatory information provided may be correct, but there may not be very much of it. It might be a true but weak proposition; one that

---

[10] The analysis of verisimilitude has been much debated. A good survey is Ilkka Niiniluoto, "Truthlikeness: Comments on Recent Discussion," *Synthese*, 38 (1978), 281–329. Some plausible analyses have failed disastrously, others conflict with one another. One conclusion that emerges is that it is probably a bad move to try to define a single virtue of verisimilitude-cum-strength. It's hard to say whether strength is a virtue in the case of false information, especially if we have no uniquely natural way of splitting the misinformation into true and false parts. Another conclusion is that even if this lumping together is avoided, verisimilitude still seems to consist of several distinguishable virtues.

excludes few (with respect to some suitable measure) of the alternative possible ways the causal history of the explanandum might be. Or the information provided might be both true and strong, but unduly disjunctive. The alternative possibilities left open might be too widely scattered, too different from one another. These defects too admit of degree. Other things being equal, it is better if more correct explanatory information is provided, and it is better if that information is less disjunctive, up to the unattainable limit in which the whole explanation is provided and there is nothing true and relevant left to add.

3. The explanatory information provided may be correct, but no thanks to the explainer. He may have said what he did not know and had no very good reason to believe. If so, the act of explaining is not fully satisfactory, even if the information provided happens to be satisfactory.

4. The information provided, even if satisfactory in itself, may be stale news. It may add little or nothing to the information the recipient possesses already.

5. The information provided may not be of the sort the recipient most wants. He may be especially interested in certain parts of the causal history, or in certain questions about its overall structure. If so, no amount of explanatory information that addresses itself to the wrong questions will satisfy his wants, even if it is correct and strong and not already in his possession.

6. Explanatory information may be provided in such a way that the recipient has difficulty in assimilating it, or in disentangling the sort of information he wants from all the rest. He may be given more than he can handle, or he may be given it in a disorganized jumble.[11] Or he may be given it in so unconvincing a way that he doesn't believe what he's told. If he is hard to convince, just telling him may not be an effective way to provide him with information. You may have to argue for what you tell him, so that he will have reason to believe you.

7. The recipient may start out with some explanatory misinformation, and the explainer may fail to set him right.

This list covers much that philosophers have said about the merits and demerits of explanations, or about what does and what doesn't deserve the name. And yet I have not been talking specifically about explanation at all! What I have been saying applies just as well to acts of providing information about *any* large and complicated structure. It might as well have been the rail and tram network of Melbourne rather than the causal

---

[11] As in the square peg example of Hilary Putnam, "Philosophy and our Mental Life," in his *Mind, Language and Reality* (Cambridge University Press, 1975), 295–7.

history of some explanandum event. The information provided, and the act of providing it, can be satisfactory or not in precisely the same ways. There is no special subject: pragmatics of explanation.

Philosophers have proposed further *desiderata*. A good explanation ought to show that the explanandum event had to happen, given the laws and the circumstances; or at least that it was highly probable, and could therefore have been expected if we had known enough ahead of time; or at least that it was less surprising than it may have seemed. A good explanation ought to show that the causal processes at work are of familiar kinds; or that they are analogous to familiar processes; or that they are governed by simple and powerful laws; or that they are not too miscellaneous. But I say that a good explanation ought to show none of these things unless they are true. If one of these things is false in a given case, and if the recipient is interested in the question of whether it is true, or mistakenly thinks that it is true, then a good explanation ought to show that it is false. But that is nothing special: it falls under points 1, 5, and 7 of my list.

It is as if someone thought that a good explanation of any current event had to be one that revealed the sinister doings of the CIA. When the CIA really does play a part in the causal history, we would do well to tell him about it: we thereby provide correct explanatory information about the part of the causal history that interests him most. But in case the CIA had nothing to do with it, we ought not to tell him that it did. Rather we ought to tell him that it didn't. Telling him what he hopes to hear is not even a merit to be balanced off against the demerit of falsehood. In itself it has no merit at all. What does have merit is addressing the right question.

This much is true. We are, and we ought to be, biased in favor of believing hypotheses according to which what happens is probable, is governed by simple laws, and so forth. That is relevant to the credibility of explanatory information. But credibility is not a separate merit alongside truth; rather, it is what we go for in seeking truth as best we can.

Another proposed *desideratum* is that a good explanation ought to produce understanding. If understanding involves seeing the causal history of the explanandum as simple, familiar, or whatnot, I have already registered my objection. But understanding why an event took place might, I think, just mean possession of explanatory information about it—the more of that you possess, the better you understand. If so, of course a good explanation produces understanding. It produces possession of that which it provides. But this *desideratum*, so construed, is empty. It adds nothing to our understanding of explanation.

6. WHY-QUESTIONS, PLAIN AND CONTRASTIVE

A why-question, I said, is a request for explanatory information. All questions are requests for information of some or other sort.[12] But there is a distinction to be made. Every question has a maximal true answer: the whole truth about the subject matter on which information is requested, to which nothing could be added without irrelevancy or error. In some cases it is feasible to provide these maximal answers. Then we can reasonably hope for them, request them, and settle for nothing less. "Who done it?—Professor Plum." There's no more to say.

In other cases it isn't feasible to provide maximal true answers. There's just too much true information of the requested sort to know or to tell. Then we do not hope for maximal answers and do not request them, and we always settle for less. The feasible answers do not divide sharply into complete and partial. They're all partial, but some are more partial than others. There's only a fuzzy line between enough and not enough of the requested information. "What's going on here?"—No need to mention that you're digesting your dinner. "Who is Bob Hawke?"—No need to write the definitive biography. Less will be a perfectly good answer. Why-questions, of course, are among the questions that inevitably get partial answers.

When partial answers are the order of the day, questioners have their ways of indicating how much information they want, or what sort. "In a word, what food do penguins eat?" "Why, in economic terms, is there no significant American socialist party?"

One way to indicate what sort of explanatory information is wanted is through the use of contrastive why-questions. Sometimes there is an explicit "rather than. . . ." Then what is wanted is information about the causal history of the explanandum event, not including information that would also have applied to the causal histories of alternative events, of the sorts indicated, if one of them had taken place instead. In other words, information is requested about the difference between the actualized causal history of the explanandum and the unactualized causal histories of its unactualized alternatives. Why did I visit Melbourne in 1979, rather than Oxford or Uppsala or Wellington? Because Monash University invited me. That is part of the causal history of my visiting Melbourne; and if I had gone to one of the other places instead, presumably that would not have been part of the causal history of my going there. It would

---

[12] Except perhaps for questions that take imperative answers: "What do I do now, Boss?"

have been wrong to answer: Because I like going to places with good friends, good philosophy, cool weather, nice scenery, and plenty of trains. That liking is also part of the causal history of my visiting Melbourne, but it would equally have been part of the causal history of my visiting any of the other places, had I done so.

The same effect can be achieved by means of contrastive stress. Why did I *fly* to Brisbane when last I went there? I had my reasons for wanting to get there, but I won't mention those because they would have been part of the causal history no matter how I'd travelled. Instead I'll say that I had too little time to go by train. If I had gone by train, my having too little time could not have been part of the causal history of my so doing.

If we distinguish plain from contrastive why-questions, we can escape a dilemma about explanation under indeterminism. On the one hand, we seem quite prepared to offer explanations of chance events. Those of us who think that chance is all-pervasive (as well as those who suspend judgment) are no less willing than the staunchest determinist to explain the events that chance to happen.[13] On the other hand, we balk at the very idea of explaining why a chance event took place—for is it not the very essence of chance that one thing happens rather than another for no reason whatsoever? Are we of two minds?

No; I think we are right to explain chance events, yet we are right also to deny that we can ever explain why a chance process yields one outcome rather than another. According to what I've already said, indeed we cannot explain why one happened *rather than the other*. (That is so regardless of the respective probabilities of the two.) The actual causal history of the actual chance outcome does not differ at all from the unactualized causal history that the other outcome would have had, if that outcome had happened. A contrastive why-question with "rather" requests information about the features that differentiate the actual causal history from its counterfactual alternative. There are no such features, so the question can have no positive answer. Thus we are right to call chance events inexplicable, if it is contrastive explanation that we have in mind.

---

[13] A treatment of explanation in daily life, or in history, dare not set aside the explanation of chance events as a peculiarity arising only in quantum physics. If current scientific theory is to be trusted, chance events are far from exceptional. The misguided hope that determinism might prevail in history if not in physics well deserves Railton's mockery: "All but the most basic regularities of the universe stand forever in peril of being interrupted or upset by intrusion of the effects of random processes. . . . The success of a social revolution might appear to be explained by its overwhelming popular support, but this is to overlook the revolutionaries' luck: if all the naturally unstable nuclides on earth had commenced spontaneous nuclear fission in rapid succession, the triumph of the people would never have come to pass." ("A Deductive-Nomological Model of Probabilistic Explanation," 223–4.) On the same point, see my Postscript B to "A Subjectivist's Guide to Objective Chance," *Philosophical Papers*, ii. 117–21.

(Likewise, we can never explain why a chance event *had* to happen, because it didn't have to.) But take away the "rather" (and the "had") and explanation becomes possible. Even a chance event has a causal history. There is information about that causal history to be provided in answer to a plain why-question. And thus we are right to proceed as we all do in explaining what we take to be chance events.

7. THE COVERING-LAW MODEL

The covering-law model of explanation has long been the leading approach. As developed in the work of Hempel and others, it is an elegant and powerful theory. How much of it is compatible with what I have said?

Proponents of the covering-law model do not give a central place to the thesis that we explain by providing information about causes. But neither do they say much against it. They may complain that the ordinary notion of causation has resisted precise analysis; they may say that mere mention of a cause provides less in the way of explanation than might be wished; they may insist that there are a few special cases in which we have good non-causal explanations of particular occurences. But when they give us their intended examples of covering-law explanation, they almost always pick examples in which—as they willingly agree—the covering-law explanation does include a list of joint causes of the explanandum event, and thereby provides information about its causal history.

The foremost version of the covering-law model is Hempel's treatment of explanation in the non-probabilistic case.[14] He proposes that an explanation of a particular event consists, ideally, of a correct deductive-nomological (henceforth D-N) argument. There are law premises and particular-fact premises and no others. The conclusion says that the explanandum event took place. The argument is valid, in the sense that the premises could not all be true and the conclusion false. (We might instead define validity in syntactic terms. If so, we should be prepared to included mathematical, and perhaps definitional, truths among the premises.) No premise could be deleted without destroying the validity of the argument. The premises are all true.

Hempel also offers a treatment for the probabilistic case; but it differs significantly from his deductive-nomological model, and also it has two

---

[14] For a full presentation of Hempel's views, see the title essay in his *Aspects of Scientific Explanation*.

unwelcome consequences. (1) An improbable event cannot be explained at all. (2) One requirement for a correct explanation—"maximal specificity"—is relative to our state of knowledge; so that our ignorance can make correct an explanation that would be incorrect if we knew more. Surely what's true is rather that ignorance can make an explanation seem to be correct when really it is not. Therefore, instead of Hempel's treatment of the probabilistic case, I prefer to consider Railton's "deductive-nomological model of probabilistic explanation".[15] This closely parallels Hempel's D-N model for the non-probabilistic case, and it avoids both the difficulties just mentioned. Admittedly, Railton's treatment is available only if we are prepared to speak of chances—single-case objective probabilities. But that is no price at all if we have to pay it anyway. And we do, if we want to respect the apparent content of science. (Which is not the same as respecting the positivist philosophy popular among scientists.) Frequencies—finite or limiting, actual or counterfactual—are fine things in their own right. So are degrees of rational belief. But they just do not fit our ordinary conception of objective chance, as exemplified when we say that any radon-222 atom at any moment has a 50% chance of decaying within the next 3.825 days. If chances are good enough for theorists of radioactive decay, they are good enough for philosophers of science.

Railton proposes that an explanation of a particular chance event consists, ideally, of two parts. The first part is a D-N argument, satisfying the same constraints that we would impose in the nonprobabilistic case, to a conclusion that the explanandum event had a certain specified chance of taking place. The chance can be anything: very high, middling, or even very low. The D-N argument will have probabilistic laws among its

[15] See Railton's paper of the same name. In what follows I shall simplify Railton's position in two respects. (1) I shall ignore his division of a D-N argument for a probabilistic conclusion into two parts, the first deriving a law of uniform chances from some broader theory and the second applying that law to the case at hand. (2) I shall pretend, until further notice, that Railton differs from Hempel only in his treatment of probabilistic explanation; in fact there are other important differences, to be noted shortly.

It is important to distinguish Railton's proposal from a different way of using single-case chances in a covering-law model of explanation, proposed in James H. Fetzer, "A Single Case Propensity Theory of Explanation," *Synthese*, 28 (1974), 171–98. For Fetzer, as for Railton, the covering laws are universal generalizations about single-case chances. But for Fetzer, as for Hempel, the explanatory argument, without any addendum, is the whole of the explanation; it is inductive, not deductive; and its conclusion says outright that the explanandum took place, not that it had a certain chance. This theory shares some of the merits of Railton's. However, it has one quite peculiar consequence. For Fetzer, as for Hempel, an explanation is an argument; however, a good explanation is not necessarily a good argument. Fetzer, like Railton, wants to have explanations even when the explanandum is extremely improbable. But in that case a good explanation is an extremely bad argument. It is an inductive argument whose premises not only fail to give us any good reason to believe the conclusion, but in fact give us very good reason to *dis*believe the conclusion.

premises—preferably, laws drawn from some powerful and general the-ory—and these laws will take the form of universal generalizations con-cerning single-case chances. The second part of the explanation is an addendum—not part of the argument—which says that the event did in fact take place. The explanation is correct if both parts are correct: if the premises of the D-N argument are all true, and the addendum also is true.

Suppose we have a D-N argument, either to the explanandum event itself or to the conclusion that it has a certain chance. And suppose that each of the particular-fact premises says, of a certain particular event, that it took place. Then those events are jointly sufficient, given the laws cited, for the event or for the chance. In a sense, they are a minimal jointly sufficient set; but a proper subset might suffice given a different selection of true law premises, and also it might be possible to carve off parts of the events and get a set of the remnants that is still sufficient under the original laws. To perform an act of explaining by producing such an argument and committing oneself to its correctness is, in effect, to make two claims: (1) that certain events are jointly sufficient, under the prevailing laws, for the explanandum event or for a certain chance of it; and (2) that only certain of the laws are needed to establish that sufficiency.

It would make for reconciliation between my account and the cover-ing-law model if we had a covering-law model of causation to go with our covering-law model of explanation. Then we could rest assured that the jointly sufficient set presented in a D-N argument was a set of causes of the explanandum event. Unfortunately, that assurance is not to be had. Often, a member of the jointly sufficient set presented in a D-N argument will indeed be one of the causes of the explanandum event. But it may not be. The counterexamples are well known; I need only list them.

1. An irrelevant non-cause might belong to a non-minimal jointly sufficient set. Requiring minimality is not an adequate remedy; we can get an artificial minimality by gratuitously citing weak laws and leaving stronger relevant laws uncited. That is the lesson of Salmon's famous example of the man who escapes pregnancy because he takes birth con-trol pills, where the only cited law says that nobody who takes the pills becomes pregnant, and hence the premise that the man takes pills cannot be left out without spoiling the validity of the argument.[16]

2. A member of a jointly sufficient set may be something other than an event. For instance, a particular-fact premise might say that something

---

[16] See Wesley C. Salmon *et al.*, *Statistical Explanation and Statistical Relevance* (Pittsburgh: University of Pittsburgh Press, 1971), 34.

has a highly extrinsic or disjunctive property. I claim that such a premise cannot specify a genuine event.[17]

3. An effect might belong to a set jointly sufficient for its cause, as when there are laws saying that a certain kind of effect can be produced in only one way. That set might be in some appropriate sense minimal, and might be a set of events. That would not suffice to make the effect be a cause of its cause.

4. Such an effect might also belong to a set jointly sufficient for another effect, perhaps a later effect, of the same cause. Suppose that, given the laws and circumstances, the appearance of a beer ad on my television could only have been caused by a broadcast which would also cause a beer ad to appear on your television. Then the first appearance may be a member of a jointly sufficient set for the second; still, these are not cause and effect. Rather they are two effects of a common cause.

5. A preempted potential cause might belong to a set jointly sufficient for the effect it would have caused, since there might be nothing that could have stopped it from causing that effect without itself causing the same effect.

In view of these examples, we must conclude that the jointly sufficient set presented in a D-N argument may or may not be a set of causes. We do not, at least not yet, have a D-N analysis of causation. All the same, a D-N argument may present causes. If it does, or rather if it appears to the explainer and audience that it does, then on my view it ought to look explanatory. That is the typical case with sample D-N arguments produced by advocates of the covering-law model.

If the D-N argument does not appear to present causes, and it looks explanatory anyway, that is a problem for me. In Section 3, I discussed three such problem cases; the alleged non-causal explanations there considered could readily have been cast as D-N arguments, and indeed I took them from Hempel's and Railton's writings on covering-law explanation. In some cases, I concluded that information was after all given about how the explanandum was caused, even if it happened in a more roundabout way than by straightforward presentation of causes. In other cases, I concluded that what was explained was not really a particular event. Either way, I'm in the clear.

If the D-N argument does not appear to present causes, and therefore fails to look explanatory, that is a problem for the covering-law theorist. He might just insist that it *ought* to look explanatory, and that our customary standards of explanation need reform. To the extent that he

---

[17] See my "Events."

takes this high-handed line, I lose interest in trying to agree with as much of his theory as I can. But a more likely response is to impose constraints designed to disqualify the offending D-N arguments. Most simply, he might say that an explanation is a D-N argument of the sort that does present a set of causes, or that provides information in some more round-about way about how the explanandum was caused. Or he might seek some other constraint to the same effect, thereby continuing the pursuit of a D-N analysis of causation itself. Railton is one covering-law theorist who acknowledges that not just any correct D-N argument (or prob-abilistic D-N argument with addendum) is explanatory; further con-straints are needed to single out the ones that are. In sketching these further constraints, he does not avoid speaking in causal terms. (He has no reason to, since he is not attempting an analysis of causation itself.) For instance, he distinguishes D-N arguments that provide an "account of the mechanism" that leads up to the explanandum event; by which he means, I take it, that there ought to be some tracing of causal chains. He does not make this an inescapable requirement, however, because he thinks that not all covering-law explanation is causal.[18]

A D-N argument may explain by presenting causes, or otherwise giving information about the causal history of the explanandum; is it also true that any causal history can be characterized completely by means of the information that can be built into D-N arguments? That would be so if every cause of an event belongs to some set of causes that are jointly sufficient for it, given the laws; or, in the probabilistic case, that are jointly sufficient under the laws for some definite chance of it. Is it so that causes fall into jointly sufficient sets of one or the other sort? That does not follow, so far as I can tell, from the counterfactual analysis of causation that I favor. It may nevertheless be true, at least in a world governed by a sufficiently powerful system of (strict or probabilistic) laws; and this may be such a world. If it is true, then the whole of a causal history could in principle be mapped by means of D-N arguments (with addenda in the probabilistic case) of the explanatory sort.

In short, if explanatory information is information about causal his-tories, as I say it is, then one way to provide it is by means of D-N arguments. Moreover, under the hypothesis just advanced, there is no explanatory information that could not in principle be provided in that way. To that extent the covering-law model is dead right.

But even when we acknowledge the need to distinguish explanatory D-N arguments from others, perhaps by means of explicitly causal con-

[18] See his *Explaining Explanation*, "A Deductive-Nomological Model of Probabilistic Explanation," and "Probability, Explanation, and Information."

straints, there is something else wrong. It is this. The D-N argument—correct, explanatory, and fully explicit—is represented as the ideal serving of explanatory information. It is the right shape and the right size. It is enough, anything less is not enough, and anything more is more than enough.

Nobody thinks that real-life explainers commonly serve up full D-N arguments which they hope are correct. We very seldom do. And we seldom could—it's not just that we save our breath by leaving out the obvious parts. We don't know enough. Just try it. Choose some event you think you understand pretty well, and produce a fully explicit D-N argument, one that you can be moderately sure is correct and not just almost correct, that provides some non-trivial explanatory information about it. Consult any science book you like. Usually the most we can do, given our limited knowledge, is to make existential claims.[19] We can venture to claim that there exists some (correct, etc.) D-N argument for the explanandum that goes more or less like this, or that includes this among its premises, or that draws its premises from this scientific theory, or that derives its conclusion from its premises with the aid of this bit of mathematics, or. . . . I would commend these existential statements as explanatory, to the extent—and only to the extent—that they do a good job of giving information about the causal history of the explanandum. But if a proper explanation is a complete and correct D-N argument (perhaps plus addendum), then these existential statements are not yet proper explanations. Just in virtue of their form, they fail to meet the standard of how much information is enough.

Hempel writes "To the extent that a statement of individual causation leaves the relevant antecedent conditions, and thus also the requisite explanatory laws, indefinite it is like a note saying that there is a treasure hidden somewhere."[20] The note will help you find the treasure provided you go on working, but so long as you have only the note you have no treasure at all; and if you find the treasure you will find it all at once. I say it is not like that. A shipwreck has spread the treasure over the bottom of the sea and you will never find it all. Every dubloon you find is one more dubloon in your pocket, and also it is a clue to where the next dubloons may be. You may or may not want to look for them, depending on how many you have so far and on how much you want to be how rich.

[19] In *Foundations of Historical Knowledge*, ch. III, Morton White suggests that "because"-statements should be seen as existential claims. You assert the existence of an explanatory argument which includes a given premise, even though you may be unable to produce the argument. This is certainly a step in the right direction. However it seems to underestimate the variety of existential statements that might be made, and also it incorporates a suspect D-N analysis of causation.

[20] *Aspects of Scientific Explanation*, 349.

If you have anything less than a full D-N argument, there is more to be found out. Your explanatory information is only partial. Yes. *And so is any serving of explanatory information we will ever get*, even if it consists of ever so many perfect D-N arguments piled one upon the other. There is always more to know. A D-N argument presents only one small part—a cross-section, so to speak—of the causal history. There are very many other causes of the explanandum that are left out. Those might be the ones we especially want to know about. We might want to know about causes earlier than those presented. Or we might want to know about causes intermediate between those presented and the explanandum. We might want to learn the mechanisms involved by tracing particular causal chains in some detail. (The premises of a D-N argument might tell us that the explanandum would come about through one or the other of two very different causal chains, but not tell us which one). A D-N argument might give us far from enough explanatory information, considering what sort of information we want and what we possess already. On the other hand, it might give us too much. Or it might be the wrong shape, and give us not enough and too much at the same time; for it might give us explanatory information of a sort we do not especially want. The cross-section it presents might tell us a lot about the side of the causal history we're content to take for granted, and nothing but stale news about the side we urgently want to know more about.

Is a (correct, etc.) D-N argument in *any* sense a complete serving of explanatory information? Yes in this sense, and this sense alone: it completes a jointly sufficient set of causes. (And other servings complete seventeen-membered sets, still others complete sets going back to the nineteenth century. . . .) The completeness of the jointly sufficient set has nothing to do with the sort of enoughness that we pursue. There is nothing ideal about it, in general. Other shapes and sizes of partial servings may be very much better—and perhaps also better within our reach.

It is not that I have some different idea about what is the unit of explanation. We should not demand a unit, and that demand has distorted the subject badly. It's not that explanations are things we may or may not have one of; rather, explanation is something we may have more or less of.

One bad effect of an unsuitable standard of enoughness is that it may foster disrespect for the explanatory knowledge of our forefathers. Suppose, as may be true, that seldom or never did they get the laws quite right. Then seldom or never did they possess complete and correct D-N arguments. Did they therefore lack explanatory knowledge? Did they have only some notes, and not yet any of the treasure? Surely not! And the reason, say I, is that whatever they may not have known about the laws, they knew a lot about how things were caused.

But once again, the covering-law model needn't have the drawback of which I have been complaining; and once again it is Railton who has proposed the remedy.[21] His picture is similar to mine. Associated with each explanandum we have a vast and complicated structure; explanatory information is information about this structure; an act of explaining is an act of conveying some of this information; more or less information may be conveyed, and in general the act of explaining may be more or less satisfactory in whatever ways any act of conveying information about a large and complicated structure may be more or less satisfactory. The only difference is that whereas for me the vast structure consists of events connected by causal dependence, for Railton it is an enormous "ideal text" consisting of D-N arguments—correct, satisfying whatever constraints need be imposed to make them explanatory, and with addenda as needed—strung together. They fit together like proofs in a mathematics text, with the conclusion of one feeding in as a premise to another, and in the end we reach arguments to the occurrence, or at least a chance, of the explanandum itself. It is unobjectionable to let the subject matter come in units of one argument each, so long as the activity of giving information about it needn't be broken artificially into corresponding units.

By now, little is left in dispute. Both sides agree that explaining is a matter of giving information, and no standard unit need be completed. The covering-law theorist has abandoned any commitment he may once have had to a D-N analysis of causation; he agrees that not just any correct D-N argument is explanatory; he goes some distance toward agreeing that the explanatory ones give information about how the explanandum is caused; and he does not claim that we normally, or even ideally, explain by producing arguments. For my part, I agree that one way to explain would be to produce explanatory D-N arguments; and further, that an explainer may have to argue for what he says in order to be believed. Explanation as argument versus explanation as information is a spurious contrast. More important, I would never deny the relevance of laws to causation, and therefore to explanation; for when we ask what would have happened in the absence of a supposed cause, a first thing to say is that the world would then have evolved lawfully. The covering-law theorist is committed, as I am not, to the thesis that all explanatory information can be incorporated into D-N arguments; however, I do not

---

[21] See *Explaining Explanation* and "Probability, Explanation, and Information."

deny it, at least not for a world like ours with a powerful system of laws. I am committed, as he is not, to the thesis that all explaining of particular events gives some or other sort of information about how they are caused; but when we see how many varieties of causal information there are, and how indirect they can get, perhaps this disagreement too will seem much diminished.

One disagreement remains, central but elusive. It can be agreed that information about the prevailing laws is at least highly relevant to causal information, and *vice versa*; so that the pursuit of explanation and the investigation of laws are inseparable in practice. But still we can ask whether information about the covering laws is itself *part* of explanatory information. The covering-law theorist says yes; I say no. But this looks like a question that would be impossible to settle, given that there is no practical prospect of seeking or gaining information about causes without information about laws, or information about laws without information about causes. We can ask whether the work of explaining would be done if we knew all the causes and none of the laws. We can ask; but there is little point trying to answer, since intuitive judgments about such preposterous situations needn't command respect.

VIII

CONTRASTIVE EXPLANATION*

PETER LIPTON

1. INTRODUCTION

According to a causal model of explanation, we explain phenomena by giving their causes or, where the phenomena are themselves causal regularities, we explain them by giving a mechanism linking cause and effect. If we explain why smoking causes cancer, we do not give the cause of this causal connection, but we do give the causal mechanism that makes it. The claim that to explain is to give a cause is not only natural and plausible, but it also avoids many of the objections to other accounts of explanation, such as the views that to explain is to give a reason to believe the phenomenon occurred, to somehow make the phenomenon familiar, or to give a deductive-nomological argument. Unlike the reason for belief account, a causal model makes a clear distinction between understanding why a phenomenon occurs and merely knowing that it does, and the model does so in a way that makes understanding unmysterious and objective. Understanding is not some sort of super-knowledge, but simply more knowledge: knowledge of the phenomenon and knowledge of its causal history. A causal model makes it clear how something can explain without itself being explained, and so avoids the regress of whys, since we can know a phenomenon's cause without knowing the cause of the cause. It also accounts for legitimate self-evidencing explanations, explanations where the phenomenon is an essential part of the evidence that the explanation is correct, so the explanation cannot supply a non-circular reason for believing the phenomenon occurred. There is no barrier to knowing a cause through its effects and also knowing that it is their cause. The speed of recession of a star

Peter Lipton, 'Contrastive Explanation', in D. Knowles (ed.), *Explanation and its Limits* (Cambridge: Cambridge University Press, 1990.) Copyright Royal Institute of Philosophy. Reprinted by permission.

* I am grateful to Philip Clayton, Trevor Hussey, Philip Pettit, David Ruben, Elliott Sober, Edward Stein, Nick Thompson, Jonathan Vogel, David Weissbord, Alan White, Tim Williamson, and Eddy Zemach for helpful discussions about contrastive explanation.

explains its observed red-shift, even though the shift is an essential part of the evidence for its speed of recession. The model also avoids the most serious objection to the familiarity view, which is that some phenomena are familiar yet not understood, since a phenomenon can be perfectly familiar, such as the blueness of the sky or the fact that the same side of the moon always faces the earth, even if we do not know its cause. Finally, a causal model avoids many of the objections to the deductive-nomological model. Ordinary explanations do not have to meet the requirements of the deductive-nomological model, because one does not need to give a law to give a cause, and one does not need to know a law to have good reason to believe that a cause is a cause. As for the notorious over-permissiveness of the deductive-nomological model, the reason recession explains red-shift but not conversely, is simply that causes explain effects but not conversely, and the reason a conjunction of laws does not explain its conjuncts is that conjunctions do not cause their conjuncts.

The most obvious objection to a causal model of explanation is that there are non-causal explanations. Mathematicians and philosophers give explanations, but mathematical explanations are never causal, and philosophical explanations seldom are. A mathematician may explain why Gödel's Theorem is true, and a philosopher may explain why there can be no inductive justification of induction, but these are not explanations that cite causes. In addition to the mathematical and philosophical cases, there are explanations of the physical world that seem non-causal. Here is a personal favourite. Suppose that some sticks are thrown into the air with a lot of 'spin', so that they separate and tumble about as they fall. Now freeze the scene at a certain point during the sticks' descent. Why are appreciably more of them near the vertical axis than near the horizontal, rather than in more or less equal numbers near each orientation, as one would have expected? The answer, roughly speaking, is that there are many more ways for a stick to be near the horizontal than near the vertical. To see this, consider purely horizontal and vertical orientations for a single stick with a fixed midpoint. There are infinitely many of the former, but only two of the latter. Less roughly, the explanation is that there are two horizontal dimensions but only one vertical one. This is a lovely explanation, but apparently not a causal one, since geometrical facts cannot be causes.

Non-causal explanations show that a causal model of explanation cannot be complete. Nevertheless, a causal model is still a good bet now, because of the backward state of alternate views of explanation, and the overwhelming preponderance of causal explanations among all explanations. Nor is it *ad hoc* to limit our attention to causal explanations. A causal model does not simply pick out a feature that certain explanations happen to have: causal explanations are explanatory *because* they are causal. Like other accounts of explanation, however, causal models face a problem of underdetermination. Most causes do not provide good explanations. This paper attempts a partial solution to this problem.

2. FACT AND FOIL

Let us focus on the causal explanation of particular events. The problem here is with the notion of explaining an event by giving *the* cause. We may explain an event by giving some information about its causal history (Lewis, 1986), but causal histories are long and wide, and most causal information does not provide a good explanation. The big bang is part of the causal history of every event, but explains only a few. The spark and the oxygen are both part of the causal history that led up to the fire, but only one of them explains it. So what makes one piece of information about the causal history of an event explanatory and another not? The short answer is that the cause that explains depends on our interests. But this does not yield a very informative model of explanation unless we can go some way towards spelling out how explanatory interests determine explanatory causes.

One way to show how we select from among causes is to reveal additional structure in the phenomenon to be explained, structure that points to a particular cause. We can account for the specificity of explanatory answers by revealing the specificity in the explanatory question. Suppose we started by construing a phenomenon to be explained simply as a concrete event, say a particular eclipse. The number of causal factors is enormous. As Hempel has observed, however, we do not explain events, only aspects of events (Hempel, 1965: 421–3). We do not explain the eclipse *tout court*, but only why it lasted as long as it did, or why it was partial, or why it was not visible from a certain place. This reduces the number of causal factors we need consider for any particular phenomenon, since there will be many causes of the eclipse that are not, for example, causes of its duration.

More recently, it has been argued that explanation is 'interest relative', and that we can analyse some of this relativity with a constrastive analysis of the phenomenon to be explained. What gets explained is not simply 'Why this', but 'Why this rather than that' (Garfinkel, 1981: 28–41; van Fraassen, 1980: 126–9). A contrastive phenomenon consists of a *fact* and a *foil*, and the same fact may have several different foils. We may not explain why the leaves turn yellow in November *tout court*, but only, for example, why they turn yellow in November rather than in January, or why they turn yellow in November rather than turning blue.

The contrastive analysis of explanation is extremely natural. We often pose our why-questions in explicitly contrastive form and it is not difficult to come up with examples where different people select different foils, requiring different explanations. When I asked my 3-year-old son why he threw his food on the floor, he told me that he was full. This may explain why he threw it on the floor rather than eating it, but I wanted to know why he threw it rather than leaving it on his plate. Similarly, an explanation of why I went to see *Jumpers* rather than *Candide* will probably not explain why I went to see *Jumpers* rather than staying at home, and an explanation of why Able rather than Baker got the philosophy job may not explain why Able rather than Charles got the job. The proposal that phenomena to be explained have a complex fact–foil structure can be seen as another step along Hempel's path of focusing explanation by adding structure to the why-question. A fact is usually not specific enough: we also need to specify a foil. Since the causes that explain a fact relative to one foil will not generally explain it relative to another, the contrastive question provides a further restriction on explanatory causes.

While the role of contrasts in explanation will not account for all the factors that determine which cause is explanatory, I believe that it does provide the central mechanism. In this essay, I want to show in some detail how contrastive questions help select explanatory causes. My discussion will fall into three parts. First, I will make three general observations about contrastive explanation. Then, I will use these observations to show why contrastive questions resist reduction to non-contrastive form. Finally, I will describe the mechanism of 'causal triangulation' by which the choice of foils in contrastive questions helps to select explanatory causes.

When we ask a contrastive why-question—'Why the fact rather than the foil?'—we presuppose that the fact occurred and that the foil did not. Often we also suppose that the fact and the foil are in some sense incompatible. When we ask why Kate rather than Frank won the Philosophy Department Prize, we suppose that they could not both have won. Similarly, when we asked about leaves, we supposed that if they turn yellow in November, they cannot turn yellow in January, and if they turn yellow in November, they cannot also turn blue then. Indeed, it is widely supposed that fact and foil are always incompatible (Garfinkel, 1981: 40;

Ruben, 1987; Temple, 1988: 144). My first observation is that this is false: many contrasts are compatible. We often ask a contrastive question when we do not understand why two apparently similar situations turned out differently. In such a case, far from supposing any incompatibility between fact and foil, we ask the question just because we expected them to turn out the same. By the time we ask the question, we realize that our expectation was disappointed, but this does not normally lead us to believe that the fact precluded the foil, and the explanation for the contrast will usually not show that it did. Consider the much discussed example of syphilis and paresis (cf. Hempel, 1965: 369–70; van Fraassen, 1980: 128). Few with syphilis contract paresis, but we can still explain why Jones rather than Smith contracted paresis by pointing out that only Jones had syphilis. In this case, there is no incompatibility. Only Jones contracted paresis, but they both could have: Jones's affliction did not protect Smith. Of course not every pair of compatible propositions would make a sensible contrast but, as we will eventually see, it is not necessary to restrict contrastive questions to incompatible contrasts to distinguish sensible questions from silly ones.

My second and third observations concern the relationship between an explanation of the contrast between a fact and foil and the explanation of the fact alone. I do not have a general account of what it takes to explain a fact on its own. As we will see, this is not necessary to give an account of what it takes to explain a contrast; indeed, this is one of the advantages of a contrastive analysis. Yet, based on our intuitive judgements of what is and what is not an acceptable explanation of a fact alone, the requirements for explaining a fact diverge from the requirements for explaining a contrast. My second observation, then, is that explaining a contrast is sometimes easier than explaining the fact alone (cf. Garfinkel, 1981: 30). An explanation of 'P rather than Q' is not always an explanation of P. This is particularly clear in examples of compatible contrasts. Jones's syphilis does not explain why he got paresis, since the vast majority of people who get syphilis do not get paresis, but it does explain why Jones rather than Smith got paresis, since Smith did not have syphilis. The relative ease with which we explain some contrasts also applies to many cases where there is an incompatibility between fact and foil. My preference for contemporary plays may not explain why I went to see *Jumpers* last night, since it does not explain why I went out, but it does explain why I went to see *Jumpers* rather than *Candide*. A particularly striking example of the relative ease with which some contrasts can be explained is the explanation that I chose A rather than B because I did not realize that B was an option. If you ask me why I ordered eggplant rather than sea bass (a 'daily special'), I may give the perfectly good answer that I did not know there were any specials, but this would not be an acceptable answer to the simple question, 'Why did you order eggplant?' One reason we can sometimes explain a contrast without explaining the fact alone seems to be that contrastive questions incorporate a presupposition that makes explanation easier. To explain '*P* rather than *Q*' is to give a certain type of explanation of *P*, *given* '*P* or *Q*', and an explanation that succeeds with the presupposition will not generally succeed without it.

My final observation is that explaining a contrast is also sometimes harder than explaining the fact alone. An explanation of *P* is not always an explanation of '*P* rather than *Q*'. This is obvious in the case of compatible contrasts: you cannot explain why Jones rather than Smith contracted paresis without saying something about Smith. But it also applies to incompatible contrasts. To explain why I went to *Jumpers* rather than *Candide*, it is not enough for me to say that I was in the mood for a philosophical play. To explain why Kate rather than Frank won the prize, it is not enough that she wrote a good essay; it must have been better than Frank's. One reason that explaining a contrast is sometimes harder than explaining the fact alone is that explaining a contrast requires giving causal information that distinguishes the fact from the foil, and information that we accept as an explanation of the fact alone may not do this.

3. FAILED REDUCTIONS

There have been a number of attempts to reduce contrastive questions to non-contrastive and generally truth-functional form. One motivation for this is to bring contrastive explanations into the fold of the deductive-nomological model since, without some reduction, it is not clear what the conclusion of a deductive explanation of '*P* rather than *Q*' ought to be. Armed with our three observations—that contrasts may be compatible, and that explaining a contrast is sometimes easier and sometimes harder than explaining the fact alone—we can show that contrastive questions resist a reduction to non-contrastive form. We have already seen that the contrastive question 'Why *P* rather than *Q*?' is not equivalent to the simple question 'Why *P*?', where two why-questions are explanatorily equivalent just in case any adequate answer to one is an adequate answer to the other. One of the questions may be easier or harder to answer than the other. Still, a proponent of the deductive-nomological model of explanation may be tempted to say that, for incompatible contrasts, the question 'Why P rather than Q?' is equivalent to 'Why P?' But it is not plausible to say that a deductive-nomological explanation of P is generally necessary to explain 'P rather than Q'. And it is even dubious that a deductive-nomological explanation of P is always sufficient to explain 'P rather than Q'. Imagine a typical deductive explanation for the rise of mercury in a thermometer. Such an explanation would explain various contrasts, for example why the mercury rose rather than fell. It may not, however, explain why the mercury rose rather than breaking the glass. A full deductive-nomological explanation of the rise will have to include a premiss saying that the glass does not break, but it does not need to explain this.

Another natural suggestion is that the contrastive question 'Why P rather than Q?' is equivalent to the conjunctive question 'Why P and not-Q?' On this view, explaining a contrast between fact and foil is tantamount to explaining the conjunction of the fact and the negation of the foil (Temple, 1988). In ordinary language, a contrastive question is often equivalent to its corresponding conjunction, simply because the 'and not' construction is often used contrastively. Instead of asking, 'Why was the prize won by Kate rather than by Frank?', the same question could be posed by asking 'Why was the prize won by Kate and not by Frank?'. But this colloquial equivalence does not seem to capture the point of the conjunctive view. So I suggest that the conjunctive view be taken to entail that explaining a conjunction at least requires explaining each conjunct; that an explanation of 'P and not-Q' must also provide an explanation of P and an explanation of not-Q. Thus, on the conjunctive view, to explain why Kate rather than Frank won the prize at least requires an explanation of why Kate won it and an explanation of why Frank did not. This account of contrastive explanation falls to the observation that explaining a contrast is sometimes easier than explaining the fact alone, since explaining P and explaining not-Q is at least as difficult as explaining P.

The observations that explaining contrasts is sometimes easier and sometimes harder than explaining the fact alone reveal another objection to the conjunctive view, on any model of explanation that is deductively closed. (A model is deductively closed if it entails that an explanation of P will also explain any logical consequence of P.) Consider cases where the fact is logically incompatible with the foil. Here P entails not-Q, so the conjunction 'P and not-Q' is logically equivalent to P alone. Furthermore, all conjunctions whose first conjunct is P and whose second conjunct is logically incompatible with P will be equivalent to each other, since they are all logically equivalent to P. Hence, for a deductively closed model of explanation, explaining 'P and not-Q' is tantamount to explaining P, whatever Q may be, so long as it is incompatible with P. We have seen, however, that explaining 'P rather than Q' is not generally tantamount to explaining P. The conjunction is explanatorily equivalent to P, and the contrast is not, so the conjunction is not equivalent to the contrast.

The failure to represent a contrastive phenomenon by the fact alone or by the conjunction of the fact and the negation of the foil suggests that, if we want a non-contrastive paraphrase, we ought instead to try something logically weaker than the fact. In some cases, it does seem that an explanation of the contrast is really an explanation of a logical consequence of the fact. This is closely related to what Hempel has to say about 'partial explanation' (1965: 415–18). He gives the example of Freud's explanation of a particular slip of the pen that resulted in writing down the wrong date. Freud explains the slip with his theory of wish-fulfilment, but Hempel objects that the explanation does not really show why that particular slip took place, but at best only why there was some wish-fulfilling slip or other. Freud gave a partial explanation of the particular slip, since he gave a full explanation of the weaker claim that there was some slip. Hempel's point fits naturally into contrastive language: Freud did not explain why it was this slip rather than another wish-fulfilling slip, though he did explain why it was this slip rather than no slip at all. And it seems natural to analyse 'Why this slip rather than no slip at all?' as 'Why some slip?'

In general, however, we cannot paraphrase contrastive questions with consequences of their facts. We cannot, for example, say that to explain why the leaves turn yellow in November rather than in January is just to explain why the leaves turn (some colour or other) in November. This attempted paraphrase fails to discriminate between the intended contrastive question and the question, 'Why do the leaves turn in November rather than falling right off?' Similarly, we cannot capture the question, 'Why did Jones rather than Smith get paresis?', by asking about some consequence of Jones's condition, such as why he contracted a disease.

A general problem with finding a paraphrase entailed by the fact P is that, as we have seen, explaining a contrast is sometimes harder than explaining P alone. There are also problems peculiar to the obvious candidates. The disjunction, 'P or Q' will not do: explaining why I went to *Jumpers* rather than *Candide* is not the same as explaining why I went to either. Indeed, this proposal gets things almost backwards: the disjunction is what the contrastive question assumes, not what calls for explanation. This suggests, instead, that the contrast is equivalent to the conditional, 'if *P* or *Q*, then *P*' or, what comes to the same thing if the conditional is truth-functional, to explaining *P* on the assumption of '*P* or *Q*'. Of all the reductions we have considered, this proposal is the most promising, but I do not think it will do. On a deductive model of explanation it would entail that any explanation of not-*Q* is also an explanation of the contrast, which is incorrect. We cannot explain why Jones rather than Smith has paresis by explaining why Smith did not get it. It would also wrongly entail that any explanation of *P* is an explanation of the contrast, since *P* entails the conditional.

4. CAUSAL TRIANGULATION

By asking a contrastive question, we can achieve a specificity that we do not seem to be able to capture either with a non-contrastive sentence that entails the fact or with one that the fact entails. But how then does a contrastive question specify the sort of information that will provide an adequate answer? It now appears that looking for a non-contrastive reduction of '*P* rather than *Q*' is not a useful way to proceed. The contrastive claim may entail no more than '*P* and not-*Q*' or perhaps better, '*P* but not-*Q*', but explaining the contrast is not the same as explaining these conjuncts. We will do better to leave the analysis of the contrastive question to one side, and instead consider directly what it takes to provide an adequate answer. David Lewis has given an interesting account of contrastive explanation that does not depend on paraphrasing the contrastive question. According to him, we explain why event *P* occurred rather than event *Q* by giving information about the causal history of *P* that would not have applied to the history of *Q*, if *Q* had occurred (Lewis, 1986: 229–30). Roughly, we cite a cause of *P* that would not have been a cause of *Q*. In Lewis's example, we can explain why he went to Monash rather than to Oxford in 1979 by pointing out that only Monash invited him, because the invitation to Monash was a cause of his trip, and that invitation would not have been a cause of a trip to Oxford, if he had taken one. On the other hand, Lewis's desire to go to places where he has good friends would not explain why he went to Monash rather than Oxford, since he has friends in both places and so the desire would have been part of either causal history.

Lewis's account, however, is too weak: it allows for unexplanatory causes. Suppose that both Oxford and Monash had invited him, but he went to Monash anyway. On Lewis's account, we can still explain this by pointing out that Monash invited him, since that invitation still would not have been a cause of a trip to Oxford. Yet the fact that he received an invitation from Monash clearly does not explain why he went there rather than to Oxford in this case, since Oxford invited him too. Similarly, Jones's syphilis satisfies Lewis's requirement even if Smith has syphilis too, yet in this case it would not explain why Jones rather than Smith contracted paresis.

It might be thought that Lewis's account could be saved by construing the causes more broadly, as types rather than tokens. In the case of the trip to Monash, we might take the cause to be receiving an invitation rather than the particular invitation to Monash he received. If we do this, we can correctly rule out the attempt to explain the trip by appeal to an invitation if Oxford also invited since, in this case, receiving an invitation would also have been a cause of going to Oxford. This, however, will not do, for two reasons. First, it does not capture Lewis's intent: he is interested in particular elements of a particular causal history, not general causal features. Secondly, and more importantly, the suggestion throws out the baby with the bath water. Now we have also ruled out the perfectly good explanation by invitation in some cases where only Monash invites. To see this, suppose that Lewis is the sort of person who only goes where he is invited. In this case, an invitation would have been part of a trip to Oxford, if he had gone there.

To improve on Lewis's account, consider John Stuart Mill's method of difference, his version of the controlled experiment (Mill, 1904, bk. 3, ch. 8, sect. 2). Mill's method rests on the principle that a cause must lie among the antecedent differences between a case where the effect occurs and an otherwise similar case where it does not. The difference in effect points back to a difference that locates a cause. Thus we might infer that contracting syphilis is a cause of paresis, since it is one of the ways Smith and Jones differed. The cause that the method of difference isolates depends on which control we use. If, instead of Smith, we have Doe, who does not have paresis but did contract syphilis and had it treated, we would be led to say that a cause of paresis is not syphilis, but the failure to treat it. The method of difference also applies to incompatible as well as to compatible contrasts. As Mill observes, the method often works particularly well with diachronic (before and after) contrasts, since these give us histories of fact and foil that are largely shared, making it easier to isolate a difference. If we want to determine the cause of a person's death, we naturally ask why he died when he did rather than at another time, and this yields an incompatible contrast, since you can only die once.

The method of difference concerns the discovery of causes rather than the explanation of effects, but the similarity to contrastive explanation is striking (cf. Garfinkel, 1981: 40). Accordingly, I propose that, for the causal explanations of events, explanatory contrasts select causes by means of what I will call the 'difference condition'. *To explain why P rather than Q, we must cite a causal difference between P and not-Q, consisting of a cause of P and the absence of a corresponding event in the history of not-Q.* Instead of pointing to a counterfactual difference, a particular cause of P that would not have been a cause of Q, as Lewis suggests, contrastive questions select as explanatory an actual difference between P and not-Q. Lewis's invitation to Monash does not explain why he went there rather than to Oxford if he was invited to both places because, while there is an invitation in the history of his trip to Monash, there is also an invitation in the history that leads him to forgo Oxford. Similarly, the difference condition correctly entails that Jones's syphilis does not explain why he rather than Smith contracted paresis if Smith had syphilis too, and that Kate's submitting an essay does not explain why she rather than Frank won the prize. Consider now some of the examples of successful contrastive explanation. If only Jones had syphilis, that explains why he rather than Smith has paresis, since having syphilis is a condition whose presence was a cause of Jones's paresis and a condition that does not appear in Smith's medical history. Writing the best essay explains why Kate rather than Frank won the prize, since that marks a causal difference between the two of them. Lastly, the fact that *Jumpers* is a contemporary play and *Candide* is not caused me both to go to one and to avoid the other.

The application of the difference condition is easiest to see in cases of compatible contrasts, since here the causal histories of P and of not-Q are generally distinct, but the condition does not require this. In cases of choice, for example, the causal histories are usually the same: the causes of my going to *Jumpers* are the same as the causes of my not going to *Candide*. The difference condition may nevertheless be satisfied if my belief that *Jumpers* is a contemporary play is a cause of going, and I do not believe that *Candide* is a contemporary play. That is why my preference for contemporary plays explains my choice. The difference condition does not require that the same event be present in the history of P but absent in the history of not-Q, a condition that could never be satisfied when the two histories are the same, but only that the cited cause of P find no corresponding event in the history of not-Q, where a corresponding event is something that would bear the same relation to Q as the cause of P bears to P.

One of the merits of the difference condition is that it brings out the way the incompatibility of fact and foil, when it obtains, is not sufficient to transform an explanation of the fact into an explanation of the contrast, even if the cause of the fact is also a cause of the foil not obtaining. Perhaps we could explain why Able got the philosophy job by pointing out that Quine wrote him a strong letter of recommendation, but this will only explain why Able rather than Baker got the job if Quine did not also write a similar letter for Baker. If he did, Quine's letter for Able does not alone explain the contrast, even though that letter is a cause of both Able's success and Baker's failure, and the former entails the latter. The letter may be a partial explanation of why Able got the job, but it does not explain why Able rather than Baker got the job. In the case where they both have strong letters from Quine, a good explanation of the contrast will have to find an actual difference, say that Baker's dossier was weaker than Able's in some other respect, or that Able's specialities were more useful to the department. There are some cases of contrastive explanation that do seem to rely on the way the fact precludes the foil, but I think these can be handled by the difference condition. For example, suppose we explain why a bomb went off prematurely at noon rather than in the evening by saying that the door hooked up to the trigger was opened at noon (I owe this example to Eddy Zemach). Here it may appear that the Difference Condition is not in play, since the explanation would stand even if the door was also opened in the evening. But the difference condition is met, if we take the cause not simply to be the opening of the door, but the opening of the door when it is rigged to an armed bomb.

My goal in this paper is to show how the choice of contrast helps to determine an explanatory cause, not to show why we choose one contrast rather than another. Still, some account of the considerations that govern our choice of why-questions would have to be a part of a full model of our explanatory practices, and it is to the credit of my model of contrastive explanation that it lends itself to this. For example, as I have already observed, not all contrasts make for sensible contrastive questions. It does not make sense, for example, to ask why Lewis went to Monash rather than Baker getting the philosophy job. One might have thought that a sensible contrast must be one where fact and foil are incompatible, but we have seen that this is not necessary, since there are many sensible compatible contrasts. There are also incompatible contrasts that do not yield reasonable contrastive questions, such as why someone died when she did rather than never having been born. The difference condition suggests instead that the central requirement for a sensible contrastive question is that the fact and the foil have a largely similar history, against which the differences stand out. When the histories are disparate, we do not know where to begin to answer the question. There are, of course, other considerations that help to determine the contrasts we actually choose. For example, in the case of incompatible contrasts, we often pick as foil the outcome we expected; in the case of compatible contrasts, as I have already mentioned, we often pick as foil a case we expected to turn out the same way as the fact. The condition of a similar history also helps to determine what will count as a corresponding event. If we were to ask why Lewis went to Monash rather than Baker getting the job, it would be difficult to see what in the history of Baker's failure would correspond to Lewis's invitation, but when we ask why Able rather than Baker got the job, the notion of a corresponding event is relatively clear.

5. FURTHER ISSUES

I will now consider three further issues connected with my analysis of contrastive explanation: the need for further principles for distinguishing explanatory from unexplanatory causes, the prospects for treating all why-questions as contrastive, and a comparison of my analysis with the deductive-nomological model. When we ask contrastive why-questions, we choose our foils to point towards the sort of causes that interest us. As we have just seen, when we ask about a surprising event, we often make the foil the thing we expected. Failed expectations are not, however, the only things that prompt us to ask why-questions. If a doctor is interested in the internal etiology of a disease, he will ask why the afflicted have it rather than other people in similar circumstances, even though the shared circumstances may be causally relevant to the disease. Again, if a machine malfunctions, the natural diagnostic contrast is its correct behaviour, since that directs our attention to the causes that we want to change. But the contrasts we construct will almost always leave multiple differences that meet the difference condition, and this raises the problem of selecting from among them. A problem of multiple differences also arises for the method of difference, in the context of inference rather than explanation. Mill tells us that we may infer that the only antecedent difference between fact and foil marks a cause, but in practice there will almost always be many such differences, not all of which will be causally relevant. Moreover, as Mill seems not to have recognized, his own deterministic assumptions entail that there will always be multiple differences as a matter of principle, since any antecedent difference itself marks an effect that must have a still earlier causal difference. (I owe this point to Trevor Hussey.)

In the case of inference, the central problem is to distinguish those differences that are causally relevant from those that are not. In the case of explanation, on the other hand, all the differences that meet the difference condition are, by definition, causally relevant. So all of them may be explanatory: the difference condition does not entail that there is only one way to explain a contrast. At the same time, however, some causally relevant differences will not be explanatory in a particular context, so while the difference condition may be necessary for the causal contrastive explanations of particular events, it is not generally sufficient. For that we need further principles of causal selection.

The considerations that govern selection from among causally relevant differences are numerous and diverse; the best I can do here is to mention what a few of them are. An obvious pragmatic consideration is that someone who asks a contrastive question may already know about some causal differences, in which case a good explanation will have to tell her something new. If she asks why Kate rather than Frank won the prize, she may assume that it was because Kate wrote the better essay, in which case we will have to tell her more about the differences between the essays that made Kate's better. A second consideration is that, when they are available, we usually prefer explanations where the foil would have occurred if the corresponding cause had occurred. Suppose that only Able had a letter from Quine, but even a strong letter from Quine would not have helped Baker much, since his specialities do not fit the department's needs. Suppose also that, had Baker's specialities been appropriate, he would have been given the job, even without a letter from Quine. In this case, the difference in specialities is a better explanation than the difference in letters. Note, however, that an explanation that does not meet this condition of counterfactual sufficiency for the occurrence of the foil may be perfectly acceptable, if we do not know of a sufficient difference. The explanation of why Jones rather than Smith contracted paresis is an example of this: even if Smith had syphilis in his medical history, he probably would not have contracted paresis. Moreover, even in cases where a set of known causes does supply a counterfactually sufficient condition, the enquirer may be much more interested in some than in others. The doctor may be particularly interested in causes he can control, the lawyer in causes that are connected with legal liability, and the accused in causes that cannot be held against him.

We also prefer differences where the cause is causally necessary for the fact in the circumstances. Consider a case of overdetermination.

Suppose that you ask me why I ordered eggplant rather than beef, when I was in the mood for eggplant and not for beef, and I am a vegetarian. My mood and my convictions are separate causes of my choice, each causally sufficient in the circumstance and neither necessary. In this case, it would be better to give both differences than just one. The difference condition could easily be modified to require necessary causes, but I think this would make the condition too strong. One problem would be cases of 'failsafe' overdetermination. Suppose we change the restaurant example so that my vegetarian convictions were not a cause of the particular choice I made: that time, it was simply my mood that was relevant. Nevertheless, even if I had been in the mood for beef, I would have resisted, because of my convictions. In this case, my explanation does not have to include my convictions, even though my mood was not a necessary cause of my choice. (Of course if I knew that you were asking me about my choice because you were planning to invite me to your house for dinner, it would be misleading for me not to mention my convictions, but this goes beyond the conditions for explaining the particular choice I made.) Again, we sometimes do not know whether a cause is necessary for the effect, and in such cases the cause still seems explanatory. But when there are differences that supply a necessary cause, and we know that they do, we prefer them. There are doubtless other pragmatic principles that play a role in determining which differences or combinations of differences yield the best explanation in a particular context. So there is more to contrastive explanation than the difference condition describes, but that condition does describe the central mechanism of causal selection.

Since contrastive questions are so common and foils play such an important role in determining explanatory causes, it is natural to wonder whether all why-questions are not at least implicitly contrastive. Often the contrast is so obvious that it is not worth mentioning. If you ask me why I was late for our appointment, the question is why I was late rather than on time, not why I was late rather than not showing up at all. Moreover, in cases where there is no specific contrast, stated or implied, we might construe 'Why P?' as 'Why P rather than not-P?', thus subsuming all causal why-questions under the contrastive analysis. But the difference condition seems to misbehave for these 'global' contrasts. It requires that we give a cause of P that finds no corresponding cause in the history of not-Q but, if the foil is simply the negation of the fact, this seems to require that we find a cause of P that finds no corresponding cause of itself, which is impossible, since it is tantamount to the requirement that we find a cause of P that is absent from the history of P.

We can, however, analyse the explanation of *P simpliciter* as the explanation of *P* rather than not-*P*. The correct way to construe the difference condition as it applies to the limiting case of the contrast, *P* rather than not-*P*, is that we must find a difference for events logically or causally incompatible with *P*, not for a single event, 'not-*P*'. Thus suppose that we ask why Jones has paresis, with no implied contrast. This would require a difference for foils where he does not have paresis. Saying that he had syphilis differentiates between the fact and the foil of a thoroughly healthy Jones, but this is not enough, since it does not differentiate between the fact and the foil of Jones with syphilis but without paresis. Excluding many incompatible foils will push us towards a sufficient cause of Jones's syphilis, since it is only by giving such a 'full cause' that we can be sure that some bit of it will be missing from the history of all the foils. To explain *P* rather than not-*P*, however, we do not need to explain every incompatible contrast. We do not, for example, need to explain why Jones contracted paresis rather than being long dead or never being born. The most we can require is that we exclude all incompatible foils with histories similar to the history of the fact.

One difficulty for this way of avoiding the pathological requirement of finding a cause of *P* that is absent from the history of *P* is that there appear to be some facts whose negation also seem to be a single fact (I owe this point to Elliot Sober). Suppose we wish to understand why there are tigers. Here the foil seems simply to be the absence of tigers, and we cannot give a cause of the existence of tigers that is not in the history of tigers. But the existence of tigers is not an event, so this example does not affect my account, which is only meant to apply to the explanation of events. So perhaps the problem does not arise for contrasts whose facts are events and whose foils are either events or sets of events. The difference condition will apply to some contrasts that are not explicitly event-contrasts, but not to all of them. Even for *P*s that are events, however, I am not certain that every apparently non-contrastive question should be analysed in contrastive form, so I am agnostic on the issue of whether all why-questions are contrastive.

Finally, let us compare my analysis of contrastive explanation to the deductive-nomological model. First, as we have already noted, a causal model of explanation has the merit of avoiding all the counter-examples to the deductive-nomological model where causes are deduced from effects. It also avoids the unhappy consequence of counting almost every explanation we give as a mere sketch, since one can give a cause of *P* that meets the difference condition for various foils without having the laws and singular premises necessary to deduce *P*. Many explanations that the deductive model counts as only very partial explanations of P are in fact reasonably complete explanations of P rather than Q. The excessive demands of the deductive model are particularly striking for cases of compatible contrasts, as least if the deductive-nomologist requires that an explanation of P rather than Q provide an explanation of P and an explanation of not-Q. In this case, the model makes explaining the contrast substantially harder than providing a deductive explanation of P, when in fact it is often substantially easier. Our inability to find a non-contrastive reduction of contrastive questions is a symptom of the inability of the deductive-nomological model to give an accurate account of this common type of explanation.

There are at least two other conspicuous advantages of a causal contrastive model of explanation over the deductive-nomological model. One odd feature of the deductive model is that it entails that an explanation cannot be ruined by adding true premises, so long as the additional premises do not render the law superfluous to the deduction, by entailing the conclusion outright. This consequence follows from the elementary logical point that additional premises can never convert a valid argument into an invalid one. In fact, however, irrelevant additions can spoil an explanation. If I say that Jones rather than Smith contracted paresis because only Jones had syphilis and only Smith was a regular church-goer, I have not simply said more than I need to, I have given an incorrect explanation, since going to church is not a prophylactic. By requiring that explanatory information be causally relevant, the causal model avoids this problem. Another related and unhappy feature of the deductive-nomological model is that it entails that explanations are virtually deductively closed: an explanation of P will also be an explanation of any logical consequence of P, so long as the consequence is not directly entailed by the initial conditions alone. (For an example of the slight non-closure in the model, notice that a deductive-nomological explanation of P will not also be a deductive-nomological explanation of the disjunction of P and one of the initial conditions of the explanation.) In practice, however, explanation seems to involve a much stronger form of non-closure. I might explain why all the men in the restaurant are wearing Paisley ties by appealing to the fashion of the times for ties to be Paisley, but this might not explain why they are all wearing ties, which is because of a rule of the restaurant. (I owe this example to Tim Williamson.) The contrastive model gives a natural account of this sort of non-closure. When we ask about Paisley ties, the implied foil is other sorts of tie; but when we ask simply about ties, the foil is not wearing ties. The fashion marks a difference in one case, but not in the other.

A defender of the deductive-nomological model may respond to some of these points by arguing that, whatever the merits of a contrastive analysis of lay explanation, the deductive model (perhaps with an additional restriction blocking 'explanations' of causes by effects) gives a better account of scientific explanation. For example, it has been claimed that since scientific explanations, unlike ordinary explanations, do not exhibit the interest relativity of foil variation that a contrastive analysis exploits, a contrastive analysis does not apply to scientific explanation (Worrall, 1984: 76–7). It is, however, a mistake to suppose that all scientific explanations even aspire to deductive-nomological status. The explanation of why Jones rather than Smith contracted paresis is presumably scientific, but it is not a deduction *manqué*. Moreover, as the example of the thermometer showed, even a full deductive-nomological explanation may exhibit interest relativity. It may explain the fact relative to some foils but not relative to others. A typical deductive-nomological explanation of the rise of mercury in a thermometer will simply assume that the glass does not break and so while it will explain, for example, why the mercury rose rather than fell, it will not explain why it rose rather than breaking the thermometer. Quite generally, a deductive-nomological explanation of a fact will not explain that fact relative to any foils that are themselves logically inconsistent with one of the premises of the explanation. Again, a Newtonian explanation of the Earth's orbit (ignoring the influence of the other planets) will explain why the Earth has its actual orbit rather than some other orbits, but it will not explain why the Earth does not have any of the other orbits that are compatible with Newton's theory. The explanation must assume information about the Earth's position and velocity at some time that will rule out the other Newtonian orbits, but it will not explain why the Earth does not travel in those paths. To explain this would require quite different information about the early history of the Earth. Similarly, an adaptionist explanation for a species's possession of a certain trait may explain why it has that trait rather than various maladaptive traits, but it may not explain why it had that trait rather than other traits that would perform the same functions equally well. To explain why an animal has one trait rather than another functionally equivalent trait requires instead appeal to the evolutionary history of the species, in so far as it can be explained at all.

With rather more justice, a deductive-nomologist might object that scientific explanations do very often essentially involve laws and theories, and that the contrastive model does not seem to account for this. For even if the fact to be explained carries no restricting contrast, the contrastive model, if it is extended to this case by analysing 'Why *P*?'

as 'Why *P* rather than not-*P*?', only requires at most that we cite a condition that is causally sufficient for the fact, not that we actually give any laws. I think, however, that the contrastive model can help to account for the undeniable role of laws in many scientific explanations. To see this, notice that scientists are often and perhaps primarily interested in explaining regularities, rather than particular events (cf. Friedman, 1974: 5; though explaining particular events is also important when, for example, scientists test their theories, since observations are of particular events). I think that the difference condition applies to many explanations of regularities, but to give a contrastive explanation of a regularity will require citing a law, or at least a generalization, since we here need some general cause (cf. Lewis, 1986: 225–6). To explain, say, why people feel the heat more when the humidity is high, we must find some general causal difference between cases where the humidity is high and cases where it is not, such as the fact that the evaporation of perspiration, upon which our cooling system depends, slows as the humidity rises. So the contrastive model, in an expanded version that applies to general facts as well as to events (a version I do not here provide), should be able to account for the role of laws in scientific explanations as a consequence of the scientific interest in general why-questions. Similarly, although the contrastive model does not require deduction for explanation, it is not mysterious that scientists should often look for explanations that do entail the phenomenon to be explained. This may not have to do with the requirements of explanation *per se*, but rather with the uses to which explanations are put. Scientists often want explanations that can be used for accurate prediction, and this requires deduction. Again, the construction of an explanation is a way to test a theory, and some tests require deduction.

Another way of seeing the compatibility of the scientific emphasis on theory and the contrastive model is by observing that scientists are not just interested in this or that explanation, but in a unified explanatory scheme. Scientists want theories, in part, because they want engines that provide many explanations. The contrastive model does not entail that a theory is necessary for any particular explanation, but a good theory is the best way to provide the many and diverse contrastive explanations that the scientist is after. This also helps to account for the familiar point that scientists are often interested in discovering causal mechanisms. The contrastive model will not require a mechanism to explain why one input into a black box causes one output, but it pushes us to specify more and more of the detailed workings of the box as we try to explain its full behaviour under diverse conditions. So I conclude that the contrastive model of explanation does not fly in the face of scientific practice.

6. CONCLUSION

The difference condition shows how contrastive questions about particular events help to determine an explanatory cause by a kind of causal triangulation. This contrastive model of causal explanation cannot be the whole story about explanation, since not all explanations are causal explanations or explanations of particular events and since, as we have seen, the choice of foil is not the only factor that affects the appropriate choice of cause. It does, however, give a natural account of much of what is going on in many explanations, and it captures some of the merits of competing accounts while avoiding some of their weaknesses. We have just seen this in some detail for the case of the deductive-nomological model. It also applies to the familiarity view. When an event surprises us, a natural foil is the outcome we had expected, and meeting the difference condition for this contrast will help to show us why our expectation went wrong. The mechanism of causal triangulation also accounts for the way a change in foil can lead to a change in explanatory cause, since a difference for one foil will not in general be a difference for another. It also shows why explaining 'P rather than Q' is sometimes harder and sometimes easier than explaining P alone. It may be harder, because it requires the absence of a corresponding event in the history of not-Q, and this is something that will not generally follow from the presence of the cause of P. Explaining the contrast may be easier, because the cause of P need not be sufficient for P, so long as it is part of a causal difference between P and not-Q. Again, causal triangulation helps to elucidate the interest relativity of explanation. We express some of our interests through our choice of foils and, by construing the phenomenon to be explained as a contrast rather than the fact alone, interest relativity reduces to the important but unsurprising point that different people are interested in explaining different phenomena. The difference condition also shows that different interests do not require incompatible explanations to satisfy them, only different but compatible causes. Moreover, my model of contrastive explanation suggests that our choice of foils is often governed by our *inferential* interests. As I argue extensively elsewhere, the structural similarity between the difference condition and the method of difference enables us to show why the inductive procedure of 'inference to the best explanation' is a reliable way of discovering causes. Because of this similarity, it can be shown that the hypothesis that would provide the best explanation of our contrastive data is also the one that is likeliest to have located an actual cause (Lipton, 1991). Finally, the mechanism of causal triangulation accounts for the failure of various attempts to reduce contrastive questions to non-contrastive form. None of these bring out the way a foil serves to select a location on the causal history leading up to the fact. Causal triangulation is the central feature of contrastive explanation that non-contrastive paraphrases suppress.

REFERENCES

Friedman, M. (1974), 'Explanation and Scientific Understanding', *The Journal of Philosophy* 71, 1–19.
Garfinkel, A. (1981), *Forms of Explanation* (New Haven: Yale University Press).
Hempel, C. (1965), *Aspects of Scientific Explanation* (New York: The Free Press).
Lewis, D. (1986), 'Causal Explanation', in *Philosophical Papers* (New York: Oxford University Press), ii. 214–40.
Lipton, P. (1991), *Inference to the Best Explanation* (London: Routledge).
Mill, J. S. (1904), *A System of Logic* (London: Longmans, Green & Co.).
Ruben, D.-H. (1987), 'Explaining Contrastive Facts', *Analysis* 47/1, 35–7.
Temple, D. (1988), 'Discussion: The Contrast Theory of Why-Questions', *Philosophy of Science* 55, 141–51.
van Fraassen, B. (1980), *The Scientific Image* (Oxford: Clarendon Press).
Worrall, J. (1988), 'An Unreal Image', *The British Journal for the Philosophy of Science*, 35, 65–80.

IX

EXPLANATORY REALISM, CAUSAL REALISM, AND EXPLANATORY EXCLUSION*

JAEGWON KIM

1

Explaining is an epistemological activity, and 'having' an explanation is, like knowing, an epistemological accomplishment. To be in need of an explanation is to be in an epistemologically imperfect state, and we look for an explanation in an attempt to remove that imperfection and thereby improve our epistemic situation. If we think in terms of the traditional divide between knowledge and reality known, explanations lie on the side of knowledge—on the side of the 'subjective' rather than that of the 'objective', on the side of 'representation' rather than that of reality represented. Our explanations are part of what we know about the world.

Knowledge implies truth: we cannot know something that is not the case. On a realist view of knowledge, every bit of knowledge has an objective counterpart, the thing that is known which is itself not part of knowledge—at least, not part of that particular bit of knowledge. But exactly what is it that we know when we have an explanation? Exactly in what does *explanatory knowledge* consist? If explanations constitute knowledge, it makes sense to ask, for each explanation that we 'have', exactly what it is that we know in virtue of having that explanation. And when we gain a new explanation, precisely what change takes place in our body of knowledge? We usually think of knowledge as consisting of a set of 'propositions', thought to represent 'facts' of the world. These propositions are discrete items, although they form a complex network of logical and evidential connections. How are we to represent explana-

Jaegwon Kim, *Midwest Studies in Philosophy*, 12 (1987), 225–39. Reprinted by permission of the author and the journal.

* I am indebted to David Benfield, Paul Boghossian, Brian McLaughlin, Joseph Mendola, and Michael Resnik for discussions of some of the issues taken up in this paper.

tory knowledge within such a picture? Where do we locate explanations in a scheme of propositions?

We can think of an explanation as a complex of propositions or statements divisible into two parts, *explanans* and *explanandum proposition*.[1] Since explanations can take a variety of linguistic forms, this division is rough; in particular, it is not to be taken to imply that an explanation is an argument or inference, with the explanans as premiss and the explanandum as conclusion. Let us focus on explanations of individual events. Such explanations typically explain an event (why a given event occurred) by reference to another event (or set of events). Let E be the explanandum, to the effect that a certain event e occurred. Let C be an explanans for this explanandum. C, let us assume, is the statement that event c occurred. Suppose then that we 'have' this explanation—that is, C and E are related as explanans to explanandum in our body of knowledge (call this the 'explanans relation'). What is the relationship between events c and e?

What I want to call *explanatory realism* takes the following position: C is an explanans for E *in virtue of* the fact that c bears to e some determinate objective relation R. Let us call R, whatever it is, an 'explanatory relation'. (The *explanans* relation relates propositions or statements; the *explanatory* relation relates events or facts in the world.) The explanatory relation is an objective relation among events that, as we might say, 'ground' the explanans relation, and constitutes its 'objective correlate'. On the realist view, our explanations are 'correct' or 'true' if they depict these relations correctly, just as our propositions or beliefs are true if they correctly depict objective facts; and explanations could be more or less 'accurate' according to how accurately they depict these relations. Thus, that c is related by explanatory relation R to e is the 'content' of the explanation consisting of C and E; it is what the explanation 'says'.

Although the attribution of truth or correctness to explanations is essential to explanatory realism, it by itself is not sufficient; those who reject explanatory realism in our sense, too, can speak of the truth or falsity of an explanation—for example, in the sense that propositions constituting the explanans are all true. What matters to realism is that the truth of an explanation requires an *objective relationship* between the events involved. By an 'objective relation', I have in mind a relation that at least meets the following condition: that it is instantiated does not entail anything about the existence or non-existence of any intentional

---

[1] I shall sometimes use 'explanandum' as short for 'explanandum proposition (or statement)'. This should cause no confusion.

psychological state—in particular, an epistemological or doxastic state—
except, of course, when it is instantiated by such states. I am not suggesting that the explanatory relation holding for events is all there is to
explanations, or to the explanans relation. Just as knowledge requires
more than truth, explanations presumably must meet further requirements
('internal' conditions—perhaps logical and epistemic ones), although
exactly what these are does not concern us here.

What could such an R be in virtue of which an event is correctly cited
in an explanation of another? The obvious first thought is this: R is the
causal relation. Perhaps there are non-causal explanations of individual
events; however, few will deny that the causal relation is at least one
important special case of R. And there are those who hold that the causal
relation is the only explanatory relation—at least the principal one.[2]

Explanatory irrealism, on the other hand, would be the view that the
relation of being an explanans for, as it relates C and E within our
epistemic corpus, is not, and need not be, 'grounded' in any objective
relation between events c and e. It is solely a matter of some 'internal'
relationship between items of knowledge. Perhaps, there are logical,
conceptual, or epistemic relationships among propositions in virtue of
which one proposition constitutes an explanans for another, and when
that happens, we could speak of the events represented as being related
by an explanatory relation. That is, given the explanans relation over
propositions, a relation over the events they represent could be defined:
c explains (is related by R to) e just in case C is an explanans for E. But
an R so defined would fail to be an objective relation, as required by
realism, for it would depend crucially on what goes on within our body
of knowledge and belief.

In the following passage Wesley Salmon gives a clear and forceful
expression to the realist view of explanation:

We need not object to [the purely psychological conception of explanation]
merely on the ground that people often invoke false beliefs and feel comfortable
with the 'explanation' thus provided. . . . We can, quite consistently with this approach, insist that adequate explanations must rest upon *true* explanatory bases. Nor
need we object on the ground that supernatural 'explanations' are often psychologically appealing. Again, we can insist that the explanation be grounded in

---

[2] E.g. Wesley C. Salmon, *Scientific Explanation and the Causal Structure of the World*
(Princeton, NJ, 1984); David Lewis, 'Causal Explanation', in *Philosophical Papers*, ii [Ch. VII
herein]. However, there are relations other than causation one might want to consider: e.g. the
relation of supervenience, the micro-reductive relation. Whether or not these possible explanatory
relations require the same explanandum as the causal relation is another question; see Robert
Cummins's distinction between 'explanation by subsumption' and 'explanation by analysis' in his
The Nature of Psychological Explanation (Cambridge, Mass., 1983), ch. 1. See also Peter
Achinstein, 'A Type of Non-Causal Explanation', *Midwest Studies in Philosophy*, 9 (1984); 221–43.

scientific fact. Even with those restrictions, however, the view that scientific explanation consists in release from psychological uneasiness is unacceptable for two reasons. First, we must surely require that there be some sort of *objective* relationship between the explanatory facts and the fact-to-be-explained.[3]

However, merely to hold that C is an explanans for E just in case c is a cause of e is not necessarily to espouse explanatory realism. Whether that is so depends on one's conception of causation. Consider, for example, Hanson, who writes:

The primary reason for referring to the cause of x is to explain x. There are as many causes of x as there are explanations of x.[4]

Causes certainly are connected with effects; but this is because our theories connect them, not because the world is held together by cosmic glue. The world *may* be glued together by imponderables, but that is irrelevant for understanding causal explanation. The notions 'the cause x' and 'the effect y' are intelligible only against a pattern of theory, namely one which puts guarantees on inference from x to y.[5]

For Hanson, causal relations essentially depend on an appropriate conceptual interlocking of our descriptions as provided by the theories we accept. He makes it evident, in the quoted passages, that he views the causal relation between x and y as derivative from an inferential relation from x to y, and the inferential relation as intimately associated with explanation; it is also evident that he does not take the dependence of causation on inference and explanation to be merely epistemological. If one accepts this view of causation and causal explanation, there is nothing realist about the position that causal explanations hold just in case the causal relation holds. For causal relations, on such an approach, depend on inferential-explanatory connections which are primary and more basic.

More generally, if one wants to *analyse* causation itself in terms of explanation,[6] one would be rejecting explanatory realism—unless one could identify an objective relation other than causation as the explanatory relation. But what could such a relation be? One might wish to propose the *nomological* relation as a candidate. The idea is this: that

---

[3] *Scientific Explanation and the Causal Structure of the World*, 13 (emphasis in the original). Explanatory realism, as I have characterized it, appears closely related to what Salmon calls 'the ontic model' of scientific explanation.

[4] Norwood Russell Hanson, *Patterns of Discovery* (Cambridge, 1958), 54.

[5] Ibid. 64. See, for a view similar to Hanson's but worked out in greater detail, William Ruddick, 'Causal Connection', *Synthese*, 18 (1968), 46–67.

[6] See, e.g. Michael Scriven, 'Causation as Explanation', *Nous*, 9 (1975); 3–16. I discuss Scriven's account in 'Causes as Explanations: A Critique', *Theory and Decision*, 13 (1981); 293–309. Some of the present material has been drawn from this paper.

two events, *c* and *e*, are 'subsumed under', or 'instantiate', an appropriate law is the objective correlate of the explanans relation for *C* and *E*. Giving an account of 'subsumption under a law' without presupposing causal notions is not an easy task, but let us not press this issue.[7] The point to consider is how we understand the notion of 'law'. If a law is taken as 'mere Humean constant conjunction', with no modal or subjunctive force intimating some tie of 'necessitation', this approach would give us realism. But it is highly dubious that a conception of an explanatory relation based on such a notion of 'law' could provide a basis for an adequate account of explanation; it is even more dubious that an analysis of causation based on such a conception of explanation will come close to capturing our concept of causation. On the other hand, if laws are endowed with sufficiently strong modal force, it is doubtful whether the nomological relation will be distinguishable, in any meaningful way, from the causal relation.[8] Indeed, the nomological account of causation is one of the more influential approaches to the analysis of the causal relation. An analysis of causation in terms of a conception of explanation that in turn is based on the nomological relation as the explanatory relation will essentially be just a nomological analysis, possibly with some psycho-epistemological embellishments. It would be difficult to see why one should not just go for a direct nomological analysis of causation, and use the causal relation as one's explanatory relation.

We must conclude that any attempt to analyse causation as explanation will result in a form of explanatory irrealism. For an analysis of causation to be a genuine explanatory analysis, the concept of explanation assumed as the basis of analysis must be a robustly epistemological and psychological notion whose core is constituted by such notions as understanding and intelligibility, not some pale, formal reconstruction of it. If, for example, the Hempelian deductive-nomological conception of explanation is used to explain causation, the result is not a genuine explanatory analysis of causation but rather the old standby, the nomological-subsumptive, or quasi-Humean, analysis. Thus, on a real explanatory approach to causation, causation will turn out to be a non-objective psycho-epistemological relation and, therefore, fail to serve as an objective correlate of the explanans relation.

---

[7] For some general difficulties in explaining 'subsumption under a law', see Donald Davidson, 'Causal Relations', *Journal of Philosophy*, 64 (1967), 691–703, and my 'Causation, Nomic Subsumption, and the Concept of Event', *Journal of Philosophy*, 70 (1973), 217–36.

[8] Where the nomological and the causal relation do not match up, the former also fails to yield the explanatory relation.

2

It is plausible to conclude, therefore, that explanatory realism requires the causal relation as an explanatory relation. As I said, we may leave open the question of whether the causal relation is the only explanatory relation. But at least this much is certain: in both everyday and most scientific contexts,[9] explanations of individual events are predominantly *causal explanations* in the sense that the events cited in the explanation of an event are its causes and, further, their explanatory efficacy is thought to stem from their causal status. And when each of a class of events can be given a similar causal explanation, we may have a causal explanation of a regularity. We shall in this paper focus exclusively on the causal relation as our explanatory relation; our general metaphysical points should be valid, *mutadis mutandis*, for other explanatory relations if any exist. Explanatory realism says this about causal explanations: a causal explanation of event e in terms of event c ('e occurred because c caused it') is *correct*, or *true*, just in case c did as a matter of objective fact cause e. That the causal relation holds between the two events constitutes the 'factual content' of the explanation. This may sound obvious and trivial.

Perhaps it sounds obvious only because we take explanatory realism for granted. But it certainly is not trivial. It requires, for its intended realist purposes, that causality itself be an objective feature of reality. This doctrine, which we may call *causal realism*, has not gone unchallenged. Hume's celebrated critique of 'necessary connection' as an objective relation characterizing events themselves was perhaps the first—clearly the most influential—expression of a systematically articulated irrealist position on causation. He wrote, 'Upon the whole, necessity is something that exists in the mind, not in objects.'[10] Hume well understood the causal realist's sentiments:

But though this be the only reasonable account we can give of necessity, the contrary notion is so riveted in the mind from the principles above-mentioned, that I doubt not but my sentiments will be treated by many as extravagant and ridiculous. What! the efficacy of causes lie in the determination of mind! As if causes did not operate entirely independent of the mind, and would not continue their operation, even though there was no mind existent to contemplate them, or reason concerning them. Thought may well depend on causes for its operation,

---

[9] By the qualification 'most', I intend to leave out consideration of what some tell us goes on at the deepest and most abstract levels of theoretical physics.

[10] *A Treatise of Human Nature*, ed. L. A. Selby-Bigge (Oxford, 1888), 165.

but not causes on thought. This is to reverse the order of nature, and make that secondary, which is really primary.[11]

Hume was understanding, but in the end dismissive:

I can only reply to all these arguments that the case here is much the same as if a blind man should pretend to find a great many absurdities in the supposition that the colour of scarlet is not the same with the sound of a trumpet, nor light the same with solidity. If we have really no idea of a power or efficacy in any object, or of any real connection betwixt causes and effects, it will be to little purpose that an efficacy is necessary in all operations.[12]

Hume regarded the other ingredients he identified in the causal relation, namely temporal precedence, spatiotemporal contiguity (or connectability), and constant conjunction, as objective and mind-independent features of causally connected events;[13] but evidently he thought that necessitation, too, was an essential element in our philosophically unenlightened (by his light) concept of causation. Most philosophers will now agree that an idea of causation devoid of some notion of necessitation is not *our* idea of causation—perhaps not an idea of causation at all. According to most conceptions of causation now current, at any rate, Hume was a causal irrealist *par excellence*.

Hume, our original causal irrealist, had some illustrious followers. Russell ridiculed causation as 'a relic of a bygone age', recommending the 'extrusion' of the word 'cause' from the philosophical vocabulary;[14] Wittgenstein said, 'Belief in causal nexus is superstition'.[15] The positivist-inspired suspicion of modalities, counterfactuals, and the like, which characterized much of analytic philosophy during the first two-thirds of this century, is of a piece with Hume's causal irrealism in their fundamental philosophical motivation, and it seems that many prominent philosophers in the analytic tradition during this period consciously avoided serious discussion of causality, making little use of it in their philosophical work.[16] More recently, Hilary Putnam has attacked the idea of 'non-Humean causation' as a physically real relation.[17] I think it is

---

[11] Ibid. 167.   [12] Ibid. 168.

[13] For discussions of this and other matters concerning Hume on causation and necessity, see Barry Stroud, *Hume* (London, 1977), chs. 3, 4, and Tom L. Beauchamp and Alexander Rosenberg, *Hume and the Problem of Causation* (New York and Oxford, 1981), especially ch. 1.

[14] Bertrand Russell, 'On the Notion of Cause', *Proceedings of the Aristotelian Society*, 13 (1913), 1–26.

[15] Ludwig Wittgenstein, *Tractatus Logico-Philosophicus* (London, 1922), 5. 1361. However, Wittgenstein may have had in mind by 'causal nexus' something much stronger than what we would now understand by 'causal necessity'.

[16] C. J. Ducasse and Hans Reichenbach were among the exceptions.

[17] 'Is the Causal Structure of the Physical Itself Something Physical?', *Midwest Studies in Philosophy*, 9 (1984); 3–16.

more difficult than one might at first suppose to find philosophers who have consciously advocated an unambiguously realist conception of causality.[18]

According to causal realism, therefore, causal connections hold independently of anyone's intentional states—in particular, epistemological or doxastic states—except, of course, when the causal connections concern such states. The realist believes, as Hume observes in the quotation above, that causal relations—the same ones—would hold even if there were no conscious beings to 'contemplate them, or reason concerning them'. This means that according to causal realism every event has a *unique and determinate causal history* whose character is entirely independent of our representation of it. We may come to know bits and pieces of an event's causal history, but whether we do, or to what extent we do, and what conceptual apparatus is used to depict it, do not in any way affect the causal relations in which events stand to other events. This entails that the existence and character of events themselves must be an objective and determinate fact; that is, causal realism makes sense only in the context of global realism.

Earlier, we raised the question of how explanatory knowledge is represented in our body of knowledge—that is, what it is that we know when we have an explanation of an event. The explanatory realist appears to have a simple answer: To 'have an explanation' of event e in terms of event c is to know, or somehow represent, that c caused e; that is, explanatory knowledge is causal knowledge, and explanations of individual events are represented by singular causal propositions. Thus, explanatory knowledge is propositional knowledge of a certain kind, and to gain an explanation of an event is to learn *a further fact about that event*.

But is there an alternative to representing explanations as additional bits of propositional knowledge? Is not explanatory knowledge a kind of knowledge, and is not all knowledge, in an epistemologically relevant sense, a matter of knowing *that*? Although I do not know whether anyone has held a view like this, it is possible to hold, I think, that explanations are essentially a matter of how a body of knowledge is organized or systematized—a matter of there being certain appropriate patterns of

---

[18] Some possibilities among recent writers: Salmon, *Scientific Explanation and the Causal Structure of the World*; J. L. Mackie, *The Cement of the Universe* (Oxford, 1974), especially ch. 8. Quine seems to have studiously avoided discussing causation or making use of it in his philosophical work. The uses to which Donald Davidson has put the concept of causation indicate a realist attitude; consider, e.g., his commitment to an event ontology and his causal criterion of event individuation in his *Essays on Actions and Events* (Oxford, 1980). But he may reject the terms in which I have formulated the positions.

coherence among items of knowledge. That is, to 'have' an explanation of why E in terms of C—that is, to 'have' C as an explanans for E—is simply for the two propositions C and E to be appropriately related within our epistemic corpus; it is not a matter of there being a further proposition within it. According to this view, therefore, explanatory knowledge supervenes on non-explanatory knowledge: if you and I know exactly the same first-order, factual propositions (roughly, propositions that can serve as explananda and elements of an explanans), we would share the same explanations. Various considerations might lead us to qualify this conclusion; for example, one might construe the notion of 'having an explanation' in such a way as to require the subject's awareness that the explanans is appropriately related to the explanandum. Thus, one might want to suggest that the presence of the two propositions in our body of knowledge is not enough, even if they in fact instantiate a required explanatory pattern, and insist that we must somehow mentally 'bring them together' and 'see' that they do so. Caution is required, however; pursuing this line may take one back to the propositional view of explanatory knowledge. At any rate, the non-propositional, 'pattern' view of explanatory knowledge differs from the propositional view on the following point: gaining a new explanation, on the pattern view but not on the propositional view, does not necessarily involve acquiring new information about facts of the world.

It seems clear that explanatory realism leads to the propositional view of explanatory knowledge; it makes 'having' an explanation a matter of knowing a certain proposition to be true. On the other hand, explanatory irrealism, although it has an affinity for the pattern view, is not committed to it; it appears consistent with the propositional view. One might hold, for example, that a certain conceptual–epistemic relation between an explanandum and its explanans is what is fundamentally constitutive of an explaining relation, there being no independent objective relation characterizing the events represented by the explanans and the explanandum that grounds it,[19] but that 'having' an explanation *is* a matter of *knowing that* this relationship does in fact hold for the explanans–explanandum pair. This, however, may not be a plausible view; it is naturally construed as requiring anyone who 'has' any explanation of anything at all to know what the explanans relation is, something that few philosophers would confidently claim to know. In any case, those who find the propositional view of explanatory knowledge too simplistic, or otherwise unpalatable, would have to settle for explanatory irrealism; explanatory realism is not an option for them.

---

[19] Recall our earlier discussion of Hanson; he seems to have held a view like this.

What difference does the choice between explanatory realism and irrealism make? We have already seen that explanatory realism plausibly entails causal realism. Does explanatory irrealism entail causal irrealism? There evidently is no strict inconsistency in holding both explanatory irrealism and causal realism. However, the combination seems somewhat incongruous and difficult to motivate: though acknowledging causation as a genuine relation in the world, the position denies it any essential role in explanation, severing the intuitive and natural tie between causality and explanation.[20] What, then, would be the point of the causal relation? The concept of causation, of course, has many roles to play, but it seems that its explanatory role is a central one, being closely tied to its other important roles. I think that the combination of explanatory irrealism and causal realism, though logically consistent, is not a plausible position.

We have also seen that explanatory realism entails the propositional account of explanatory knowledge, whereas explanatory irrealism, again, seems consistent with each of the two alternatives, the propositional view and the non-propositional, pattern view. I think that the issue of causal realism versus irrealism and that concerning the nature of explanatory knowledge are significant issues, both interesting in themselves and important in what they imply for other philosophical problems. Problems about what explanatory knowledge consists in—that is, what 'understanding' something amounts to—have been almost entirely neglected within traditional epistemology; this is surprising in view of the centrality of explanation in philosophy of science, which, by and large, is the epistemology of scientific knowledge. The issue of causal realism is obviously important: whether causal relations are real and objective, or mere projections of the cognizing mind, is an issue that directly affects the significance of causation within both science and philosophy. If it is an objective relation characterizing physical events in the world, is it physically reducible, or physically based in some sense, as we expect of other physical properties and relations? If not, what accounts for its special status? Which of the special sciences are responsible for investigating the properties of the causal relation itself?[21]

As for the philosophical implications of the choice between causal realism and irrealism, it is an interesting question, for example, whether any of the so-called causal theories (of perception, memory, knowledge, action, event-identity, reference, time, persistence, properties, and no

---

[20] Peter Achinstein's views in *The Nature of Explanation* (New York and Oxford, 1983), ch. 7, seem to approximate this position.

[21] See Putnam, 'Is the Causal Structure of the Physical Itself Something Physical?'

doubt many others) will be able to retain, under an irrealist conception of causation, what plausibility it enjoys. It is also an interesting question whether a substantive version of global realism can be combined with causal irrealism. I suspect that if all causal facts are taken away from the world, not much of interest may remain—the world would become so impoverished, a pale imitation of a world, that we may not care much whether it is real or only 'ideal'. (If all those 'causal theories' mentioned above are true, a world devoid of causal relations would be one in which there is no perception, no knowledge, no naming or referring, no intentional action, no time, no persisting object, and none of the rest. It would also be a world in which there are no killings, no breakings, no pushings or pullings, and so on.)

Some may consider it a disadvantage of explanatory realism that it comes only in a package with causal realism, whereas explanatory irrealism can in principle be purchased separately. However, others may consider that an advantage: causal realism gives more content to explanatory realism, and as a result explanatory realism can do work that its rival cannot. Moreover, there is a certain satisfying unity in the combination of explanatory and causal realism. In any case, there seems to be some incongruity, as we saw, in combining explanatory irrealism and causal realism, so that an explanatory irrealist may in effect have no real choice but to embrace causal irrealism as well. In what follows, I will explore the implications of the realist view of explanation for the issue of 'explanatory exclusion' and the irrealist (or 'internalist') implications of the Hempelian inferential view of explanation.

3

I have argued elsewhere[22] that proffered explanations of a single event, with mutually consistent explanantia, can exclude one another in the following sense: there can be no more than a single *complete* and *independent* explanation of any one event, and we may not accept two (or more) explanations of a single event unless we know, or have reason to believe, that they are appropriately related—that is, related in such a way that one of the explanations is either not complete in itself or dependent on the other. This constraint on explanations, which we may call *the principle of explanatory exclusion*, has two clauses: the first is about the

[22] 'Mechanism, Purpose, and Explanatory Exclusion', in J. Tomberlin (ed.), *Philosophical Perspectives* (Atascadero, Calif.: Ridgeview, 1989), iii. 77–108.

existence of explanations, the second about *acceptance* of explanations. The first clause, I shall argue, can be seen as a plausible thesis if we assume explanatory realism. We shall not discuss the second clause here.

Suppose, then, that each of C_1 and C_2 is claimed to be a causal explanans for E. Let c_1, c_2 and e be the events represented by C_1, C_2, and E. According to explanatory realism, it follows that c_1 caused e and also that c_2 caused e. How are we to understand this situation? There are various possibilities:

1. It turns out that $c_1 = c_2$. A single event is picked out by non-equivalent descriptions. Here, there is in reality only one pair of events related by the explanatory relation (that is, the causal relation), and this gives sense to the claim that there is, here, one explanation, not two. The exclusion principle makes sense only if a criterion of individuation is assumed for explanations—that is, only if we can make sense of 'same' and 'different' as applied to explanations. Now, explanatory realism yields a natural way of individuating explanations: explanations are individuated in terms of the events related by the explanatory relation (the causal relation, for explanations of events).[23] For on realism it is the objective relationship between events that ultimately grounds explanations and constitutes their objective content. This provides us with a basis for regarding explanations that appeal to the same events standing in the same relation as giving, or stating, one explanation, not two—just as two inequivalent descriptions can represent the same fact. Thus, on explanatory realism, we can make good sense of the idea that logically inequivalent explanations can represent the same explanatory relation, and therefore state the same explanation. To the explanatory irrealist, this way of individuating explanations would be unmotivated: explanations would be more appropriately individuated in terms of descriptions or propositions and their internal logical, conceptual, and epistemic relationships. Nothing needs to prevent the explanatory realist from accepting this 'internal' individuation criterion as well, as defining *another* useful sense in which we can count explanations. The point is only that explanatory realism motivates an 'objective' individuation of explanations, which is both intuitively plausible and well suited for the exclusion principle.

2. C_1 is reducible to, or supervenient on, c_2. This sort of relationship might obtain, for example, on some accounts of the mind–body relation, which, though eschewing an outright psychophysical identification, none

---

[23] If relations other than the causal relation can serve as explanatory relation, they can also be considered as a basis for individuation; however, that probably would be redundant. It is unlikely that when the explanatory relation is different, exactly the same events would be involved.

the less recognizes the reductive or supervenient dependency of the mental on the physical. In such a case, the causal relation involving the supervenient or reduced event must itself be thought of as supervenient or reducible to the causal or nomological relation involving the 'base' event.[24] In this sense, the two explanations are not independent; for the one involving the reduced causal (that is, explanatory) relation is dependent on the one representing the 'base' causal relation. This, again, is an example of realist thinking: dependency between explanations is understood in terms of the dependency between the objective explanatory relations that they represent.

3. c_1 and c_2 are only partial causes, being constituents in a single sufficient set of causal conditions. Example: You push the stalled car and I pull it, and the car moves. In this case, neither explanation is complete: each gives only a partial picture of the causal conditions that made up a sufficient cause of the effect. This sense of *explanatory completeness*, understood in terms of *sufficient cause*, is again entirely natural within the realist picture. For, according to the realist view, the causal relation between events constitutes the objective correlate, or content, of the explanans relation; where a particular causal relation gives us a cause event that is only a partial cause, or one among the many constituents of a sufficient cause, the corresponding explanans, too, can be thought to be only partial and incomplete. Conversely, when the causal relation provides a sufficient cause, the explanans can also be said to be complete and sufficient. The realist scheme also yields a more global sense of 'complete explanation', one in which a complete explanation of an event specifies its entire causal history in every detail (as we noted earlier, under explanatory realism each event has a unique determinate causal history). This is an idealized sense of completeness, and no explanation can be complete in that sense (the notion of an ideally complete explanation, however, may be useful in explicating the concept of explanation).[25] Obviously, in this idealized sense there is at most one complete explanation of any given event; again, though obvious and uninteresting, this is not trivial, unless causal realism is trivial.

4. c_1 and c_2 are different links in the same causal chain leading to event e. But then they are not independent: the later event is causally

---

[24] For further discussion see my 'Epiphenomenal and Supervenient Causation', *Midwest Studies in Philosophy*, 9 (1984), 257–70. I believe that the case in which c_1 'generates' c_2 in Alvin Goldman's sense (see his *A Theory of Human Action* (Englewood Cliffs, NJ, 1970)) can be handled in a similar way, although the details may have to be somewhat different.

[25] Compare Peter Railton's notions of 'ideal explanatory text' and 'ideal causal D-N text' in his 'Probability, Explanation, and Information', *Synthese*, 45 (1981), 233–56 [Ch. VI herein]; see also David Lewis, 'Causal Explanation', *Philosophical Papers*, ii [Ch. VII herein].

dependent on the earlier one, and, therefore, the two explanations are not independent. This, too, reflects realist thinking: two explanations are thought to be non-independent because the explanatory relations represented by them are not independent.

5. c_1 is part of c_2.[26] The explanations, then, are not independent; nor can they both be complete.

6. c_1 and c_2 are independent, each a sufficient cause of e. This, then, is a standard case of 'causal overdetermination'. Do we in this case have a counter-example to the explanatory exclusion principle? Why are both explanations, 'e happened because c_1 caused it' and 'e happened because c_2 caused it', not sufficient and independent explanations? This is an interesting case from the point of view of both explanatory exclusion and the question of explanatory realism versus irrealism, and we shall discuss this in some detail.

Hempel has called cases like this 'explanatory overdetermination':[27] suppose that a copper rod is heated while simultaneously being subjected to longitudinal stress. As a result, its length increases. Two deductive-nomological (hereafter 'D-N') arguments can be formulated: the first would invoke the lawlike premiss that copper rods lengthen when they are heated, and the 'initial condition' that this particular copper rod was heated on this occasion; the second would appeal to the law stating that copper rods increase in length when subjected to longitudinal stress, and the initial condition that this copper rod was subjected to that kind of stress. The two arguments share the same conclusion, the statement that the rod's length increased on this occasion. According to the standard D-N account of explanation, therefore, each argument counts as an explanation.

It is not surprising that Hempel rejects the view that these D-N arguments are not 'complete' as explanations. He writes:

It might be objected that—even granting the truth of all the premisses—both accounts are unacceptable since they are 'incomplete': each neglects one of the two factors that contributed to the lengthening. In appraising the force of this objection it is again important to be clear about just what is to be explained. If as in our example, this is simply the fact that Lr, i.e., that r lengthened, or that there was *some* increase in the length of r, then, I think, either of the two arguments conclusively does *that*, and the charge of incompleteness is groundless.[28]

Here he seems simply to affirm that, as an explanation of why the rod lengthened, 'each of the two arguments conclusively does *that*'. But why

---

[26] Karl Pfeifer brought this case to my attention.

[27] Carl G. Hempel, *Aspects of Scientific Explanation* (New York, 1965), 418–20.

[28] Ibid. 418–20.

does he say this? The use of the term 'conclusively' suggests that he was moved by the consideration that each D-N argument provides a premiss-set that is *deductively conclusive* for the truth of the explanandum statement. This is not surprising. For, fundamental to the D-N conception of explanation is the idea that explanations are *inferences* or *arguments* of a certain form. Given this assumption, a natural sense of 'completeness' or 'sufficiency' emerges for explanations: when an argument has the correct D-N form, it is *complete* and *sufficient*. Hempel writes:

I think it is important and illuminating to distinguish such partial explanations ... from what might be called *deductively complete explanations*, i.e., those in which the explanandum as stated is logically implied by the explanans; for the latter do, whereas the former do not, account for the explanandum phenomenon in the specificity with which the explanandum sentence describes it. An explanation that conforms to the D-N model is, therefore, automatically complete in this sense; and a partial explanation as we have characterized it always falls short of being a D-N explanation.[29]

As explanation is conceived under the D-N model, there is nothing one can do to a D-N argument to improve it in regard to its 'completeness' as an explanation. One may be able to make it deeper, more perspicuous, more systematic, and so on; but what could one possibly do to make it 'more complete' ? The D-N conception of explanation does not seem to leave room for any other sense of explanatory completeness than deductive conclusiveness.

These considerations suggest that a preoccupation with the deductive or inferential character of explanation leads to a form of explanatory irrealism ('explanatory internalism', perhaps, is more appropriate), and this is certainly what we see in Hempel. This internalist tendency is evident also in Hempel's well-known emphasis on the predictive character of explanation, and in one of his two conditions of adequacy on explanations, that is, the requirement of 'explanatory relevance' to the effect that 'explanatory information adduced affords good grounds for believing that the phenomenon to be explained did, or does, indeed occur'.[30] Hempel's idea that explanations are arguments, his condition of 'explanatory relevance', and his emphasis on the predictive aspect of explanations go hand in hand: they all point to explanatory irrealism—at least, point away from explanatory realism with the causal relation serving as objective correlate of the explanatory relation.

[29] Ibid. 416–17.

[30] *Philosophy of Natural Science* (Englewood Cliffs, NJ, 1966), 47–9. See also his *Aspects of Scientific Explanation*, 367–8. The other adequacy condition is the unexceptionable requirement of 'testability', to the effect that the explanatory premisses must be capable of empirical test in a broad sense.

Hempel's primary focus in analysing the structure of explanation is on the logical and conceptual characteristics of statements making up an explanation (the 'internal' properties, as I have called them), not on the events or other entities these statements describe and their interrelations. In fact, we get from Hempel a precise and elaborately constructed definition of what an explanation is, but only a very intuitive and unanalysed idea of what it is that a given explanation is an explanation *of*.[31] Hempel's treatment of causal explanation and causation is also symptomatic of this attitude:[32] the idea of a D-N argument, an essentially internal notion, is primary in the characterization of explanations, and the idea of causal explanation falls out of this characterization as a not-so-clearly-defined special case. Hempel evidently does not regard the concept of causal explanation, or that of causation, as at all crucial to a theory of explanation; his discussion of causal explanation often comes across as a concession to the popular practice of referring to causes and causal explanations, not something that he sees as essential to the development of his theory. From such an internalist perspective, it is entirely natural that each of the two D-N arguments about the expanding copper rod is regarded as complete and sufficient in itself as an explanation.

What does explanatory realism say about the expanding copper rod? If the heating and the stress are each an independent sufficient cause of the rod's lengthening, we have a standard case of causal overdetermination. Moreover, if, as explanatory realism seems to suggest, explanatory completeness is to be understood in terms of sufficient cause, it follows that in the present case we have two independent and complete explanations. Thus, explanatory realism seems to yield the same result as Hempel's irrealism: both seem to contradict the explanatory exclusion principle.

The explanatory realist who wants to save explanatory exclusion might deny that the rising temperature and the stress were each a sufficient cause of the event to be explained, and deny, more generally, that genuine instances of causal overdetermination exist. Peter Unger has claimed that each event has a single unique cause (at most),[33] and if this is right, then not both the heating and the stress can be a cause of the lengthening. Therefore, there could be at most one causal explanation here. But Unger's thesis is a radical one, too strong to be plausible: he construes

---

[31] For further elaboration of this point and some suggestions, see my 'Events and Their Descriptions: Some Considerations', in Nicholas Rescher *et al.* (eds.), *Essays in Honor of Carl G. Hempel* (Dordrecht, 1969).

[32] Hempel, *Aspects of Scientific Explanation*, 347–54.

[33] 'The Uniqueness in Causation', *American Philosophical Quarterly*, 14 (1977), 177–88.

it to entail the denial of transitivity of causation, and hence the impossibility of causal chains with more than two links. And, his arguments rely exlusively on a certain kind of linguistic evidence whose point I find difficult to evaluate.

Martin Bunzl, too, has argued that there are no genuine cases of causal overdetermination.[34] His basic point is that the usual examples, when closely scrutinized, turn out to be either cases of causal pre-emption or of joint cause. That is, one of the two alleged overdetermining causes pre-empts the other (by 'getting there first') so that the second, in fact, is not a cause of the effect in question, or else the two causes together make up a single sufficient cause, neither of them alone being sufficient. I think Bunzl's arguments, on the whole, are plausible, though not conclusive.[35] Thus, when applied to the case of the copper rod, his analysis would probably give this diagnosis: the particular lengthening that took place was caused by the single joint cause made up of the heating and the stress. Neither of the two events was, in itself, a sufficient cause of it. Thus, a complete explanation of the lengthening must refer to both the heating and the stress as a single sufficient cause.

We must set aside the question whether genuine instances of causal overdetermination exist. What is of interest to us here is that under explanatory realism, the causal relation can be made to do some real work, in characterizing and constraining explanations. As we saw, the association between causation and explanation, underwritten by explanatory realism, yields a principle of individuation for explanations and a notion of 'complete explanation', both essential to interpreting the principle of explanatory exclusion. We also saw that if causal overdetermination is not possible, that takes away one potential case of explanatory overdetermination. It seems to me that we are inclined to take these considerations involving causation as both natural and relevant in discussing the nature of explanation. What accounts for this inclination, I think, is our tacit acceptance of explanatory realism: for a causal explanation to hold, the explaining event must be a cause of the event explained. Given this

---

[34] 'Causal Overdetermination', *Journal of Philosophy*, 76 (1979), 134–50.

[35] Bunzl, however, says that his considerations depend essentially on a certain view of the nature and individuation of events associated with Donald Davidson, and that they are ineffective if we assume the sort of view of events that I myself have advocated, namely, one that takes events as property-exemplifications. See Bunzl, 'Causal Overdetermination', 150. However, I am not convinced of this; I think Bunzl may have been misled by just the kind of consideration that led Hempel to believe in explanatory overdetermination. It is interesting to note that Bunzl accepts explanatory overdetermination in Hempel's sense (p. 145). On causal overdetermination, see also Louis E. Loeb, 'Causal Theories and Causal Overdetermination', *Journal of Philosophy*, 71 (1974), 525–44.

connection between causal explanations and causal relations, we are able to use facts about the latter to say something about the former.

To return briefly to the matter of explanatory exclusion: if our considerations are generally right (especially in the treatment of the six cases in which two causal explanations are offered for one event), explanatory realism is seen to provide a sense, as well as support, for the explanatory exclusion principle—except, perhaps, in the case of causal overdetermination, which we set aside without a clear-cut resolution.[36] I believe it is more difficult, though not impossible, to interpret and argue for explanatory exclusion if by embracing explanatory irrealism we lose the causal handle on explanation.[37]

[36] From the point of view of explanatory exclusion, causal overdetermination is not crucial; the exclusion principle has content of sufficient interest even if causal overdetermination is simply exempted.

[37] I think explanatory exclusion can hold under explanatory irrealism as well; however, unlike explanatory realism, irrealism does not, I think, provide a positive basis for explanatory exclusion. For some considerations favouring explanatory exclusion that are not based on explanatory realism, see my 'Mechanism, Purpose, and Explanatory Exclusion'.

X

A THEORY OF SINGULAR CAUSAL EXPLANATION*

JAMES WOODWARD

My interest in this essay is in singular causal explanations like

(1) The short-circuit caused the fire; or

(2) The blow of the hammer caused the chestnut to shatter

I call such sentences 'explanations' deliberately, for it is one of my central claims that we may legitimately think of them as constituting, in appropriate contexts, explanations of why the fire occurred or of why the chestnut shattered. In this essay I shall be interested in the structure and explanatory import of such sentences—in what they tell us and why it is that we are entitled to take them as explanatory. The position I shall defend is in many respects the mirror image of the most commonly accepted philosophical view of singular causal explanations. The commonly accepted view[1] is that there is a sharp distinction to be drawn between singular causal explanations and those sentences (call them singular causal sentences) which simply report causal connections. While singular causal sentences are said to typically relate events, and to be extensional, singular causal explanations are held to relate items which are sentential or factlike in structure and to be non-extensional. Singular

James Woodward, 'A Theory of Singular Causal Explanation', *Erkenntnis*, 21 (1984), 231–62. Copyright, © 1984 by D. Reidel Publishing Company, Dordrecht. Reprinted by permission of Kluwer Academic Publishers.

* Portions of this paper were written while I was a visiting fellow at the Center for Philosophy of Science at the University of Pittsburgh during Spring, 1983. I am grateful to the Center and the University of Pittsburgh for their support. [Note added in 1992: I have corrected a number of typographical errors that appeared in the original (1984) version of this paper. A feature of this paper which is conspicuous upon re-reading (and about which I feel very apologetic) is its failure to refer at all to the excellent discussions of contrastive explanation in van Fraassen's *The Scientific Image* (1980) and Garfinkel's *Forms of Explanation* (1981), both of which appeared in print several years before my discussion. The explanation (not justification) for this is that the bulk of 'A Theory of Singular Causal Explanation' was actually written in 1979 and not updated, as it should have been, when these books appeared.]

[1] Views of this sort are defended by, for example, J. L. Mackie (1974), Donald Davidson (1967), Tom Beauchamp and Alexander Rosenberg (1981), and Michael Levin (1976).

causal explanations are held on this view to always be explanations of why events or other objects belong to certain kinds or possess certain properties.

According to the view I shall defend, singular causal sentences and singular causal explanations do not differ fundamentally in structure in this way. A singular explanation will simply consist of a sentence which reports a causal connection like (1) or (2) above. Singular causal explanations are extensional on both the cause-side and the effect-side and (subject to various qualifications and refinements to be introduced below) many such explanations relate terms which are eventlike rather than factlike in structure. Many singular causal explanations are not explanations of why events or other particulars possess certain properties, but are rather, as we shall see, something quite different—explanations of the occurrence of individual events.[2]

While the conventional view is that singular causal explanations are in some way truncated or implicit covering-law explanations (or at least are to be understood by reference to a covering-law paradigm), the view I shall develop treats singular causal explanations as a distinct genre of explanation, which does not possess anything remotely like a covering-law structure.

There are two basic keys to understanding the structure of singular causal explanations. The first consists in taking seriously our naïve pre-analytic view about sentences like (1) and (2)—a view which tells us that such sentences are explanatory and that what they explain is simply the occurrence of a particular event (the fire, the shattering of the chestnut) rather than some more complicated explanandum. The second key consists in rethinking the question of how other kinds of why-explanation—scientific explanation and statistical explanation—work. According to one very influential view, to provide a why-explanation is, in the paradigmatic case, to provide a nomologically sufficient condition. This view leads to the covering-law model of scientific explanation, to the view that statistical explanations proceed by showing that their explananda are highly probable, and to the claim that singular causal explanations explain in virtue of approximating, or pointing, to an implicit covering-

---

[2] The intuition that sometimes what is explained by a singular causal explanation is the occurrence of an individual event, where this is something different from explaining why that event has some property or is an event of a certain kind, is nicely captured in Thomas Nickles's essay 'Davidson on Explanation'. Nickles writes 'It certainly is intuitively appealing to think that we can explain event occurrences period and that if we have explained the occurrence of e and know that $e = f$, then we can also explain why f occurred.' (1977: 144). Nickles goes on to express doubts that this idea of explaining the occurrence of an individual event can be made clear, despite its intuitive attractiveness. Much of this essay is an attempt to show that the notion of explaining the occurrence of an individual event described by Nickles in the above passage can be made clear, once we reject the claim that singular causal explanation is a variety of implicit covering-law explanation.

law structure. I believe that if one is guided by this conception of what it is to give a why-explanation one will miss or misconstrue those features of singular causal explanation which are responsible for their explanatory import. My discussion will draw attention to features of scientific and statistical explanation which contribute to their explanatory import and which are quite different from those emphasized by the covering-law paradigm. I shall attempt to show that it is the presence of analogous features in singular causal explanations and not the fact that such explanations possess or point to anything like an implicit covering-law structure which accounts for their explanatory efficacy.[3]

This essay falls into three major parts. Section 1 describes some general desiderata which a theory of singular causal explanation ought to satisfy. Section 2 attempts to place singular causal explanation within the context of a general theory of why-explanation. Section 3 explores the structure of singular causal explanation in detail.

1

I begin by setting down, somewhat arbitrarily, some general adequacy conditions which, in my view, an account of singular causal explanation ought to satisfy. These conditions will help to guide my subsequent discussion, but I will make little independent attempt to motivate them here.[4] Whatever justification they possess will emerge in the course of my discussion.

First, an acceptable theory of singular causal explanation must make it clear how (that is, identify the structural features in virtue of which) such explanations explain and it presumably ought to do this by setting such explanations within the context of a general theory of why-explanation. Analyses that take singular causal explanation to represent an entirely *sui generis* form of explanation, possessing features that are not shared at all by other varieties of explanation (such as scientific explanation), are *ad hoc* and unsatisfying.

Secondly, the features in virtue of which singular causal explanations explain must satisfy certain epistemological requirements. Users of singular causal explanations must be able to recognize and appreciate these

---

[3] For reasons of space, I shall not explore in this essay the suggestion that singular causal explanations may be construed as implicit IS explanations (or, for that matter, implicit SR explanations). My reasons for rejecting this suggestion are developed in detail in my 'Causal Explanation in History' (unpublished).

[4] I say more to motivate these conditions in my 'Are Singular Causal Explanations Implicit Covering-Law Explanations?' (1986).

features and to readily ascertain whether proffered explanations provide them. Consider in the light of this second requirement the familiar suggestion that to explain an explanandum is to deduce it from a law of nature and a statement of initial conditions and that a singular causal explanation like

(1) The short-circuit caused the fire

is explanatory in virtue of (implicitly) possessing such a deductive-nomological structure. Whatever the merits of this proposal, it becomes implausible as an account of singular causal explanation when combined with the contention that ordinary users of singular causal explanations do not know, even in very rough outline, what the DN structure so instantiated looks like. (A position very much like this seems to be adopted by Davidson (1967).) A theory of singular causal explanation ought to identify the structural features of such explanations which function so as to produce understanding in the ordinary user. Features which are entirely unknown to the ordinary user are not plausible candidates for what produces such understanding. If, as Davidson quite reasonably suggests, ordinary speakers typically do not know the laws which singular causal explanations like (1) instantiate, it is implausible to suppose that explanations like (1) provide understanding by tacitly invoking such laws.

Third, an acceptable theory of singular causal explanation must provide a basis for distinguishing what such explanations explain from what they merely presuppose. Consider, for example, the difference between

(3) The short-circuit caused the purple fire (causally explains why the purple fire occurred)[5]

and

(4) The presence of potassium salts caused the fire to be purple (causally explains why the fire was purple)

The difference between (3) and (4), about which I shall say more below, seems to come to something like this. What (4) attempts to provide is not some factor which explains, among other things, why the fire occurred, but rather some factor which explains why, given that the fire occurred, it was purple rather than some other colour. By contrast (3) does not even purport to explain why the fire was purple. Rather what (3) attempts to provide is some factor which explains why a particular fire (the purple one) occurred. In (4) the occurrence of the fire is pre-

[5] As I indicated in the introduction to this paper, I take the claims 'The short-circuit caused the purple fire' (3), and 'The short-circuit causally explains why the purple fire occurred' (3*) to be, in many circumstances, equivalent in explanatory import. A justification for this claim is given in Sections 2 and 3 below.

supposed, rather than explained, while in (3) the purple colour of the fire is presupposed rather than explained. Intuitively, it looks as though the difference between these two explananda depends in part on the fact that an expression which plays the role of a definite description in the explanandum of (3) plays the role of ascribing a property, the possession of which is to be explained, in the explanandum of (4). In (3) the function of 'the purple fire' is apparently to indicate or pick out the particular event the occurrence of which we wish to explain. In (4) what we wish to explain is rather why a certain particular (the fire) possesses the property of being purple.

A similar distinction should be kept in mind with regard to a claim like

(1) The short-circuit caused the fire

The expression 'the fire' occurring in (1) is a definite description and its correct use requires that there be only one fire in the vicinity of interest. However, it would be a mistake to suppose that (1) purports to explain *why* this is the case. It *presupposes*, rather than explains why, only one fire occurs in the vicinity of interest in essentially the same way that (3) presupposes, rather than explains why, the fire is purple.

I have stressed this point because capturing the distinction between what singular causal explanations explain and what they merely presuppose is a non-trivial matter. Suppose, for example, we take singular causal explanation to have an implicit DN structure and represent the explananda of (3) and (1) respectively as $(\exists!x)\,(Fx \cdot Px)$ and $(\exists!x)\,Fx$.[6] (A suggestion along these lines occurs in Davidson (1967).) Such an analysis appears inadequate, for it collapses the difference between (3) and (4), the explanandum of which is presumably also to be represented as $(\exists!x)\,(Fx \cdot Px)$. Moreover, it also in effect mistakenly construes (1) as an explanation of why one and only one fire occurs in the vicinity of interest. Thus, there seems to be a clear difference between explaining why an event occurs, where this event happens to have certain properties and we pick the event out by means of a definite description which mentions those properties, and explaining why the event has those properties. We require an account of singular causal explanation which is sensitive to this difference.[7]

[6] I do not mean that this is the only way of construing the explananda of (3) and (1), given that these explanations are taken as possessing a covering-law structure; merely, that it is a natural way of so construing them, particularly when one considers the hopelessness of the obvious alternatives (such as, e.g., construing the explanandum of (1) as attributing to some fire the 'property' of being a unique occurrence).

[7] In my 'Do Singular Causal Explanations Possess an Implicit Covering-Law Structure?' I argue that no plausible analysis which attributes an implicit covering-law structure to singular causal explanations will be sensitive to this difference. My argument is somewhat complex, and I shall not rehearse it here.

Lastly (and more controversially), once we agree that we need to distinguish what singular causal explanations explain from what they merely presuppose, it becomes plausible to think that those singular causal explanations which explain the occurrence of individual events will obey a substitution principle concerning individual event identities on the effect side, i.e. that they will be extensional. Suppose that I have a singular causal explanation for the occurrence of a fire, e.g.,

(1) The short-circuit caused the fire

Suppose this particular fire happens to be purple and odd-shaped:

(5) The fire = the purple, odd-shaped fire

Then it seems reasonable to infer that we also have a singular causal explanation for the occurrence of the purple, odd-shaped fire:

(6) The short-circuit caused the purple, odd-shaped fire

After all, the occurrence of the purple, odd-shaped fire is the very same occurrence which is explained by (1). It is true, of course, that (6) does not represent an explanation of why the fire is purple and odd-shaped, but as we noted above, it is a mistake to construe (6) as even purporting to be such an explanation.

Indeed, I would maintain that we may regard the suggestion that singular causal explanations like (1) obey such a substitution principle as explicating part of what is meant by the claim that such explanations genuinely explain the occurrence of particular events rather than why those events possess certain properties.[8] We require an account of singular causal explanation which makes it plausible that they will obey such a substitution principle.[9]

---

[8] I hope to defend this claim in more detail in a future paper. The underlying idea is that there are certain broad theoretical considerations which show that explanatory contexts must be extensional. We may attribute structure to an explanatory context by determining what sorts of substitution principles it obeys. If an explanation permits substitutions in its explanandum on the basis of individual event-identities *salva explanatione*, this indicates that what is being explained is the occurrence of an individual event. Similarly if the structure of an explanation is such that what is being explained is the possession of a property by a particular, we should expect the explanation to obey a properly formulated principle concerning the substitution of identical properties. Indeed, I would think that this is the only clear sense that can be given to such claims as (some) singular causal explanations explain the occurrence of individual events or covering-law explanations explain why certain particulars possess certain properties.

[9] There are passages in Davidson (1967) which seem to endorse the claim that singular causal explanations will obey such a substitution principle, although much of what Davidson says in this and other essays is inconsistent with his claim, or so I argue in my 'Do Singular Causal Explanations Possess an Implicit Covering-Law Structure?' For example, Davidson (ibid.) suggests that we can explain the occurrence of an individual event a, by finding a 'redescription' of a, such that the claim that a, so redescribed, occurs is deducible from an appropriate statement of laws and initial conditions. If these remarks are taken literally—if what Davidson intends is that by so proceeding

2

What does a sentence like (1) 'The short-circuit caused the fire' tell us? The natural, pre-analytic answer is that (1) makes a counterfactual claim. (1) tells us that, putting aside complications having to do with overdetermination, if the short-circuit had not occurred and matters had otherwise remained unaltered, the fire would not have occurred and that given that the short-circuit did occur, the probability of the fire was higher than the zero value it would have had if the short-circuit had not occurred.[10] (1) does *not* claim, at least explicitly, that given the short-circuit and the circumstances in which it occurred, the fire was certain or even likely. Rather what (1) claims is that the short-circuit 'made a difference for' the occurrence of the fire—that without the short-circuit there would have been no fire and that given the short-circuit, the fire was at least made more likely than it would have been in the absence of the short-circuit. That is, (1) claims that, given the circumstances, $P(^F/\bar{S}) = 0$ and $P(^F/S) > 0$, where S = the occurrence of a short-circuit, F = the occurrence of fire, and these conditional probabilities are understood as claims about the hypothetical relative frequencies for F that are expected when S and $\bar{S}$ are realized.

We might think of (1) as implicitly identifying (in a sense which I shall try to make more precise in Section 3) two possible outcomes—the occurrence or non-occurrence of the fire—and two possible conditions in its explanans—the occurrence or non-occurrence of the short-circuit—and then asserting that there is a pattern of dependence of these outcomes on these conditions. We might express this by saying that (1) answers a 'What-if-things-had-been-different' question (a 'w-question'). Sentence (1) shows us how a certain kind of change in the conditions identified in its explananans (from a short-circuit to no short-circuit and vice versa) would favour a change in the outcomes identified in its explanandum (from a situation in which the fire has non-zero probability to one in which it does not or vice versa).

we can actually explain the occurrence of *a* and not just why *a* satisfies such-and-such a description—then they seem to commit Davidson to something very like the above substitution principle. As I note (n. 2), Thomas Nickles (1977*a*) also admits the plausibility of, but does not endorse, such a substitution principle.

[10] I do not claim that this is *all* that (1) tells us. Clearly not every case of counterfactual dependence is a case of causal dependence, and relatedly, not all claims of counterfactual dependence can be used to explain in the way that (1) can. It does not follow from this, however, that it is incorrect to say that singular causal explanations explain in virtue of answering w-questions.

Writers like Davidson and Mackie have noted this point and have then passed on to what they have thought of as the very different issue of the structure of singular causal explanation. They have done so because they have worked with paradigms of explanation—most notably the covering-law model—according to which it was difficult to see why the fact that (1) answers a w-question or provides the sort of counter-factual information described above should make any difference to its status as an explanation.

I want to suggest, on the contrary, that we think of (1) as explanatory precisely because it contains the sort of counter-factual information and answers the w-question described above. Singular causal explanations wear the source of their explanatory efficacy on their face—they explain not because they tacitly invoke a 'hidden' law or statement of sufficient conditions, but because they identify conditions such that changes in these conditions would make a difference for whether the explanandum-phenomenon or some specified alternatives to it would ensue.

But why should the fact that singular causal explanations provide this sort of information lead us to take them as explanatory? This question obviously gets its bite from the assumption that other varieties of why-explanation do not explain in virtue of providing similar information. If scientific and statistical explanations explain simply in virtue of showing that their explananda were to be expected, given certain laws and initial conditions, then it seems that singular causal explanations will explain (if they explain at all) in virtue of providing similar information. It is this assumption that I wish to challenge. I shall attempt to show that both scientific (Section 2.1) and statistical (Section 2.2) explanations explain in part in virtue of answering a what-if-things-had-been-different question which is a generalized analogue to the w-question answered by a singular causal explanation. As we shall see, *all* why explanations (whether scientific, statistical, or singular causal) proceed at least in part by indicating a range of alternatives to the explanandum-phenomenon and then identifying conditions in their explanans such that if these conditions had obtained, these alternatives rather than the explanandum-phenomenon would have been favoured and such that, given the condition that actually obtained, the actual outcome was favoured over these alternatives.

2.1 Consider an explanation we might naturally describe as 'scientific'—an explanation (Ex. 7) in terms of Coulomb's law, ($L7$), of why the magnitude of the electric intensity (force per unit charge) at a perpendicular distance r from a very long fine wire having a positive charge uniformly distributed along its length, is given by the expression

$$E = \frac{1}{2\pi\varepsilon_0} \frac{\lambda}{r}$$

(where λ is the charge per unit length on the wire) and is at right angles to the wire. The usual textbook[11] explanation for this explanandum proceeds as follows:

(7) One thinks of the wire as divided into short segments of length dx, each of which may be treated as a point charge dq. The resultant intensity at any point will then be the vector sum of the fields set up by all these point charges. By Coulomb's law, the element dq will set up a field of magnitude.

$$(L7) \quad dE = \frac{1}{4\pi\varepsilon_0} \frac{dq}{s^2}$$

at a point P a distances s from the element. When we integrate the x and y components of dE separately and take the wire to be infinitely long, we obtain

$$E_x = 0$$

$$E_y = \frac{1}{2\pi\varepsilon_0} \frac{\lambda}{r}$$

This shows that the resultant field will be at right angles to the wire and that its intensity is given by

$$E = \frac{1}{2\pi\varepsilon_0} \frac{\lambda}{r}$$

The first thing to note about this explanation is it does indeed involve a derivation of (an idealized approximation to) its explanandum from a law and a statement of initial conditions. In this respect (7) does not differ significantly from an explanation like

(8) (L8) All ravens are black
Arthur is a raven
Arthur is black

which covering-law theorists take to be paradigmatic of all scientific explanation. It is clear, however, that there are other respects in which (7) differs rather strikingly from (8). Unlike the law occurring in (8), the version of Coulomb's law which occurs in (7) is stated in terms of variables (charge, distance, field intensity or force per unit charge) which are such that a whole range of different states or conditions can be characterized in terms of variations of their values. Coulomb's law for-

---

[11] For more details see, e.g. Sears and Zemansky (1970: 343–4).

mulates a systematic relation between these variables. It shows us how a range of different changes in certain of these variables will be linked to changes in others of these variables. In consequence, Coulomb's law is such that when the variables in it assume one set of values (when we make certain assumptions about boundary and initial conditions) the explanandum in the above explanation is derivable, and when the variables in it assume other sets of values, a range of other explananda are derivable. For example, Coulomb's law can be used, in conjuction with other initial and boundary conditions, to derive expressions for the electrical intensity between two equally and uniformly charged plates or inside and outside a uniformly charged hollow sphere. Coulomb's law can be used to show how the explanandum of (7) would alter if the infinitely long wire in (7) were twisted into a circle of finite diameter or coiled into a solenoid. While (L8) can also be used, in conjunction with different initial conditions, to explain a set of different explananda (e.g. raven a is black, raven b is black) the different possible explananda of (L7) differ from one another in a much more radical and fundamental way than the different possible explananda of (L8).[12]

My suggestion is that we think of (7) as explanatory not simply because it shows that its explanandum 'had' to obtain, but because it possesses a sort of generalized analogue of the explanation-making features which we attributed above to singular causal explanations. A singular causal explanation in effect specifies two possible conditions such that a change from one of these to the other would make a difference for which of two possible outcomes obtains. A good scientific explanation like (7) explains in part because it does something resembling this on a larger, more detailed scale. It permits the specification in terms of variations in the values of some small set of variables of both the explanandum phenomenon and a whole range of significantly different alternatives to the explanandum-phenomenon, any one of which might have occurred had initial and boundary conditions been different in various ways from what they actually were. We can think of the law occurring in a scientific explanation like (7) as implicitly providing answers to a whole range of what-if-things-had-been-different questions, with respect to these

---

[12] For an attempt to state more precisely the difference between the relatively trivial way in which the potential explananda of (7) differ from each other and the more fundamental way in which the potential explananda of Coulomb's law differ from each other, as well as a more detailed defence of the account of scientific explanation adumbrated here, see my (1979) and (1980). One way of seeing the difference between the range of potential explananda of (7) and the range of potential explananda is to note that with the appropriate initial conditions, the potential explananda of (7) are naturally thought of as different generalizations, while the potential explananda (7) are most naturally thought of as different instances of the same generalization.

alternatives. This law makes it clear how if the various conditions identified in the explanans of (7) were to change in various ways, (if, for example, the wire was coiled into a solenoid, or 'stretched' into a hollow sphere), one of these alternatives rather than the explanandum of (7) would ensue. In showing us how various alterations in the inital conditions cited in (7) would affect the resulting expression for the electrical field, (7) enables us to see how those conditions are relevant to or make a difference for this explanandum, just as, in answering a w-question a singular causal explanation identifies a condition relevant to its explanandum. We might say that scientific explanation, like singular causal explanation (and, as we shall see, statistical explanation), is always explanation of why the actual outcome *rather than* certain other possible outcomes obtains; such explanations always proceed by showing that given the conditions that actually obtained, the actual outcome was more favoured over these alternatives than it otherwise would have been.

2.2. It seems to me that the account of statistical explanation recently developed by Wesley Salmon (1971) is, in some respects, intuitively satisfying precisely because it suggests that successful statistical explanations will exhibit similar features.[13] Salmon thinks of statistical explanations as answers to questions of the form 'Why does this which is a member of class A have property B?' The answer to such a question will on Salmon's view consist of a partition of the class A into a number of sub-classes, which are the broadest possible homogeneous subclasses with respect to B, along with the probabilities of B within each of these sub-classes, and a statement identifying the subclass to which X belongs. That is, an explanation of why this X which is A is a B will take the following form.

$$P(A \cdot C_1, \ B) = P_1$$
$$P(A \cdot C_2, \ B) = P_2$$
$$P(A \cdot C_n, \ B) = P_n$$

where $A \cdot C_1, A \cdot C_2, \ldots, A \cdot C_n$ is a homogeneous partition with respect to B,

[13] I should say, however, that it is not clear to me whether Salmon would subscribe to the following reconstruction of the intuitive rationale of his model. In attempting to explain why his S-R model deserves to be described as a model of explanation Salmon appeals both to (a) the fact that the model captures intuitives notions of explanatory relevance and weight, and to (b) the fact that when the requirements of his model are satisfied we know what sort of betting behaviour is rational with respect to the question of whether any A is B. It seems that (a) but not (b) fits comfortably with my reconstruction. I might add that my argument is simply that there is a parallel between Salmon's multiple homogeneity requirement and my emphasis on the role of w-questions in explanation. I do not claim that every feature of the S-R model can be reconstructed within the framework of a 'w-question' conception of explanation; in fact I think that several important features of the S-R model are inconsistent with my conception.

$$P_i = P_j \text{ only if } i = j, \text{ and}$$

$$X \varepsilon A \cdot C_k$$

(Salmon employs the Reichenbachian notation for probability statements, according to which $P(A, B)$ is the probability of an object X in A having B. A is homogeneous with respect to B if every property that determines a 'place selection' is statistically irrelevant to B in A. A property C is said to be statistically irrelevant to B within A if and only if $P(A \cdot C, B) = P(A, B)$.)

The feature of Salmon's account on which I wish to focus is his 'multiple homogeneity' requirement—his requirement that all of the subclasses into which A is partitioned be homogeneous and that the probabilities of B within each of these subclasses be given. What is the rationale for insisting on this requirement rather than the simpler requirement that a statistical explanation must yield only a homogeneous partition of the particular subclass to which X belongs (and must give the probability of occurrence of B within that subclass)? If $X \varepsilon A \cdot C_k$, why should it matter whether the subclasses $A \cdot C_i$ for $i \neq k$ are homogeneous?

To focus our intuitions, let us suppose that President Carter secures his party's nomination for the 1980 presidential election and (implausibly) that the following (actually obtaining) factors are the only factors statistically relevant to this outcome.

C_1 = being an incumbent president
C_2 = being perceived as decent and honest
C_3 = having the use of the patronage powers of the presidency

Suppose that the following represents a partition of A into the broadest possible homogeneous reference classes together with the associated probabilities

   (i) $P(A \cdot C_1 \cdot C_2 \cdot C_3, B) = r_1$
   (ii) $P(A \cdot C_1 \cdot C_2 \cdot \sim C_3, B) = r_2$
   (iii) $P(A \cdot C_1 \cdot \sim C_2 \cdot C_3, B) = r_3$
   (iv) $P(A \cdot \sim C_1 \cdot C_2 \cdot C_3, B) = r_4$

where A = class of persons attempting to obtain their parties' nomination for president and B = class of persons succeeding in doing so. (We suppose that President Carter = $X \varepsilon A \cdot C_1 \cdot C_2 \cdot C_3$.) This partition does not simply tell us (i) how likely it was, given the conditions which actually obtain ($C_1 \cdot C_2 \cdot C_3$), that Carter would be renominated but also tells us (ii) how, if those conditions were to change in various ways (if, for example, Carter were no longer perceived as decent and honest or if Carter stopped making use of the patronage powers of the presidency), the probability of his nomination would be altered. As in the case of scientific and singular causal explanation, it is natural to think of the additional information represented by (ii) as telling us about the weight of relative significance of the various factors relevant to Carter's renomination, as telling us which factors made a difference for Carter's renomination and how much of a difference they made. Suppose, in the example above, $r_1 = .70$, $r_2 = .68$ and $r_3 = .15$. In learning this, we learn that the presence of C_3 had only a small effect on the probability of Carter's being renominated, while the presence of C_2 made a great deal of difference for the likelihood of Carter's being renominated. There is an obvious and natural preanalytic sense in which such information is relevant to an understanding of why Carter was renominated. In insisting on the multiple homogeneity requirement, and denying the Hempelian claim that we have an adequate explanation of why Carter was renominated if we merely know (i) and r_1 is sufficiently high, Salmon is in effect suggesting that a statistical explanation does not explain by showing that its explanandum was to be expected, and that the ability of a statistical explanation to answer a w-question is essential to its status as an explanation.

2.3. My general claim, then, is that scientific and statistical explanations have explanatory import at least in part because they show how, if the conditions cited in their explanans were to alter in various ways, their explananda would also alter. I claim that singular causal explanations explain because they possess an analogous feature, and not because they tacitly invoke a law or show that their explananda-phenomena were to be expected. So where Hempel would claim that the common thread running through all why explanations is that such explanations must show that their explananda were to be expected, I claim that the common thread is that such explanations must answer a 'what-if-things-had-been-different' question.

Perhaps all of this will become clearer if we briefly consider a (purported) explanation which fails to answer a w-question. Suppose I claim, to use Wesley Salmon's example (1971: 34), that

(9) John's taking birth-control pills explains his failure to get pregnant and that John is a male.

(9) fails to answer a (relevant) w-question. It identifies a condition (John's taking birth-control pills) which is nearly sufficient for John's failure to get pregnant, but this condition is not such that, if it were to change, a change in the explanandum-phenomenon cited in (9) would be found. Intuitively, a change in John's taking birth-control pills is not relevant to or does not make a difference for whether or not he gets

pregnant in the way in which the occurrence or non-occurrence of the short-circuit makes a difference for whether or not the fire occurs or changes in the various factors cited in the above statistical explanation make a difference for whether or not Carter is renominated. This failure of (9) to answer a w-question is reflected in the fact that (9) does not seem explanatory.

2.4. I want to conclude this section with some brief remarks on the distinctive picture of singular causal explanations that emerges when we locate the explanatory import of such explanations in their ability to answer a w-question. First, on this view, a sentence reporting a causal relationship between events may constitute a singular causal explanation, for such a sentence answers a what-if-things-had-been-different question. Many recent writers on singular causal explanation have denied this. They have claimed that only sentences having a form like 'X's being A causes Y's being B' are potentially explanatory. For example, J. L. Mackie apparently holds that

(10) The giving way of the bolt caused the collapse of the bridge

is not potentially explanatory, while

(11) The fact that the bolt gave way so suddenly and unexpectedly caused the bridge to collapse

is potentially explanatory (1974: xv, 257–65). For Mackie, as for most writers who hold similar views, the motivation for this distinction between (10) and (11) largely derives from the idea that (to put it very roughly) singular causal explanation, to the extent it is genuinely explanatory, always relates properties that are linked by law. (This idea presumably derives in turn from the idea that the paradigm of a successful why-explanation is a covering-law explanation.) In contrast, on my account of how singular causal explanation works there is no motivation for drawing a sharp contrast of this kind between (10) and (11). A sentence like (10) is seen as explaining in just the way that (11) explains—by identifying conditions in its explanans such that changes in these would make a difference for whether the explanandum-phenomenon or a specified alternative to it occurred. Sentence (11) is seen as explanatory not because, as Mackie seems to suppose, the predicate 'gave way so suddenly and unexpectedly' figures in a law of nature specifying the conditions under which bridges collapse (it is implausible to think that there is such a law) but because (11) answers a w-question—because it tells us that if the bolt had given way, but not so suddenly and unexpectedly, the bridge collapse would not have occurred and that given that the bolt did give way so suddenly and unexpectedly, the collapse was made more likely than it would have been if the giving way had not been so sudden and unexpected. The 'caused' of *both* (10) and (11) can, in the appropriate context,[14] mean 'causally explains'. I submit that this unified treatment of (10) and (11) is desirable on both theoretical and intuitive grounds. Sentences like (1) and (10) reporting causal relationships between events *are* taken to be explanatory by ordinary speakers and my proposal leads to a far more plausible account of the kind of understanding provided by a sentence like (11) than Mackie's.

Finally, if it is correct that singular causal explanations explain because they identify conditions under which the actual outcome or some specified alternative to it would have been favoured, it seems reasonable to expect that they will continue to be explanatory if we substitute some alternative way of referring to the actual outcome *and* this specified alternative. If (1) explains because it tells us (truly) that a certain kind of change would make a difference for whether the fire or an alternative situation in which there is no fire ensues and if the phrase 'the fire on the north side of the house' picks out the same fire, it ought to be true that this same change would make a difference (the same difference) for whether the fire on the north side of the house or the specified alternative situation in which there is no fire ensues. As we shall see in Section 3 this expectation is correct—singular causal explanations are extensional on the effect side. (Put differently, if the explanans of (1) explains why the fire occurs rather than fails to occur, it will also explain why the fire on the north side of the house occurs rather than fails to occur, it will also explain why the fire on the north side of the house occurs rather than fails to occur. However, this explanans need not, to anticipate our discussion in Section 3, explain why the fire occurs on the north side of the house rather than some place else.)

3

I have suggested that singular causal explanations explain in virtue of answering a what-if-things-had-been-different question but I have said relatively little about the structure of such explanations. It is to this task that I now turn. For reasons of space I shall concentrate mainly on the

---

[14] It is no part of my argument that sentences like (10) and (11) are always used to explain. Sometimes they are used for other purposes, e.g. to attribute responsibility or to satisfy practical interests in control and manipulability.

explanandum side of singular causal explanations but much of what I say will transfer neatly to the explanans side of such explanations.[15]

One of the keys to understanding singular causal explanations is to recognize that the explananda of such explanations possess what Fred Dretske calls 'contrastive focus' (Dretske, 1973). Consider, to use Dretske's example, the question

(12) Why did Clyde lend Alex $300?

There are a number of different ways of understanding this question and, correspondingly, a number of different explanations one may give by way of answer to it. In asking this question one may want to know

(13) Why did Clyde lend Alex *$300* (rather than some other amount)?

and the answer given by way of explanation may be that this was all Alex wanted or all that Clyde had available to lend. Or one may want to know

(14) Why did Clyde *lend* (rather than, e.g., give) Alex $300?

and the answer might be that although Clyde could well afford to give Alex $300 he is stingy and so insists on full repayment. Or one may, in asking (12), want to know

(15) Why did *Clyde* (rather than someone else) lend Alex $300?

and the answer may be that Clyde was the only friend Alex had who could afford to lend him the money.

In each of these cases the task of explaining why Clyde lent Alex $300 is a matter of accounting for a contrast, of explaining why the actual situation rather than some specific alternative obtained. Thus, if one

---

[15] That is, if it is correct that singular causal explanations explain simply in virtue of identifying two possible conditions such that a change from one to the other would make a difference for whether the explanandum-phenomenon or some specified alternative to it is favoured, then it seems reasonable to expect that such explanations will continue to be explanatory when we substitute an alternative way of describing or referring to the two conditions cited in its explanans. In particular it seems to me that most apparent counter-examples to the thesis that singular causal explanation is extensional on the cause-side arise because it is easy and natural to interpret some substitutions of apparently co-referential phrases in sentences like (1) as picking out different possibilities than those picked out in the explanans of (1). Suppose (1) is understood to mean 'The occurrence of short-circuit (rather than its failure to occur) causally explains the fire' (1*). Then if the short-circuit = the short-circuit occurring at dusk, it will also be true that 'The occurrence of the short-circuit at dusk (rather than its failure to occur at all) causally explains the fire' (1**). Of course it presumably need not be true that 'The occurrence of the short-circuit at dusk (rather than at some other time) causally explains the fire'(1***). But if we take seriously my remarks above it should be clear that (1**) and (1***) simply have different explanans. (1**) focuses on what would happen if there was a change from a case in which the short-circuit occurs to a case in which the short-circuit fails to occur, while (1***) focuses on the quite different question of what would happen if there was a change from the case in which the fire occurs at dusk to a case in which the fire occurs at some other time. So (1**) and (1***) do not really represent a case in which the same explanans, differently described, both explains and fails to explain the same explanandum. (See Section 3 for a similar way of handling counter-examples to the thesis of extensionality on the effect-side).

attempts to explain (12) construed as (13), one is interested in the contrast between Clyde's lending Alex $300 and an alternative possible case in which Clyde lends Alex a different amount of money—what one undertakes to explain is the limited question of why the first alternative rather than the second alternative is realized. The answer one gives does not purport to account for the contrast between Clyde's lending Alex $300 and other conceivable alternatives—it does not, for example, purport to answer (12) construed as (14) or (15).

Given that one's interest is in explaining (12) construed as (13), the facts that Clyde lent Alex some amount of money or other, that the money was lent rather than given, or that it was Clyde rather than someone else who lent the money are all, in Dretske's suggestive phrase, 'idle elements'—they are treated as assumed or presupposed by (13), as not standing in need of explanation. And because the explanatory task associated with (13) is limited in this way—because explaining (13) is a matter of accounting for the contrast between the actual situation and one distinctive alternative rather than accounting for the contrast between the actual situation and all other possible alternatives—we may successfully explain (13) (by showing, for example, that this was all the money Clyde had available) without also explaining (14) or (15). Indeed we may successfully explain (13) without committing ourselves on the question of whether it is even possible to explain (14) or (15). The explanation we give by way of answer to (13) is not shown to be defective or incomplete (as an answer to (13)) by the fact that it is not also an answer to (14) or (15), or even by the fact that it may be epistemologically impossible to answer (14) or (15). Following Dretske, we may express these features of (13), and similar features of (14) and (15) by saying that '$300', 'lent', and 'Clyde' are respectively, the 'contrastive foci' of (13), and (14), and (15).

3.1.   While the explanations provided above by way of answer to (13), (14), and (15) are perhaps not singular causal explanations, it seems to me that the explananda of singular causal explanations will also always possess a distinctive contrastive focus. Failure to appreciate this has led to many misunderstandings regarding the structure of such explanations.

When one asks 'What caused b?' one typically has in mind a contrast between the actual situation, in which b occurs and some otherwise similar situation in which b does not occur. (We shall call this hypothetical alternative a contrast state; it indicates the contrastive focus of the above question.) This contrast state and the actual outcome constitute the two possible outcomes which, I claimed in Section 2, a singular causal explanation will implicitly identify. We can say, following the usage introduced above, that a singular causal explanation explains (in the manner described in Section 2), why the actual outcome *rather than* this specific alternative eventuated. Just as in Dretske's example, what one undertakes to explain in citing the cause of *b* (in giving a singular causal explanation for the occurrence of *b*) is the limited question of why *b* rather than this specific alternative occurred and not the more general question of why *b* rather than any possible alternative to *b* occurred.

Consider again the singular causal explanation

(1) The short-circuit caused the fire

Ordinarily when one makes this claim one is contrasting the situation in which the fire occurs with some otherwise similar situation (containing the same background conditions—the presence of oxygen, inflammable materials, and so forth) in which no fire at all occurs. The presence of the short-circuit is taken to be the factor which accounts, in the manner described in Section 2, for the contrast between the actual situation and this alternative. Just as a satisfactory answer to question (13) above need not also be a satisfactory answer to (14) or (15), so (1) may be a perfectly satisfactory singular causal explanation even though there are many causal questions which (1) does not purport to answer, even by implication. For example, the fire referred to in (1) presumably had a certain colour (e.g. purple), and reached a certain maximum temperature (e.g. 500°). But (1) does not purport to tell us why the fire was purple, or reached a maximum temperature of 500°. These matters involve different questions and different contrast states than those which (1) involves. Thus, if one is interested in a singular causal explanation of why the fire is purple one will ordinarily be interested in accounting for the contrast between the actual situation, and an otherwise similar situation in which the fire also occurs but without a purple flame. In this case, what one undertakes to explain is the limited question of why the fire occurs with a purple flame and not the quite different question of why the fire occurs rather than fails to occur. (This is the question to which (1) addresses itself.) One may be able to identify a factor—for example, the presence of a potassium salt—which explains why the fire has flames which are purple in colour without being also able to explain why the fire rather than no fire at all has occurred. In a situation in which one undertakes to give a singular causal explanation of why the fire has purple flames, the fact that the fire occurs and has flames of some colour or other is simply assumed—these are 'idle elements' which are not ordinarily thought of as standing in need of explanation. And in a similar way one can give a singular causal explanation of why the fire occurred (rather than failed to occur) without undertaking to explain why it had the purple colour which it in fact possessed.

Indeed, I would make a stronger claim—not only does a singular causal explanation like (1) not purport to answer (even implicitly) all the causal questions which may arise in connection with the fire, but (1) does not even carry with it the implication that it must be possible in principle to answer these questions. That is to say, (1) does not carry with it the implication that it must be possible to explain every feature of the fire (its exact colour, shape, time of occurrence, etc.) in terms of the short-circuit and the circumstances in which it occurs, or, indeed, in terms of any other explanans. To suppose otherwise is to overlook the point that (the warranted use of) a singular causal explanation like (1) requires simply that we be able to account for the contrast between the actual situation and some one specific alternative and not that we be able to account for the contrast between the actual situation and every conceivable alternative (i.e. that we be able to explain why the fire is this particular colour rather than some slightly different colour, this particular shape rather than some slightly different shape and so forth). Sentence (1) does *not* claim, even by implication, that there is a sufficient condition for every aspect of the fire, exactly as it occurred.

We are now in a position to discuss in more detail some of the adequacy conditions developed in Section 1. First of all, we can now give content to the intuition that singular causal explanations may be used to explain the occurrence of individual events. To give a singular causal explanation of an individual event will not of course be to vainly attempt to account for every feature of that event, but will rather be to account for the contrast between the situation in which that event occurs and some specific contrast situation in which that event fails to occur (by finding conditions such that a change in these would make a difference for which of these two possible outcome situations occurs). So what is explained by a sentence like (1) is indeed the occurrence of *this* individual fire (rather than the occurrence of no fire at all) and not why some event has the property of being the one and only fire in the vicinity, but of course explaining why this particular fire occurred does not require explaining why this fire has the properties it possesses.[16]

Secondly, we should now be able to see, as suggested in Section 1, that there is indeed a clear difference between explaining why some

---

[16] In 'Aspects of Scientific Explanation', Hempel asks whether what he calls 'concrete events' such as the eruption of Mt Vesuvius in AD 79—particulars that can be referred to by means of noun phrases—can ever be 'completely explained'. He writes, 'What could be meant, in this case by a complete explanation? Presumably, one that accounts for every aspect of the given event. If that is

particular event, which has a certain feature, occurred, and explaining why that event has that feature. A request for a singular causal explanation of the purple, odd-shaped fire will be ordinarily understood as a request for a factor which distinguishes between the situation in which that fire occurs and a contrasting situation in which no fire occurs, while a request for an explanation of why the fire is odd-shaped or purple will ordinarily be understood as a request for a factor which distinguishes between the actual situation in which an otherwise similar fire occurs, but is not odd-shaped or purple.

3.2. There are many ways in which the contrastive focus of a singular causal explanation may be indicated. One common device is emphasis. Consider, to use Peter Achinstein's (1975) examples, the singular causal explanations

(16) Socrates' drinking hemlock caused his death *at dusk*

and

(17) Socrates' drinking hemlock caused his *death* at dusk

On the analysis offered above, (16) claims (presumably falsely) that Socrates' drinking hemlock is a factor which accounts for the contrast between the actual situation in which Socrates dies at some other time. What is explained is (it is claimed) why Socrates died at dusk rather than

the idea, then indeed no concrete event has finitely many different aspects and thus cannot even be completely described, let alone completely explained. For example, a complete description of the eruption of Mt Vesuvius in AD 79 would have to specify the exact time of its occurrence; the path of the lava stream as well as its physical and chemical characteristics—including temperatures, pressures, densities, at every point—and their changes in the course of time; the most minute details of the destruction wreaked upon Pompeii and Herculaneum. Indeed, there seems to be no clear and satisfactory way at all of separating off some class of facts that do not constitute aspects of the concrete event here referred to. Clearly then, it is quite pointless to ask for a complete explanation of an individual event thus understood' (1965: 421–2).

Given this difficulty in attaching sense to the notion of explaining a particular event like an eruption, Hempel concludes that explaining a particular event must always be a matter of explaining why a particular event has a certain property or is of a certain kind.

But the difficulty Hempel finds in the notion of explaining a particular eruption or abdication, where this is something different from explaining why these events had certain properties, only arises because of Hempel's prior commitment to the covering-law model. If one thinks that all why-explanations must be covering-law explanations, then of course it will look as though if explaining an eruption is not a matter of explaining why the eruption had some particular feature, it can only be a matter of explaining why the eruption has all of the features it has. My account of singular causal explanation avoids Hempel's dilemma because it rejects the claim that singular causal explanation is covering-law explanation. On my account, we explain an eruption or abdication (or better, the occurrence of these events) by finding a factor which accounts (in the manner described above) for the contrast between the actual situation and a specific alternative situation in which no eruption or abdication occurs. So what we explain is the occurrence of the particular event—the eruption of Mt Vesuvius in AD 79—and not why this event lasted for so many hours or had some other property, but of course explaining this does not involve explaining why this event has all of the properties it has.

at some other time. On the other hand, (17) claims (presumably truly) that Socrates' drinking hemlock is a factor which accounts for the contrast between the actual situation in which Socrates dies (his death being a death which happens to occur at dusk) and an otherwise similar situation in which Socrates does not die at all.

Constrastive focus may also be suggested by syntax. For example, it seems to me that usually the most natural way of understanding a request for a singular causal explanation of why $(^{\imath}x)$ (Hx) is G is as a request for a factor which distinguishes between the actual situation and one in which there is an otherwise similar particular which is also H but not G. And the most natural way of understanding a request for singular causal explanation of why $(^{\imath}x)$ (Hx) verbed χly (why the tall man talked loudly) is as a request for a factor which distinguishes the actual situation from an otherwise similar situation in which $(^{\imath}x)$ (Hx) verbs, but not χly (an otherwise similar situation in which the tall man talks, but not loudly).

In many cases a request for a singular causal explanation will have several natural readings and we must rely on background information about the interest of the speaker or await further clarification in order to determine what was meant. Suppose (to use a hackneyed example) that A who has an ulcerated stomach eats parsnips and develops indigestion. A may (truly) say

(18) Eating parsnips caused my indigestion

and mean that his eating parsnips accounts for the contrast between the actual situation and an otherwise similar situation in which his stomach is ulcerated but in which he develops no indigestion. Here A attempts to answer the following question 'Given that my stomach is ulcerated why did I get indigestion on this particular occasion and not on others?' A's doctor, on the other hand, may understand the question 'Why did A get indigestion?' in a different way—as a request for a factor which distinguishes the actual situation from an otherwise similar situation in which A's digestive system functions normally and he does not develop indigestion while eating parsnips. If so, A's doctor will offer a different explanation. He will say

(19) A's ulcerated stomach caused his indigestion

and mean that this factor explains why A developed indigestion while other people (or A at an earlier time) do not when they eat parsnips. So there are two plausible readings of the question 'What caused A's indigestion?' and corresponding to these, a sense in which (18) is correct and a sense in which it is not, and similarly for (19). What this shows is that as they stand (18) and (19) are incomplete or ambiguous—to fully specify what they mean we must specify the contrast state to which they make implicit reference. Singular causal explanation is always, on the explanandum side, a relational matter—what is explained is always why this alternative rather than that alternative was realized—and until we specify both of the terms of this relation we have not fully specified what a singular causal claim means.[17]

3.3. I can further bring out the distinctive features of the above account of singular causal explanation by briefly considering a recent argument that singular causal explanations are not extensional in the effect-position. In his paper, 'Doing Without Events', Russell Trenholme (1978: 197) asks us to consider the sentences

(20) Stubbing his toe caused John to walk slowly

(21) Stubbing his toe caused John to walk

Trenholme contends that although it seems plausible to hold that

(22) the event of John's walking slowly = the event of John's walking

(20) may very well be true on some particular occasion and (21) false. He concludes that (20) and (21) are not extensional in the effect position. In a similar vein, Peter Achinstein (1975) asks us to imagine a crooked roulette wheel in which each number appears twice, once below a black space and once below a red one. If I push button 1 the ball will land on a red number. If I push button 2 the ball will land on one of the sevens. Then it may be true that

(23) My pushing button 1 causes the ball to land on the *red* seven

---

[17] My point is *not* that whether or not a given sentence is a successful singular causal explanation depends upon the expectations and interests of the person to whom it is addressed. Once we specify the contrastive focus of a given explanandum it is entirely an objective matter whether a certain singular causal explanation explains that explanandum. Of course someone who asks 'Why did *A* develop indigestion' may wish to know what caused *A*'s indigestion when other people with ulcerated stomachs do not develop indigestion. Such a person will not be satisfied if he is told (19) rather than (18). But this does not show that (19) is not an adequate explanation of the explanandum at which it is directed. (In just the same way, Einstein's explanation of the photoelectric effect is not shown to be an incorrect explanation if it is addressed to someone who wants an explanation of some different phenomenon.)

In insisting that singular causal explanation is always explanation of why an explanandum phenomenon rather than some specific alternative occurs, I mean to make a point about the semantics (and not just the pragmatics) of singular causal explanation. If a singular causal claim does not make explicit reference to a contrast state and is thus susceptible of several different interpretations, we may be guided by our knowledge of the interest of the speaker and his audience in deciding which interpretation (which contrastive focus) is intended, but once we settle on a certain interpretation (a certain contrastive focus) whether or not the singular causal explanation adequately explains that explanandum, thus interpreted, is independent of anyone's interests.

and false that

> (24) My pushing button 1 causes the ball to land on the red *seven*

On the apparently plausible assumption that changing the emphasis in a referring expression does not alter its referent so that

> (25) The event of the ball's landing on the *red* seven = the event of the ball's landing on the red *seven*

it again seems to follow that (23) and (24) are not extensional in the effect position.

One possible response to Trenholme's argument is to deny (22). This approach is in effect taken by philosophers like Kim (1968, 1973) and Goldman (1970), who employ examples like Trenholme's together with the assumption that sentences like (20) and (21) are extensional, to argue in favour of a finer-grained account of event identity than Trenholme's. But such an approach is not a terribly plausible way of dealing with (23) and (24), and is at any event unnecessary, as we shall see. Even if we accept (22) and (25), as I suspect we should, and even if we agree that (20) and (21), and (23) and (24) differ in truth-value, it does not follow that these sentences fail to be extensional in the explanandum position.

On the account of singular causal explanation sketched above the true form of a sentence like

> (26) *a* caused (causally explains) *b*

is perspicuously represented as

> (27) *a* caused (causally explains) *b* rather than *c*

If this is correct, it seems clear that the question of whether singular causal explanations are referentially transparent on the effect side is properly understood as the question of whether the substitution of co-referential terms for '*b*' and '*c*' in (27) will preserve truth or alternatively, as the question of whether those substitutions on the effect side which preserve and do not shift contrastive focus also preserve truth. If we compare the contexts

> (27) *a* caused *b* rather than *c*

and

> (28) *a* caused *b'* rather than *c'*

where $b = b'$ but $c \neq c'$ and determine that (27) and (28) differ in truth-value, this does not establish that (27) and (28) are non-extensional. Instead it establishes only the unsurprising conclusion that when we substitute into an explanatory context expressions that describe or refer to different explananda-phenomena we may alter truth-value, or move from an acceptable explanation to a non-acceptable explanation. It seems to me that arguments like Trenholme's and Achinstein's do just this.

On the most natural interpretation of (20), it is to be understood as

(20*) Stubbing his toe caused John to walk slowly rather than to walk at a normal pace

That is to say, (20) understood as (20*) claims that the event of John's stubbing his toe accounts for the contrast between the actual situation in which John walks slowly and the specific alternative situation in which John also walks, but at a more normal pace. On this interpretation, (20*) does *not* purport to explain why John walked rather than did not walk— that John walks at some pace or other is assumed or presupposed in (20). On the other hand, the most natural interpretation of (21) is

(21*) Stubbing his toe caused John to walk (rather than not to walk at all)

Here the claim is that the event of John's stubbing his toe accounts for the contrast between the actual situation in which John walks and an otherwise similar situation in which he fails to walk.

Thus on their most natural interpretation (20) and (21) have entirely different explananda and are directed towards entirely different explanatory tasks—towards accounting for very different contrasts. (20) and (21) do not represent a situation in which an explanans explains an explanandum-phenomenon described or referred to in one way, and fails to explain that same phenomenon described or referred to in a different way and thus do not show that singular causal explanations are non-extensional in the effect position. The fact that (20) and (21) may differ in truth-value even if (22) is true shows simply that the structure of the explanans of (20) and (21) is more complicated than one might otherwise suppose.[18]

A similar point can be made about Achinstein's example. To avoid the conclusion that (23) and (24) are non-extensional in the effect position, we need not take the heroic course of denying (25). Instead we need only

---

[18] If my account of singular causal explanation is correct, it also follows of course that arguments parelleling Trenholme's cannot be used to defend a 'fine-grained' theory of event individuation. In *A Theory of Human Action* (1970: 3), Alvin Goldman argues that because (36), 'John's being in a tense emotional state caused him to say "hello" loudly', may be true and (37), 'John's being in a tense emotional state caused him to say "hello" ', false, it follows (on the assumption that (36) and (37) represent extensional contexts) that the event of John's saying 'hello' ≠ the event of John's saying 'hello' loudly. On my view (36) and (37) have different explananda (because they involve reference to different contrast states) and so Goldman's conclusion does not follow. Even if (36) and (37) differ in truth value on their most natural interpretation, and even if they are extensional contexts, it still might be true the event of John's saying 'hello' = the event of John's saying 'hello' loudly.

note that (23) and (24) purport to explain quite different explananda. Sentence (23) claims in effect that pushing button 1 caused (or causally explains) the ball landing on a red square (which happens to be a seven) rather than a black square. By contrast, (24) claims that pushing button 1 caused (or causally explains) the ball landing on a square marked seven (which happens to be red) rather than a square marked with some other number.[19]

3.4. We can deepen our appreciation of the relation between extensionality and contrastive focus by returning to an observation I stressed earlier—that there is a difference between

(3) The short-circuit caused the purple fire

and

(29) The short-circuit caused the fire to be purple

As we noted above, (3) claims to cite a factor which distinguishes between the actual situation in which a certain fire (the purple one) occurs and an otherwise similar situation in which no fire occurs. That is to say the expression 'the purple fire' functions referentially, as a definite description, in (3). (A sentence like (3) is perhaps most naturally used in a situation in which there is some uncertainty regarding which fire is being explained. If we imagine a situation in which several fires occur, only one of which is purple, (3) may be used to indicate that it is the occurrence of the purple fire (rather than some other fire) which is being explained.) It seems plausible to hold that if the function of the definite description 'the purple fire' in (3) is simply to indicate which fire is being explained (that is, to indicate that it is the contrast between the occurrence of this particular fire and a situation in which no fire occurs which is being explained), then a definite description that can be understood as having the same referent and preserving the same contrastive focus ought to serve this purpose equally well and thus will be substitutable in (3). If the purple fire was the only fire within the relevant vicinity which occurred at dusk, or which occurred on the north side of the church, then the expression 'the fire which occurred at dusk' or 'the fire which occurred on the north side of the church' will be substitutable for 'the purple fire' in (3). So we can truly say

(30) The short-circuit caused the fire which occurred at dusk

---

[19] The reader who is still inclined to attempt to deal with (20) and (21), (23) and (24) by denying (22) and (25) might consider how he would deal with the analogous argument which may be constructed in connection with (18), 'Eating parsnips caused *A*'s indigestion.'

and

(31) The short-circuit caused the fire which occurred on the north side of the church

In making the substitutions which allow us to move from (3) to (30) and (31) we must, however, be careful to interpret (30) and (31) in such a way that the contrastive focus implicit in (3) is preserved. (It is a failure to do this that makes it seem plausible that (3) is non-extensional in the effect position.) That is to say, (30) and (31) must be understood as

(30*) The short-circuit caused the fire which occurred at dusk (rather than no fire at all)

and

(31*) The short-circuit caused the fire which occurred on the north side of the church (rather than no fire at all)

It is only when we understand (30) and (31) in this way that the phrases 'the fire which occurred at dusk' and 'the fire on the north side of the church' function as 'the purple fire' does in (3)—that is, as definite descriptions. If on the other hand, we take the results of substituting into (3) to mean

(32) The short-circuit caused the fire to occur at dusk (rather than at some other time)

and

(33) The short-circuit caused the fire to occur on the north side of the church (rather than to occur at some other place)

the resulting sentences are presumably false. With (32) and (33) we have shifted the contrastive focus from that present in (3), (30), and (31). Sentences (32) and (33) purport to account for a different explanandum from the explanandum of (3) and so it is not surprising that (30) and (33) can be false while (30) and (31) are true. Like 'to be purple' in (29), the role of 'to occur at dusk' in (32) and 'to occur on the north side of the church' in (33) is not referential. Rather these phrases function so as to attribute a property, the possession of which by the fire is to be explained. We might put the point this way: (30) and (31) presuppose, but do not explain why, the fire occurred at dusk and on the north side of the church, while (32) and (33) do purport to explain why the fire has just these features. So while it is indeed true that a singular causal explanation of why the fire, which was at dusk, occurred is not a singular causal explanation of why the fire occurred at dusk, this has to do with the fact that singular causal explanations possess contrastive focus, and not with their non-extensionality.[20]

Just as an account of singular causal explanation in terms of contrastive focus helps to support the claim that sentences like (3) are extensional in the effect position, it seems also to support the claim that sentences like

(4) The presence of potassium salts caused the fire to be purple

are extensional in the effect position, provided we take care to read the sentences which result from the substitution of identicals in (4) as possessing the same contrastive focus as (4). Suppose, for example, the colour purple is the colour which reflects light of wave length λ or the colour which is Mrs Jones's favourite colour. Then if (4) is true, it will also be true that

(34) The presence of potassium salts caused the fire to be the colour which reflects light of wave length λ

and

(35) The presence of potassium salts caused the fire to be the colour which is Mrs Jones's favourite

provided these explanations are understood as possessing the same contrastive focus possessed by (4). Since (4) is most naturally interpreted as

> The presence of potassium salts accounts for the contrast between the actual situation in which the fire occurred and is purple and a contrasting situation in which the fire occurs and is some other colour

(34) and (35) will be true if we construe them as

(34*) The presence of potassium salts accounts for the contrast be-
(35*) tween the actual situation in which the fire occurs and is the colour which reflects light of wavelength λ (is the colour which is Mrs Jones's favourite) and a contrasting situation in which the fire occurs and is some other colour

To be sure, a singular causal explanation of why the fire is purple is not also a singular causal explanation of why the fire reflects light (of some wavelength or other) or of why purple light is light of wavelength λ or of why purple, which happens to be the colour of the fire, is Mrs Jones's favourite colour or of why Mrs Jones has a favourite colour at all and if

---

[20] Those familiar with Kim's work on events will recognize that the different roles that 'is purple' and 'the purple fire' play in (29) and (3) correspond, respectively, to cases in which, in Kim's language, purpleness is, and is not, a constitutive property of the event of the fire (cf. (1968) and (1973)).

we wish to preserve truth as we move from (4) to (34) and (35) we must construe (34) and (35) so that they do not suggest otherwise. However, the reasons for this and its irrelevance to the question of extensionality should by now be familiar to the reader.

Let us take stock. At the end of Section 1, I suggested a number of desiderata which an adequate theory of singular causal explanation should meet. How does our theory fare with respect to these desiderata?

First our account has, I would argue, identified the structural features in virtue of which singular causal explanations explain and it has done so by placing such explanations within the context of a general theory of why-explanation. Secondly, these structural features are features which are readily epistemologically accessible to users of singular causal explanations. We have not claimed, as many theories of singular causal explanation do, that such explanations explain in virtue of satisfying criteria which are such that most users are in no position to recognize whether those criteria are satisfied or not. Thirdly, the notion of contrastive focus has provided a basis for distinguishing what singular causal explanations explain from what they merely presuppose. We now have an account of singular causal explanations which identifies a structural difference between, e.g., an explanation of the occurrence of a purple fire and an explanation of why that fire was purple. Fourth, we have provided an account of what it might mean to explain the occurrence of an individual event, while avoiding Hempel's worry that such an explanation can only be an explanation of why that event has all the properties it does (cf. Hempel, 1965). Lastly, on our account, singular causal explanations do indeed obey a substitution principle like that advocated in Section 1. A singular causal explanation of the occurrence of a will be a singular causal explanation of the occurrence of b, when $a = b$.

REFERENCES

Achinstein, P. (1975), 'Causation, Transparency, and Emphasis', *Canadian Journal of Philosophy*, 5, 1–23.

Beauchamp, T., and Rosenberg, A. (1981), *Hume and the Problem of Causation* (New York: Oxford University Press).

Causey, R. (1977), *The Unity of Science* (Dordrecht: D. Reidel).

Davidson, D. (1967), 'Causal Relations', *Journal of Philosophy*, 64, 691–703. Reprinted in Donald Davidson, *Actions and Events* (Oxford: Oxford University Press, 1981), 149–62.

—— (1969), 'The Individuation of Events', in N. Rescher *et al.* (eds.), *Essays in Honor of Carl G. Hempel* (Dordrecht: D. Reidel), 216–34.

Davidson, D. (1970), 'Mental Events', in L. Foster and J. W. Swanson (eds.), *Actions and Events* (Amherst, Mass.: University of Massachusetts Press); reprinted in Davidson, *Actions and Events* (Oxford: Oxford University Press, 1981), 207–25.

Dretske, F. (1973), 'Contrastive Statements', *Philosophical Review*, 82, 411–37.

Goldman, A. (1970), *A Theory of Human Action* (Englewood Cliffs, NJ: Prentice-Hall).

Hempel, C. (1965), 'Aspects of Scientific Explanation', in *Aspects of Scientific Explanation* (New York: Free Press), 331–496.

Kim, J. (1968), 'Events and Their Descriptions: Some Considerations', in N. Rescher (ed.), *Essays in Honor of Carl Hempel* (Dordrecht: Reidel), 198–215.

—— (1973), 'Causation, Nomic Subsumption and the Concept of Event', *Journal of Philosophy*, 70, 212–36.

Levin, M. (1976), 'Extensionality of Causation and Causal Explanatory Contexts', *Philosophy of Science*, 43, 266–77.

Mackie, J. L. (1974), *The Cement of the Universe* (Oxford: Oxford University Press).

Nickles, T. (1977a), 'On the Independence of Singular Causal Explanation in Social Science: Archaeology', *Philosophy of the Social Sciences*, 1963–87.

—— (1977b), 'Davidson on Explanation', *Philosophical Studies*, 31, 141–5.

Salmon, W. (1971), 'Statistical Explanation and Statistical Relevance', in Salmon (ed.) *Statistical Explanation and Statistical Relevance* (Pittsburgh: University of Pittsburgh Press).

—— (1975), 'Theoretical Explanation', in S. Korner (ed.), *Explanation* (Oxford: Blackwell), 118–45.

—— (1978), 'Why Ask "Why?"?', Presidential Address, *Proceedings of the American Philosophical Association*, 51, 683–705.

Sears, F., and Zemansky, M. (1970), *University Physics* (Reading, Mass.: Addison-Wesley).

Trenholme, R. (1978), 'Doing Without Events', *Canadian Journal of Philosophy*, 8, 173–85.

Woodward, J. (1979), 'Scientific Explanation', *The British Journal for the Philosophy of Science*, 30, 41–67.

—— (1980), 'Developmental Explanation', *Synthese*, 44, 443–66.

—— (1986), 'Are Singular Causal Explanations Implicit Covering-Law Explanations?', *Canadian Journal of Philosophy*, 16, 253–80.

—— 'Causal Explanation in History' (unpublished).

XI

THE PRAGMATICS OF EXPLANATION*

BAS C. VAN FRAASSEN

> If cause were non-existent everything would have been produced by
> everything and at random. Horses, for instance, might be born,
> perchance, of flies, and elephants of ants: and there would have been
> severe rains and snow in Egyptian Thebes, while the southern dis-
> tricts would have had no rain, unless there had been a cause which
> makes the southern parts stormy, the eastern dry.
>
> Sextus Empiricus, *Outlines of Pyrrhonism*, 3. 5. 1

A theory is said to have explanatory power if it allows us to explain; and
this is a virtue. It is a pragmatic virtue, albeit a complex one that includes
other virtues as its own preconditions. After some preliminaries in Section 1, I shall give a frankly selective history of philosophical attempts
to explain explanation. Then I shall offer a model of this aspect of
scientific activity in terms of why-questions, their presuppositions, and
their context-dependence. This will account for the puzzling features
(especially asymmetries and rejections) that have been found in the
phenomenon of explanation, while remaining compatible with empiricism.

1. THE LANGUAGE OF EXPLANATION

One view of scientific explanation is encapsulated in this argument:
science aims to find explanations, but nothing is an explanation unless
it is true (explanation requires true premisses); so science aims to find
true theories about what the world is like. Hence scientific realism is
correct. Attention to other uses of the term 'explanation' will show that
this argument trades on an ambiguity.

* This chapter is based in part on my paper by the same title, *American Philosophical Quarterly*,
14 (1977), 143–50, presented to the American Philosophical Association, Portland, March 1977,
with commentary by Kit Fine and Clark Glymour.

1.1. Truth and Grammar

It is necessary first of all to distinguish between the locutions 'we have an explanation' and 'this theory explains'. The former can be paraphrased 'we have a theory that explains'—but then 'have' needs to be understood in a special way. It does not mean, in this case, 'have on the books', or 'have formulated', but carries the conversational implicature that the theory tacitly referred to is acceptable. That is, you are not warranted in saying 'I have an explanation' unless you are warranted in the assertion 'I have a theory *which is acceptable* and which explains'. The important point is that the mere statement 'theory *T* explains fact *E*' does not carry any such implication: not that the theory is true, not that it is empirically adequate, and not that it is acceptable.

There are many examples, taken from actual usage, which show that truth is not presupposed by the assertion that a theory explains something. Lavoisier said of the phlogiston hypothesis that it is too vague and consequently 's'adapte à toutes les explications dans lesquelles on veut le faire entrer'.[1] Darwin explicitly allows explanations by false theories when he says 'It can hardly be supposed that a false theory would explain, in so satisfactory a manner as does the theory of natural selection, the several large classes of facts above specified.'[2] Gilbert Harman, we recall, has argued similarly: that a theory explains certain phenomena is part of the evidence that leads us to accept it. But that means that the explanation-relation is visible before we believe that the theory is true. Finally, we criticize theories selectively: a discussion of celestial mechanics around the turn of the century could surely contain the assertion that Newton's theory does explain many planetary phenomena. Yet it was also agreed that the advance in the perihelion of Mercury seems to be inconsistent with the theory, suggesting therefore that the theory is not empirically adequate—and hence, is false—without this agreement undermining the previous assertion. Examples can be multiplied: Newton's theory explained the tides, Huygens's theory explained the diffraction of light, Rutherford's theory of the atom explained the scattering of alpha particles, Bohr's theory explained the hydrogen spectrum, Lorentz's theory explained clock retardation. We are quite willing to say all this, although we will add that, for each of these theories, phenomena were discovered which they could not only not explain, but could not

---

[1] A. Lavoisier, *Œuvres* (Paris: Imp. Imperiale, 1862), ii. 640. I owe this and the other historical references below to my former student, Paul Thagard.

[2] C. Darwin, *On the Origin of the Species* (text of 6th edn., New York: Collier, 1962), 476.

even accommodate in the minimal fashion required for empirical adequacy.

Hence, to say that a theory explains some fact or other, is to assert a relationship between this theory and that fact, which is independent of the question of whether the real world, as a whole, fits that theory.

Let us relieve the tedium of terminological discussion for a moment and return to the argument displayed at the beginning. In view of the distinctions shown, we can try to revise it as follows: science tries to place us in a position in which we have explanations, and are warranted in saying that we do have. But to have such warrant, we must first be able to assert with equal warrant that the theories we use to provide premisses in our explanations are true. Hence science tries to place us in a position where we have theories which we are entitled to believe to be true.

The conclusion may be harmless of course if 'entitled' means here only that one cannot be convicted of irrationality on the basis of such a belief. That is compatible with the idea that we have warrant to believe a theory only because, and in so far as, we have warrant to believe that it is empirically adequate. In that case it is left open that one is at least as rational in believing merely that the theory is empirically adequate.

But even if the conclusion were construed in this harmless way, the second premiss will have to be disputed, for it entails that someone who merely accepts the theory as empirically adequate, is not in a position to explain. In this second premiss, the conviction is perhaps expressed that having an explanation is not to be equated with having an acceptable theory that explains, but with having a true theory that explains.

That conviction runs afoul of the examples I gave. I say that Newton could explain the tides, that he had an explanation of the tides, that he did explain the tides. In the same breath I can add that this theory is, after all, not correct. Hence I would be inconsistent if by the former I meant that Newton had a true theory which explained the tides—for if it was true then, it is true now. If what I meant was that it was true *then* to say that Newton had an acceptable theory which explains the tides, that would be correct.

A realist can of course give his own version: to have an explanation means to have 'on the books' a theory which explains and which one is entitled to believe to be true. If he does so, he will agree that to have an explanation does not require a true theory, while maintaining his contention that science aims to place us in a position to give *true* explanations. That would bring us back, I suppose, to our initial disagreement, the detour through explanation having brought no benefits. If you can only be entitled to assert that the theory is true because, and in so far as, you are entitled to assert that it is empirically adequate, then the distinction drawn makes no practical difference. There would of course be a difference between *believe* (to-be-true) and *accept* (believe-to-be-empirically-adequate) but no real difference between be-entitled-to-believe and be-entitled-to-accept. A realist might well dispute this by saying that if the theory explains facts then that gives you an *extra* good reason (over and above any evidence that it is empirically adequate) to believe that the theory is true. But I shall argue that this is quite impossible, since explanation is not a special additional feature that can give you good reasons for belief in addition to evidence that the theory fits the observable phenomena. For 'what more there is to' explanation is something quite pragmatic, related to the concerns of the user of the theory and not something new about the correspondence between theory and fact.

So I conclude that (*a*) the assertion that theory *T* explains, or provides an explanation for, fact *E* does not presuppose or imply that *T* is true or even empirically adequate, and (*b*) the assertion that we have an explanation is most simply construed as meaning that we have 'on the books' an acceptable theory which explains. I shall henceforth adopt this construal.

To round off the discussion of the terminology, let us clarify what sorts of terms can be the grammatical subjects, or grammatical objects, of the verb 'to explain'. Usage is not regimented: when we say 'There is the explanation!', we may be pointing to a fact, or to a theory, or to a thing. In addition, it is often possible to point to more than one thing which can be called 'the explanation'. And, finally, whereas one person may say that Newton's theory of gravitation explained the tides, another may say that Newton used that theory to explain the tides. (I suppose no one would say that the hammer drove the nail through the wood; only that the carpenter did so, using the hammer. But today people do sometimes say that the computer calculated the value of a function, or solved the equations, which is perhaps similar to saying that the theory explained the tides.)

This bewildering variety of modes of speech is common to scientists as well as philosophers and laymen. In Huygens and Young the typical phrasing seemed to be that phenomena may be explained *by means of* principles, laws, and hypotheses, or *according* to a view.[3] On the other hand, Fresnel writes to Arago in 1815 'tous ces phénomènes . . . sont réunis et expliqués par la même théorie des vibrations', and Lavoisier says that the oxygen hypothesis he proposes *explains* the phenomena of combustion.[4]

---

[3] Cf. Christiaan Huygens, *Treatise on Light*, trans. S. P. Thompson (New York: Dover, 1962), 19f., 22, 63; Thomas Young, *Miscellaneous Works*, ed. G. Peacock (London: John Murray, 1855), i. 168, 170.

[4] A. Fresnel, *Œuvres complètes* (Paris: Imp. Imperiale, 1866), i. 36 (see also 254, 355); Lavoisier, *Œuvres*, 233.

Darwin also speaks in the latter idiom: 'In scientific investigations it is permitted to invent any hypothesis, and if it explains various large and independent classes of facts it rises to the rank of a well-grounded theory'; though elsewhere he says that the facts of geographical distribution are *explicable on* the theory of migration.[5]

In other cases yet, the theory assumed is left tacit, and we just say that one fact explains another. For example: the fact that water is a chemical compound of oxygen and hydrogen explains why oxygen and hydrogen appear when an electric current is passed through (impure) water.

To put some order into this terminology, and in keeping with previous conclusions, we can regiment the language as follows. The word 'explain' can have its basic role in expressions of form 'fact E explains fact F relative to theory T'. The other expressions can then be parsed as: 'T explains F' is equivalent to: 'there are facts which explain F relative to T'; 'T was used to explain F' equivalent to 'it was shown that there are facts which explain F relative to T'; and so forth. Instead of 'relative to T' we can sometimes also say 'in'; for example, 'the gravitational pull of the moon explains the ebb and flow of the tides in Newton's theory'.

After this, my concern will no longer be with the derivative type of assertion that we *have* an explanation. After this point, the topic of concern will be that basic relation of explanation, which may be said to hold between facts relative to a theory, quite independently of whether the theory is true or false, believed, accepted, or totally rejected.

2.8. Why-Questions

Another approach to explanation was initiated by Sylvain Bromberger in his study of why-questions.[6] After all, a why-question is a request for explanation. Consider the question:

(1) Why did the conductor become warped during the short-circuit?

This has the general form

(2) Why (is it the case that) P?

where P is a statement. So we can think of 'Why' as a function that turns statements into questions.

5 C. Darwin, *The Variations of Animals and Plants* (London: John Murray, 1868), i. 9; *On the Origin of the Species* (Facs. of the first edn., Cambridge, Mass.: Harvard, 1964), 408.

6 S. Bromberger, 'Why-Questions', in R. G. Colodny (ed.), *Mind and Cosmos* (Pittsburgh: University of Pittsburgh Press, 1966), 86–108.

Question 1 *arises*, or *is in order*, only if the conductor did indeed become warped then. If that is not so, we do not try to answer the question, but say something like: 'You are under a false impression, the conductor became warped much earlier,' or whatever. Hence Bromberger calls the statement that *P* the *presupposition* of the question '*Why P?*' One form of the rejection of explanation requests is clearly the denial of the presupposition of the corresponding why-question.

I will not discuss Bromberger's theory further here, but turn instead to a criticism of it. The following point about why-questions has been made in recent literature by Alan Garfinkel and Jon Dorling, but I think it was first made, and discussed in detail, in unpublished work by Bengt Hannson circulated in 1974.[7] Consider the question

(3) Why did Adam eat the apple?

This same question can be construed in various ways, as is shown by the variants:

(3*a*) Why was it Adam who ate the apple?

(3*b*) Why was it the apple Adam ate?

(3*c*) Why did Adam *eat* the apple?

In each case, the canonical form prescribed by Bromberger (as in (2) above) would be the same, namely

(4) Why (is it the case that) (Adam ate the apple)?

yet there are three different explanation requests here.

The difference between these various requests is that they point to different contrasting alternatives. For example, (3*b*) may ask why Adam ate *the apple* rather than some other fruit in the garden, while (3*c*) asks perhaps why Adam *ate* the apple rather than give it back to Eve untouched. So to (3*b*), 'because he was hungry' is not a good answer, whereas to (3*c*) it is. The correct general, underlying structure of a why-question is therefore

(5) Why (is it the case that) *P in contrast to* (other members of) *X*?

where *X*, the *contrast-class*, is a set of alternatives. *P* may belong to *X* or not; further examples are:

Why did the sample burn green (rather than some other colour)?

---

[7] 'Explanations-of-What?', mimeographed and circulated, Stanford University, 1974. The idea was independently developed, by Jon Dorling in a paper circulated in 1976, and by Alan Garfinkel in *Forms of Explanation* (New Haven, Conn., and London: Yale University Press, 1981). I wish to express my debt to Bengt Hannson for discussion and correspondence in the autumn of 1975 which clarified these issues considerably for me.

Why did the water and copper reach equilibrium temperature 22.5°C (rather than some other temperature)?

In these cases the contrast-classes (colours, temperatures) are 'obvious'. In general, the contrast-class is not explicitly described because, *in context*, it is clear to all discussants what the intended alternatives are.

This observation explains the tension we feel in the paresis example. If a mother asks why her eldest son, a pillar of the community, mayor of his town, and best beloved of all her sons, has this dread disease, we answer: because he had latent untreated syphilis. But if that question is asked about this same person, immediately after a discussion of the fact that everyone in his country club has a history of untreated syphilis, *there is no answer*. The reason for the difference is that in the first case the contrast-class is the mother's sons, and in the second, the members of the country club, contracting paresis. Clearly, an answer to a question of form (5) must adduce information that *favours P in contrast to* other members of X. Sometimes the availability of such information depends strongly on the choice of X.

These reflections have great intuitive force. The distinction made is clearly crucial to the paresis example and explains the sense of ambiguity and tension felt in earlier discussion of such examples. It also gives us the right way to explicate such assertions as: individual events are never explained, we only explain a particular event *qua* event of a certain kind. (We can explain *this* decay of a uranium atom *qua* decay of a uranium atom, but not *qua* decay of a uranium atom at *this* time.)

But the explication of what it is for an answer to favour one alternative over another proves difficult. Hannson proposed: answer A is a good answer to (Why P in contrast to X?) exactly if the probability of P given A is higher than the average probability of members of X given A. But this proposal runs into most of the old difficulties. Recall Salmon's examples of irrelevancy: the probability of recovery from a cold *given* administration of vitamin C is nearly one, while the probability of not recovering *given* the vitamins is nearly zero. So by Hannson's criterion it would be a good answer—even if taking vitamin C has no effect on recovery from colds one way or the other.

Also, the asymmetries are as worrisome as ever. By Hannson's criterion, the length of the shadow automatically provides a good explanation of the height of the flagpole. And 'Because the barometer fell' is a good answer to 'Why is there a storm?' (upon selection of the 'obvious' contrast-classes, of course). Thus it seems that reflection on the contrast-class serves to solve some of our problems, but not all.

2.9. The Clues Elaborated

The discussions of causality and of why-questions seem to me to provide essential clues to the correct account of explanation. In the former we found that an explanation often consists in listing salient factors, which point to a complete story of how the event happened. The effect of this is to eliminate various alternative hypotheses about how this event did come about, and/or eliminate puzzlement concerning how the event could have come about. But salience is context-dependent, and the selection of the correct 'important' factor depends on the range of alternatives contemplated in that context. In N. R. Hanson's example, the barrister wants this sort of weeding out of hypotheses about the death relevant to the question of legal accountability; the carriage-builder, a weeding out of hypotheses about structural defects or structural limitations under various sorts of strain. *The context*, in other words, *determines relevance* in a way that goes well beyond the statistical relevance about which our scientific theories give information.

This might not be important if we were not concerned to find out exactly how having an explanation goes beyond merely having an acceptable theory about the domain of phenomena in question. But that is exactly the topic of our concern.

In the discussion of why-questions, we have discovered a further contextually determined factor. The range of hypotheses about the event which the explanation must 'weed out' or 'cut down' is not determined solely by the interests of the discussants (legal, mechanical, medical) but also by a range of contrasting alternatives to the event. This *contrast-class* is also determined by context.

It might be thought that when we request a *scientific* explanation, the relevance of possible hypotheses, and also the contrast-class are automatically determined. But this is not so, for both the physician and the motor mechanic are asked for a scientific explanation. The physician explains the fatality *qua* death of a human organism, and the mechanic explains it *qua* automobile crash fatality. To ask that their explanations be scientific is only to demand that they rely on scientific theories and experimentation, not on old wives' tales. Since any explanation of an individual event must be an explanation of that event *qua* instance of a certain kind of event, nothing more can be asked.

The two clues must be put together. The description of some account as an explanation of a given fact or event, is incomplete. It can only be an explanation with respect to a certain *relevance relation* and a certain *contrast-class.* These are contextual factors, in that they are determined neither by the totality of accepted scientific theories, nor by the event or fact for which an explanation is requested. It is sometimes said that an Omniscient Being would have a complete explanation, whereas these contextual factors only bespeak our limitations due to which we can only grasp one part or aspect of the complete explanation at any given time. But this is a mistake. If the Omniscient Being has no specific interests (legal, medical, economic; or just an interest in optics or thermodynamics rather than chemistry) and does not abstract (so that he never thinks of Caesar's death *qua* multiple stabbing, or *qua* assassination), then no why-questions ever arise for him in any way at all—and he does not have any explanation in the sense that we have explanations. If he does have interests, and does abstract from individual peculiarities in his thinking about the world, then his why-questions are as essentially context-dependent as ours. In either case, his advantage is that he always has all the information needed to answer any specific explanation request. But that information is, in and by itself, not an explanation; just as a person cannot be said to be older, or a neighbour, except in relation to others.

3. ASYMMETRIES OF EXPLANATION: A SHORT STORY

3.1. Asymmetry and Context: the Aristotelian Sieve

That vexing problem about paresis, where we seem both to have and not to have an explanation, was solved by reflection on the contextually supplied contrast-class. The equally vexing, and much older, problem of the asymmetries of explanation, is illuminated by reflection on the other main contextual factor: contextual relevance.

If that is correct, if the asymmetries of explanation result from a contextually determined relation of relevance, then it must be the case that these asymmetries can at least sometimes be reversed by a change in context. In addition, it should then also be possible to account for specific asymmetries in terms of the interests of questioner and audience that determine this relevance. These considerations provide a crucial test for the account of explanation which I propose.

Fortunately, there is a precedent for this sort of account of the asymmetries, namely in Aristotle's theory of science. It is traditional to understand this part of his theory in relation to his metaphysics; but I

maintain that the central aspects of his solution to the problem of asymmetry of explanations are independently usable.[8]

Aristotle gave examples of this problem in the *Posterior Analytics* 1.
13; and he developed a typology of explanatory factors ('the four
causes'). The solution is then simply this. Suppose there is a definite
number (e.g. four) of types of explanatory factors (i.e. of relevance
relations for why-questions). Suppose also that relative to our background information and accepted theories, the propositions *A* and *B* are
equivalent. It may then still be that these two propositions describe
factors of different types. Suppose that in a certain context, our interest
is in the mode of production of an event, and 'Because *B*' is an acceptable
answer to 'Why *A*?' Then it may well be that *A* does not describe any
mode of production of anything, so that, *in this same context*, 'Because
A' would not be an acceptable answer to 'Why *B*?'

Aristotle's lantern example (*Posterior Analytics* 2. 11) shows that he
recognized that in different contexts, verbally the same why-question
may be a request for different types of explanatory factors. In modern
dress the example would run as follows. Suppose a father asks his teenage
son, 'Why is the porch light on?' and the son replies 'The porch switch
is closed and the electricity is reaching the bulb through that switch.' At
this point you are most likely to feel that the son is being impudent. This
is because you are most likely to think that the sort of answer the father
needed was something like: 'Because we are expecting company.' But it
is easy to imagine a less likely question context: the father and son are
rewiring the house and the father, unexpectedly seeing the porch light
on, fears that he has caused a short-circuit that bypasses the porch light
switch. In the second case, he is *not* interested in the human expectations
or desires that led to the depressing of the switch.

Aristotle's fourfold typology of causes is probably an over-simplifica-
tion of the variety of interests that can determine the selection of a range
of relevant factors for a why-question. But in my opinion, appeal to some
such typology will successfully illuminate the asymmetries (and also the
rejections, since no factor of a *particular* type may lead to a telling
answer to the why-question). If that is so then, as I said before, asymmetries must be at least sometimes reversible through a change in context.
The story which follows is meant to illustrate this. As in the lantern (or
porch light) example, the relevance changes from one sort of efficient
cause to another, the second being a person's desires. As in all explana-

---

[8] For a fuller account of Aristotle's solution of the asymmetries, see my 'A Re-examination of
Aristotle's Philosophy of Science', *Dialogue*, 1980. The story was written in reply to searching
questions and comments by Professor J. J. C. Smart, and circulated in Nov. 1976.

tions, the correct answer consists in the exhibition of a single factor in the causal net, which is made salient in that context by factors not overtly appearing in the words of the question.

3.2. 'The Tower and the Shadow'

During my travels along the Saône and Rhône last year, I spent a day and night at the ancestral home of the Chevalier de St X, an old friend of my father's. The Chevalier had in fact been the French liaison officer attached to my father's brigade in the First War, which had—if their reminiscences are to be trusted—played a not insignificant part in the battles of the Somme and Marne.

The old gentleman always had *thé à l'Anglaise* on the terrace at five o'clock in the evening, he told me. It was at this meal that a strange incident occurred; though its ramifications were of course not yet perceptible when I heard the Chevalier give his simple explanation of the length of the shadow which encroached upon us there on the terrace. I had just eaten my fifth piece of bread and butter and had begun my third cup of tea when I chanced to look up. In the dying light of that late afternoon, his profile was sharply etched against the granite background of the wall behind him, the great aquiline nose thrust forward and his eyes fixed on some point behind my left shoulder. Not understanding the situation at first, I must admit that to begin with, I was merely fascinated by the sight of that great hooked nose, recalling my father's claim that this had once served as an effective weapon in close combat with a German grenadier. But I was roused from this brown study by the Chevalier's voice.

'The shadow of the tower will soon reach us, and the terrace will turn chilly. I suggest we finish our tea and go inside.'

I looked around, and the shadow of the rather curious tower I had earlier noticed in the grounds, had indeed approached to within a yard from my chair. The news rather displeased me, for it was a fine evening; I wished to remonstrate but did not well know how, without overstepping the bounds of hospitality. I exclaimed,

'Why must that tower have such a long shadow? This terrace is so pleasant!'

His eyes turned to rest on me. My question had been rhetorical, but he did not take it so.

'As you may already know, one of my ancestors mounted the scaffold with Louis XVI and Marie Antoinette. I had that tower erected in 1930 to mark the exact spot where it is said that he greeted the Queen when she first visited this house, and presented her with a peacock made of soap, then a rare substance. Since the Queen would have been one hundred and seventy-five old in 1930, had she lived, I had the tower made exactly that many feet high.'

It took me a moment to see the relevance of all this. Never quick at sums, I was at first merely puzzled as to why the measurement should have been in feet; but of course I already knew him for an Anglophile. He added drily, 'The sun not being alterable in its course, light travelling in straight lines, and the laws of trigonometry being immutable, you will perceive that the length of the shadow is determined by the height of the tower.' We rose and went inside.

I was still reading at eleven that evening when there was a knock at my door. Opening it I found the housemaid, dressed in a somewhat old-fashioned black dress and white cap, whom I had perceived hovering in the background on several occasions that day. Courtseying prettily, she asked, 'Would the gentleman like to have his bed turned down for the night?'

I stepped aside, not wishing to refuse, but remarked that it was very late—was she kept on duty to such hours? No, indeed, she answered, as she deftly turned my bed covers, but it had occurred to her that some duties might be pleasures as well. In such and similar philosophical reflections we spent a few pleasant hours together, until eventually I mentioned casually how silly it seemed to me that the tower's shadow ruined the terrace for a prolonged, leisurely tea.

At this, her brow clouded. She sat up sharply. 'What exactly did he tell you about this?' I replied lightly, repeating the story about Marie Antoinette, which now sounded a bit far-fetched even to my credulous ears.

'The *servants* have a different account', she said with a sneer that was not at all becoming, it seemed to me, on such a young and pretty face. 'The truth is quite different, and has nothing to do with ancestors. That tower marks the spot where he killed the maid with whom he had been in love to the point of madness. And the height of the tower? He vowed that shadow would cover the terrace where he first proclaimed his love, with every setting sun—that is why the tower had to be so high.'

I took this in but slowly. It is never easy to assimilate unexpected truths about people we think we know—and I have had occasion to notice this again and again.

'Why did he kill her?' I asked finally.

'Because, sir, she dallied with an English brigadier, an overnight guest in this house.' With these words she arose, collected her bodice and cap, and faded through the wall beside the doorway.

I left early the next morning, making my excuses as well as I could.

4. A MODEL FOR EXPLANATION

I shall now propose a new theory of explanation. An explanation is not the same as a proposition, or an argument, or list of propositions; it is an *answer*. (Analogously, a son is not the same as a man, even if all sons are men, and every man is a son.) An explanation is an answer to a why-question. So, a theory of explanation must be a theory of why-questions.

To develop this theory, whose elements can all be gleaned, more or less directly, from the preceeding discussion, I must first say more about some topics in formal pragmatics (which deals with context-dependence) and in the logic of questions. Both have only recently become active areas in logical research, but there is general agreement on the basic aspects to which I limit the discussion.

4.1. Contexts and Propositions [9]

Logicians have been constructing a series of models of our language, of increasing complexity and sophistication. The phenomena they aim to save are the surface grammar of our assertions and the inference patterns detectable in our arguments. (The distinction between logic and theoretical linguistics is becoming vague, though logicians' interests focus on special parts of our language, and require a less faithful fit to surface grammar, their interests remaining in any case highly theoretical.) Theoretical entities introduced by logicians in their models of language (also called 'formal languages') include domains of discourse ('universes'), possible worlds, accessibility ('relative possibility') relations, facts, and propositions, truth-values, and, lately, contexts. As might be guessed, I take it to be part of empiricism to insist that the adequacy of these models does not require all their elements to have counterparts in reality. They will be good if they fit those phenomena to be saved.

Elementary logic courses introduce one to the simplest models, the languages of sentential and quantificational logic which, being the simplest, are of course the most clearly inadequate. Most logic teachers being somewhat defensive about this, many logic students, and other philo-

---

[9] At the end of my 'The Only Necessity is Verbal Necessity', *Journal of Philosophy*, 74 (1977), 71–85 (itself an application of formal pragmatics to a philosophical problem), there is a short account of the development of these ideas, and references to the literature. The paper 'Demonstratives' by David Kaplan which was mentioned there as forthcoming, was completed and circulated in mimeo'd form in the spring of 1977; it is at present the most important source for the concepts and applications of formal pragmatics, although some aspects of the form in which he develops this theory are still controversial.

sophers, have come away with the impression that the over-simplifications make the subject useless. Others, impressed with such uses as elementary logic does have (in elucidating classical mathematics, for example), conclude that we shall not understand natural language until we have seen how it can be regimented so as to fit that simple model of horseshoes and truth tables.

In elementary logic, each sentence corresponds to exactly one proposition, and the truth-value of that sentence depends on whether the proposition in question is true in the actual world. This is also true of such extensions of elementary logic as free logic (in which not all terms need have an actual referent), and normal modal logic (in which non-truth functional connectives appear), and indeed of almost all the logics studied until quite recently.

But, of course, sentences in natural language are typically context-dependent; that is, which proposition a given sentence expresses will vary with the context and occasion of use. This point was made early on by Strawson, and examples are many:

> 'I am happy now' is true in context x exactly if the speaker in context x is happy at the time of context x

where a context of use is an actual occasion, which happened at a definite time and place, and in which are identified the speaker (referent of 'I'), addressee (referent of 'you'), person discussed (referent of 'he'), and so on. That contexts so conceived are idealizations from real contexts is obvious, but the degree of idealization may be decreased in various ways, depending on one's purposes of study, at the cost of greater complexity in the model constructed.

What must the context specify? The answer depends on the sentence being analysed. If that sentence is

> Twenty years ago it was still possible to prevent the threatened population explosion in that country, but now it is too late

the model will contain a number of factors. First, there is a set of possible worlds, and a set of contexts, with a specification for each context of the world of which it is a part. Then there will be for each world a set of entities that exist in that world, and also various relations of relative possibility among these worlds. In addition there is time, and each context must have a time of occurrence. When we evaluate the above sentence we do so relative to a context and a world. Varying with the context will be the referents of 'that country' and 'now', and perhaps also the relative possibility relation used to interpret 'possible', since the speaker may have intended one of several senses of possibility.

This sort of interpretation of a sentence can be put in a simple general form. We first identify certain entities (mathematical constructs) called propositions, each of which has a truth-value in each possible world. Then we give the context as main task the job of selecting, for each sentence, the proposition it expresses 'in that context'. Assume as simplification that when a sentence contains no indexical terms (like 'I', 'that', 'here', etc.), then all contexts select the same proposition for it. This gives us an easy intuitive handle on what is going on. If A is a sentence in which no indexical terms occur, let us designate as $|A|$ the proposition which it expresses in every context. Then we can generally (though not necessarily always) identify the proposition expressed by any sentence in a given context as the proposition expressed by some index-ical-free sentence. For example:

> In context x, 'Twenty years ago it was still possible to prevent the population explosion in that country' expresses the proposition 'In 1958, it is (tenseless) possible to prevent the population explosion in India'

To give another example, in the context of my present writing, 'I am here now' expresses the proposition that Bas van Fraassen is in Vancouver, in July 1978.

This approach has thrown light on some delicate conceptual issues in philosophy of language. Note for example that 'I am here' is a sentence which is true no matter what the facts are and no matter what the world is like, and no matter what context of usage we consider. Its truth is ascertainable *a priori*. But the proposition expressed, that van Fraassen is in Vancouver (or whatever else it is) is not at all a necessary one: I might not have been here. Hence, a clear distinction between *a priori* ascertainability and necessity appears.

The context will generally select the proposition expressed by a given sentence A via a selection of referents for the terms, extensions for the predicates, and functions for the functors (i.e. syncategorematic words such as 'and' or 'most'). But intervening contextual variables may occur at any point in these selections. Among such variables there will be the assumptions taken for granted, theories accepted, world-pictures or para-digms adhered to, in that context. A simple example would be the range of conceivable worlds admitted as possible by the speaker; this variable plays a role in determining the truth-value of his modal statements in that context, relative to the 'pragmatic presuppositions'. For example, if the actual world is really the only possible world there is (which exists) then the truth-values of modal statements in that context but *tout court* will be very different from their truth-values relative to those pragmatic pre-suppositions—and only the latter will play a significant role in our under-standing of what is being said or argued in that context.

Since such a central role is played by propositions, the family of propositions has to have a fairly complex structure. Here a simplifying hypothesis enters the fray: propositions can be uniquely identified through the worlds in which they are true. This simplifies the model considerably, for it allows us to identify a proposition with a set of possible worlds, namely, the set of worlds in which it is true. It allows the family of propositions to be a complex structure, admitting of inter-esting operations, while keeping the structure of each individual prop-osition very simple.

Such simplicity has a cost. Only if the phenomena are simple enough, will simple models fit them. And sometimes, to keep one part of a model simple, we have to complicate another part. In a number of areas in philosophical logic it has already been proposed to discard that simpli-fying hypothesis, and to give propositions more 'internal structure'. As will be seen below, problems in the logic of explanation provide further reasons for doing so.

4.2. Questions

We must now look further into the general logic of questions. There are of course a number of approaches; I shall mainly follow that of Nuel Belnap, though without committing myself to the details of his theory.[10]

A theory of questions must needs be based on a theory of propositions, which I shall assume given. A *question* is an abstract entity; it is ex-pressed by an *interrogative* (a piece of language) in the same sense that a proposition is expressed by a declarative sentence. Almost anything can be an appropriate response to a question, in one situation or another; as 'Peccavi' was the reply telegraphed by a British commander in India to the question as to how the battle was going (he had been sent to attack the province of Sind).[11] But not every response is, properly speaking, an answer. Of course, there are degrees; and one response may be more or less of an answer than another. The first task of a theory of questions is

[10] Belnap's theory was first presented in *An analysis of questions: preliminary report* (Santa Monica, Cal.: System Development Corporation, technical memorandum 7–1287–1000/00, 1963), and is now more accessible in N. D. Belnap Jun. and J. B. Steel, Jun., *The Logic of Questions and Answers* (New Haven, Conn.: Yale University Press, 1976).

[11] I heard the example from my former student Gerald Charlwood. Ian Hacking and J. J. C. Smart told me that the officer was Sir Charles Napier.

to provide some typology of answers. As an example, consider the following question, and a series of responses:

Can you get to Victoria both by ferry and by plane?

(a) Yes

(b) You can get to Victoria both by ferry and by plane

(c) You can get to Victoria by ferry

(d) You can get to Victoria both by ferry and by plane, but the ferry ride is not to be missed

(e) You can certainly get to Victoria by ferry, and that is something not to be missed

Here (b) is the 'purest' example of an answer: it gives enough information to answer the question completely, but no more. Hence it is called a *direct answer*. The word 'Yes' (a) is a *code* for this answer.

Responses (c) and (d) depart from that direct answer in opposite directions: (c) says properly less than (b)—it is implied by (b)—while (d), which implies (b), says more. Any proposition implied by a direct answer is called a *partial answer* and one which implies a direct answer is a *complete answer*. We must resist the temptation to say that therefore an answer, *tout court*, is any combination of a partial answer with further information, for in that case, every proposition would be an answer to any question. So let us leave (e) unclassified for now, while noting it is still 'more of an answer' than such responses as 'Gorilla!' (which is a response given to various questions in the film *Ich bin ein Elephant, Madam*, and hence, I suppose, still more of an answer than some). There may be some quantitative notion in the background (a measure of the extent to which a response really 'bears on' the question) or at least a much more complete typology (some more of it is given below), so it is probably better not to try and define the general term 'answer' too soon.

The basic notion so far is that of direct answer. In 1958, C. L. Hamblin introduced the thesis that a question is uniquely identifiable through its answers.[12] This can be regarded as a simplifying hypothesis of the sort we came across for propositions, for it would allow us to identify a question with the set of its direct answers. Note that this does not preclude a good deal of complexity in the determination of exactly what question is expressed by a given interrogative. Also, the hypothesis does not identify the question with the disjunction of its direct answers. If that were done, the clearly distinct questions

12 C. L. Hamblin, 'Questions', *Australasian Journal of Philosophy*, 36 (1958), 159–68.

Is the cat on the mat?
>*Direct answers*: The cat is on the mat
>>The cat is not on the mat

Is the theory of relativity true?
>*Direct answers*: The theory of relativity is true
>>The theory of relativity is not true

would be the same (identified with the tautology) if the logic of propositions adopted were classical logic. Although this simplifying hypothesis
is therefore not to be rejected immediately, and has in fact guided much
of the research on questions, it is still advisable to remain somewhat
tentative towards it.

Meanwhile we can still use the notion of direct answer to define some
basic concepts. One question Q may be said to *contain* another, Q', if
Q' is answered as soon as Q is—that is, every complete answer to Q is
also a complete answer to Q'. A question is *empty* if all its direct answers
are necessarily true, and *foolish* if none of them is even possibly true. A
special case is the *dumb* question, which has no direct answers. Here are
examples:

(1) Did you wear the black hat yesterday or did you wear the white
one?

(2) Did you wear a hat which is both black and not black, or did you
wear one which is both white and not white?

(3) What are three distinct examples of primes among the following
numbers: 3, 5?

Clearly (3) is dumb and (2) is foolish. If we correspondingly call a
necessarily false statement foolish too, we obtain the theorem *Ask a
foolish question and get a foolish answer*. This was first proved by
Belnap, but attributed by him to an early Indian philosopher mentioned
in Plutarch's *Lives* who had the additional distinction of being an early
nudist. Note that a foolish question contains all questions, and an empty
one is contained in all.

Example (1) is there partly to introduce the question form used in (2),
but also partly to introduce the most important semantic concept after
that of direct answer, namely presupposition. It is easy to see that the
two direct answers to (1) ('I wore the black hat', 'I wore the white one')
could both be false. If that were so, the respondent would presumably
say 'Neither', which is an answer not yet captured by our typology.
Following Belnap who clarified this subject completely, let us introduce
the relevant concepts as follows:

A *presupposition*[13] of question Q is any proposition which is implied by all direct answers to Q

A *correction* (or *corrective answer*) to Q is any denial of any presupposition of Q

The (*basic*) *presupposition* of Q is the proposition which is true if and only if some direct answer to Q is true

In this last notion, I presuppose the simplifying hypothesis which identifies a proposition through the set of worlds in which it is true; if that hypothesis is rejected, a more complex definition needs to be given. For example (1), 'the' presupposition is clearly the proposition that the addressee wore either the black hat or the white one. Indeed, in any case in which the number of direct answers is finite, 'the' presupposition is the disjunction of those answers.

Let us return momentarily to the typology of answers. One important family is that of the partial answers (which includes direct and complete answers). A second important family is that of the corrective answer. But there are still more. Suppose the addressee of question (1) answers 'I did not wear the white one.' This is not even a partial answer, by the definition given: neither direct answer implies it, since she might have worn both hats yesterday, one in the afternoon and one in the evening, say. However, since the questioner is presupposing that she wore at least one of the two, the response is *to him* a complete answer. For the response plus the presupposition together entail the direct answer that she wore the black hat. Let us therefore add:

A *relatively complete answer* to Q is any proposition which, together with the presupposition of Q, implies some direct answer to Q

We can generalize this still further: a complete answer to Q, relative to theory T, is something which together with T, implies some direct answer to Q—and so forth. The important point is, I think, that we should regard the introduced typology of answers as open-ended, to be extended as needs be when specific sorts of questions are studied.

Finally, which question is expressed by a given interrogative? This is highly context-dependent, in part because all the usual indexical terms appear in interrogatives. If I say, 'Which one do you want?' the context determines a range of objects over which my 'which one' ranges—for example, the set of apples in the basket on my arm. If we adopt the

---

[13] The defining clause is equivalent to 'any proposition which is true if any direct answer to Q is true'. This includes, of course, propositions which would normally be expressed by means of 'metalinguistic' sentences—a distinction which, being language-relative, is unimportant.

simplifying hypothesis discussed above, then the main task of the context is to delineate the set of direct answers. In the 'elementary questions' of Belnap's theory ('whether-questions' and 'which-questions') this set of direct answers is specified through two factors: a *set of alternatives* (called the *subject* of the question) and *request* for a selection among these alternatives and, possibly, for certain information about the selection made ('distinctness and completeness claims'). What those two factors are may not be made explicit in the words used to frame the interrogative, but the context has to determine them exactly if it is to yield an interpretation of those words as expressing a unique question.

4.3. A Theory of Why-Questions

There are several respects in which why-questions introduce genuinely new elements into the theory of questions.[14] Let us focus first on the determination of exactly what question is asked, that is, the contextual specification of factors needed to understand a why-interrogative. After that is done (a task which ends with the delineation of the set of direct answers) and as an independent enterprise, we must turn to the evaluation of those answers as good or better. This evaluation proceeds with reference to the part of science accepted as 'background theory' in that context.

As example, consider the question 'Why is this conductor warped?' The questioner implies that the conductor is warped, and is asking for a reason. Let us call the proposition that the conductor is warped the *topic* of the question (following Henry Leonard's terminology, 'topic of concern'). Next, this question has a *contrast-class*, as we saw, that is, a set of alternatives. I shall take this contrast-class, call it X, to be a class of propositions which includes the topic. For this particular interrogative, the contrast could be that it is *this* conductor rather than *that* one, or that this conductor has warped rather than retained its shape. If the question is 'Why does this material burn yellow' the contrast-class could be the set of propositions: this material burned (with a flame of) colour x.

Finally, there is the respect-in-which a reason is requested, which determines what shall count as a possible explanatory factor, the relation

---

[14] In Belnap and Steel, *Logic of Questions and Answers*, Bromberger's theory of why-questions is cast in the general form common to elementary questions. I think that Bromberger arrived at his concept of 'abnormic law' (and the form of answer exhibited by ' "Grünbaum" is spelled with an umlaut because it is an English word borrowed from German, and no English words are spelled with an umlaut except those borrowed from another language in which they are so spelled'), because he ignored the tacit *rather than* (contrast-class) in why-interrogatives, and then had to make up for this deficiency in the account of what the answers are like.

of *explanatory relevance*. In the first example, the request might be *for events 'leading up to' the warping*. That allows as relevant an account of human error, of switches being closed or moisture condensing in those switches, even spells cast by witches (since the evaluation of what is a good answer comes later). On the other hand, the events leading up to the warping might be well known, in which case the request is likely to be for the standing conditions that made it possible for those events to lead to this warping: the presence of a magnetic field of a certain strength, say. Finally, it might already be known, or considered immaterial, exactly how the warping is produced, and the question (possibly based on a misunderstanding) may be about exactly what function this warping fulfils in the operation of the power station. Compare 'Why does the blood circulate through the body?' answered (1) 'Because the heart pumps the blood through the arteries,' and (2) 'To bring oxygen to every part of the body tissue.'

In a given context, several questions agreeing in topic but differing in contrast-class, or conversely, may differ further in what counts as explanatorily relevant. Hence we cannot properly ask what is relevant to this topic, or what is relevant to this contrast-class. Instead we must say of a given proposition that it is or is not relevant (in this context) to the topic with respect to that contrast-class. For example, in the same context one might be curious about the circumstances that led Adam to eat the apple rather than the pear (Eve offered him an apple) and also about the motives that led him to eat it rather than refuse it. What is 'kept constant' or 'taken as given' (that he ate the fruit; that what he did, he did to the apple) which is to say, the contrast-class, is not to be dissociated entirely from the respect-in-which we want a reason.

Summing up then, the why-question Q expressed by an interrogative in a given context will be determined by three factors:

The *topic* P_k
The *contrast-class* $X = \{P_1, \ldots, P_k, \ldots\}$
The *relevance relation* R

and, in a preliminary way, we may identify the abstract why-question with the triple consisting of these three:

$$Q = \langle P_k, X, R \rangle$$

A proposition A is called *relevant to* Q exactly if A bears relation R to the couple $\langle P_k, X \rangle$.

We must now define what are the direct answers to this question. As a beginning let us inspect the form of words that will express such an answer:

P_k *in contrast to* (the rest of) X *because A*   (*)

This sentence must express a proposition. What proposition it expresses, however, depends on the same context that selected Q as the proposition expressed by the corresponding interrogative ('Why P_k?'). So some of the same contextual factors, and specifically R, may appear in the determination of the proposition expressed by (*).

What is claimed in answer (*)? First of all, that P_k is true. Secondly, (*) claims that the other members of the contrast-class are not true. So much is surely conveyed already by the question—it does not make sense to ask why Peter rather than Paul has paresis if they both have it. Thirdly, (*) says that A is true. And finally, there is that word 'because': (*) claims that A is a *reason*.

This fourth point we have awaited with bated breath. Is this not where the inextricably modal or counterfactual element comes in? But not at all; in my opinion, the word 'because' here signifies only that A is relevant, in this context, to this question. Hence the claim is merely that A bears relation R to $\langle P_k, X \rangle$. For example, suppose you ask why I got up at seven o'clock this morning, and I say 'Because I was woken up by the clatter the milkman made.' In that case I have interpreted your question as asking for a sort of reason that at least includes events-leading-up-to my getting out of bed, and my word 'because' indicates that the milkman's clatter was that sort of reason, that is, one of the events in what Salmon would call the causal process. Contrast this with the case in which I construe your request as being specifically for a motive. In that case I would have answered 'No reason, really. I could easily have stayed in bed, for I don't particularly want to do anything today. But the milkman's clatter had woken me up, and I just got up from force of habit I suppose.' In this case, I do not say 'because' for the milkman's clatter does not belong to the relevant range of events, as I understand your question.

It may be objected that 'because A' does not only indicate that A is *a* reason, but that it is *the* reason, or at least that it is a good reason. I think that this point can be accommodated in two ways. The first is that the relevance relation, which specifies what sort of thing is being requested as answer, may be construed quite strongly: 'give me a motive strong enough to account for murder', 'give me a statistically relevant preceding event not screened off by other events', 'give me a common cause', etc. In that case the claim that the proposition expressed by A falls in the relevant range, is already a claim that it provides a telling reason. But more likely, I think, the request need not be construed that strongly; the point is rather that anyone who answers a question is in some sense tacitly claiming to be giving a good answer. In either case, the determination of whether the answer is indeed good, or telling, or better than other answers that might have been given, must still be carried out, and I shall discuss that under the heading of 'evaluation'.

As a matter of regimentation I propose that we count (*) as a direct answer *only if A* is relevant.[15] In that case, we do not have to understand the claim that *A* is relevant as explicit part of the answer either, but may regard the word 'because' solely as a linguistic signal that the words uttered are intended to provide an answer to the why-question just asked. (There is, as always, the tacit claim of the respondent that what he is giving is a good, and hence a relevant answer—we just do not need to make this claim part of the answer.) The definition is then:

B is a *direct answer* to question $Q = \langle P_k, X, R \rangle$ exactly if there is some proposition *A* such that *A* bears relation *R* to $\langle P_k, X \rangle$ and *B* is the proposition which is true exactly if (P_k; *and* for all $i \neq k$, *not* P_i; and *A*) is true

where, as before, $X = \{P_1, \ldots, P_k, \ldots\}$. Given this proposed definition of the direct answer, what does a why-question presuppose? Using Belnap's general definition we deduce:

A why-question *presupposes* exactly that
 (*a*) its topic is true
 (*b*) in its contrast-class, only its topic is true
 (*c*) at least one of the propositions that bears its relevance relation to its topic and contrast-class, is also true

However, as we shall see, if all three of these presuppositions are true, the question may still not have a *telling* answer.

Before turning to the evaluation of answers, however, we must consider one related topic: when does a why-question arise? In the general theory of questions, the following were equated: question *Q* arises, all the presuppositions of *Q* are true. The former means that *Q* is not to be rejected as mistaken, the latter that *Q* has some true answer.

In the case of why-questions, we evaluate answers in the light of accepted background theory (as well as background information) and it seems to me that this drives a wedge between the two concepts. Of course, sometimes we reject a why-question because we think that it has no true answer. But as long as we do not think that, the question does arise, and is not mistaken, regardless of what is true.

---

[15] I call this a matter of regimentation, because the theory could clearly be developed differently at this point, by building the claim of relevance into the answer as an explicit conjunct. The result would be an alternative theory of why-questions which, I think, would equally save the phenomena of explanation or why-question asking and answering.

To make this precise, and to simplify further discussion, let us introduce two more special terms. In the above definition of 'direct answer', let us call proposition A *the core* of answer B (since the answer can be abbreviated to '*Because A*'), and let us call the proposition that (P_k *and for all* $i \neq k$, *not* P_i) the *central presupposition* of question Q. Finally, if proposition A is relevant to $\langle P_k, X \rangle$ let us also call it relevant to Q.

In the context in which the question is posed, there is a certain body K of accepted background theory and factual information. This is a factor in the context, since it depends on who the questioner and audience are. It is this background which determines whether or not the question arises; hence a question may arise (or conversely, be rightly rejected) in one context and not in another.

To begin, whether or not the question genuinely *arises*, depends on whether or not K implies the central presupposition. As long as the central presupposition is not part of what is assumed or agreed to in this context, the why-question does not arise at all.

Secondly, Q presupposes *in addition* that one of the propositions A, relevant to its topic and contrast-class, is true. Perhaps K does not imply that. In this case, the question will still arise, provided K does not imply that all those propositions are false.

So I propose that we use the phrase 'the question arises in this context' to mean exactly this: K implies the central presupposition, and K does not imply the denial of any presupposition. Notice that this is very different from 'all the presuppositions are true', and we may emphasize this difference by saying 'arises in context'. The reason we must draw this distinction is that K may not tell us which of the possible answers is true, but this *lacuna* in K clearly does not eliminate the question.

4.4. Evaluation of Answers

The main problems of the philosophical theory of explanation are to account for legitimate rejections of explanation requests, and for the asymmetries of explanation. These problems are successfully solved, in my opinion, by the theory of why-questions as developed so far.

But that theory is not yet complete, since it does not tell us how answers are evaluated as telling, good, or better. I shall try to give an account of this too, and show along the way how much of the work by previous writers on explanation is best regarded as addressed to this very point. But I must emphasize, first, that this section is not meant to help in the solution of the traditional problems of explanation; and second, that I believe the theory of why-questions to be basically correct as developed so far, and have rather less confidence in what follows.

Let us suppose that we are in a context with background K of accepted theory plus information, and the question Q arises here. Let Q have topic B, and contrast-class $X = \{B, C, \ldots, N\}$. How good is the answer *Because A*?

There are at least three ways in which this answer is evaluated. The first concerns the evaluation of A itself, as acceptable or as likely to be true. The second concerns the extent to which A *favours* the topic B as against the other members of the contrast-class. (This is where Hempel's criterion of giving reasons to expect, and Salmon's criterion of statistical relevance may find application.) The third concerns the comparison of *Because A* with other possible answers to the same question; and this has three aspects. The first is whether A is more probable (in view of K); the second whether it favours the topic to a greater extent; and the third, whether it is made wholly or partially irrelevant by other answers that could be given. (To this third aspect, Salmon's considerations about *screening off* apply.) Each of these three main ways of evaluation needs to be made more precise.

The first is of course the simplest: we rule out *Because A* altogether if K implies the denial of A; and otherwise ask what probability K bestows on A. Later we compare this with the probability which K bestows on the cores of other possible answers. We turn then to favouring.

If the question why B rather than $C, \ldots, N$ arises here, K must imply B and imply the falsity of $C, \ldots, N$. However, it is exactly the information that the topic is true, and the alternatives to it not true, which is irrelevant to how favourable the answer is to the topic. The evaluation uses only that part of the background information which constitutes the general theory about these phenomena, plus other 'auxiliary' facts which are known but which do not imply the fact to be explained. This point is germane to all the accounts of explanation we have seen, even if it is not always emphasized. For example, in Salmon's first account, A explains B only if the probability of B given A does not equal the probability of A *simpliciter*. However, if I know that A and that B (as is often the case when I say that B because A), then my *personal probability* (that is, the probability given all the information I have) of A equals that of B and that of B given A, namely 1. Hence the probability to be used in evaluating answers is not at all the probability given all my background information, but rather, the probability given some of the general theories

I accept plus some selection from my data.[16] So the evaluation of the answer *Because A* to question *Q* proceeds with reference only to a certain part $K(Q)$ of K. How that part is selected is equally important to all the theories of explanation I have discussed. Neither the other authors nor I can say much about it. Therefore the selection of the part $K(Q)$ of K that is to be used in the further evaluation of *A*, must be a further contextual factor.[17]

If $K(Q)$ plus *A* implies *B*, and implies the falsity of $C, \ldots, N$, then *A* receives in this context the highest marks for favouring the topic *B*.

Supposing that *A* is not thus, we must award marks on the basis of how well *A* redistributes the probabilities on the contrast-class so as to favour *B* against its alternatives. Let us call the probability in the light of $K(Q)$ alone the *prior* probability (in this context) and the probability given $K(Q)$ plus *A* the *posterior* probability. Then *A* will do best here if the posterior probability of *B* equals 1. If *A* is not thus, it may still do well provided it shifts the mass of the probability function toward *B*; for example, if it raises the probability of *B* while lowering that of $C, \ldots, N$; or if it does not lower the probability of *B* while lowering that of some of its closest competitors.

I will not propose a precise function to measure the extent to which the posterior probability distribution favours *B* against its alternatives, as compared to the prior. Two facts matter: the minimum odds of *B* against $C, \ldots, N$, *and* the number of alternatives in $C, \ldots, N$ to which *B* bears

[16] I mention Salmon because he does explicitly discuss this problem, which he calls *the problem of the reference class*. For him this is linked with the (frequency) interpretation of probability. But it is a much more general problem. In deterministic, non-statistical (what Hempel called a deductive-nomological) explanation, the adduced information implies the fact explained. This implication is relative to our background assumptions, or else those assumptions are part of the adduced information. But clearly, our information that the fact to be explained is actually the case, and all its consequences, must carefully be kept out of those background assumptions if the account of explanation is not to be trivialized. *Mutatis mutandis* for statistical explanations given by a Bayesian, as is pointed out by Glymour in his *Theory and Evidence*.

[17] I chose the notation $K(Q)$ deliberately to indicate the connection with models of rational belief, conditionals, and hypothetical reasoning, as discussed for example by William Harper. There is, for example, something called the Ramsey test: to see whether a person with total beliefs K accepts that if *A* then *B*, he must check whether $K(A)$ implies *B*, where $K(A)$ is the 'least revision' of K that implies *A*. In order to 'open the question' for *A*, such a person must similarly shift his beliefs from K to $K?A$, the 'least revision' of K that is consistent with *A*; and we may conjecture that $K(A)$ is the same as $(K?A)\&A$. What I have called $K(Q)$ would, in a similar vein, be a revision of K that is compatible with every member of the contrast class of Q, and also with the denial of the topic of Q. I do not know whether the 'least revision' picture is the right one, but these suggestive similarities may point to important connections; it does seem, surely, that explanation involves hypothetical reasoning. Cf. W. Harper, 'Ramsey Test Conditionals and Iterated Belief Change', in W. Harper and C. A. Hooker, *Foundations of Probability Theory, Statistical Inference, and Statistical Theories of Science* (Dordrecht: Reidel, 1976), 117–35, and his 'Rational Conceptual Change', in F. Suppe and P. Asquith (eds.), *PSA 1976* (East Lansing, Mich.: Philosophy of Science Association, 1977).

these minimum odds. The first should increase, the second decrease. Such an increased favouring of the topic against its alternatives is quite compatible with a decrease in the probability of the topic. Imagining a curve which depicts the probability distribution, you can easily see how it could be changed quite dramatically so as to single out the topic—as the tree that stands out from the wood, so to say—even though the new advantage is only a relative one. Here is a schematic example:

Why E_1 rather than $E_2, \ldots, E_{1000}$?
Because A
$Prob\,(E_1) = \ldots = Prob\,(E_{10}) = 99/1000 = 0.099$
$Prob\,(E_{11}) = \ldots = Prob\,(E_{1000}) = 1/99,000 \doteq 0.00001$
$Prob\,(E_1/A) = 90/1000 = 0.090$
$Prob\,(E_2/A) = \ldots = Prob\,(E_{1000}/A) = 10/999,000 \doteq 0.00001$

Before the answer, E_1 was a good candidate, but in no way distinguished from nine others; afterwards, it is head and shoulders above all its alternatives, but has itself lower probability than it had before.

I think this will remove some of the puzzlement felt in connection with Salmon's examples of explanations that lower the probability of what is explained. In Nancy Cartwright's example of the poison ivy ('Why is this plant alive?') the answer ('It was sprayed with defoliant') was stat-istically relevant, but did not redistribute the probabilities so as to favour the topic. The mere fact that the probability was lowered is, however, not enough to disqualify the answer as a telling one.

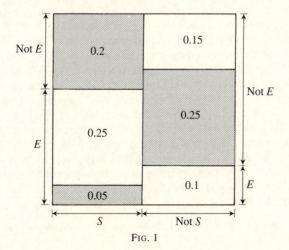

FIG. 1

There is a further way in which A can provide information which favours the topic. This has to do with what is called Simpson's Paradox; it is again Nancy Cartwright who has emphasized the importance of this for the theory of explanation.[18] Here is an example she made up to illustrate it. Let H be 'Tom has heart disease'; S be 'Tom smokes'; and E, 'Tom does exercise'. Let us suppose the probabilities to be as Fig. 1.

Shaded areas represent the cases in which H is true, and numbers the probabilities. By the standard calculation, the conditional probabilities are

$Prob\,(H/S) = Prob\,(H) = 1/2$
$Prob\,(H/S\,\&\,E) = 1/6$
$Prob\,(H/E) = 1/8$
$Prob\,(H/S\,\&\,\text{not}\,E) = 1$
$Prob\,(H/\,\text{not}\,E) = 3/4$

In this example, the answer 'Because Tom smokes' does favour the topic that Tom has heart disease, in a straightforward (though derivative) sense. For as we would say it, the odds of heart disease increase with smoking regardless of whether he is an exerciser or a non-exerciser, and he must be one or the other.

Thus we should add to the account of what it is for A to favour B as against $C, \ldots, N$ that: if $Z = \{Z_1, \ldots, Z_n\}$ is a logical partition of explanatorily relevant alternatives, and A favours B as against $C, \ldots, N$ if any member of Z is added to our background information, then A does favour B as against $C, \ldots, N$.

We have now considered two sorts of evaluation: how probable is A itself? and, how much does A favour B as against $C, \ldots, N$? These are independent questions. In the second case, we know what aspects to consider, but do not have a precise formula that 'adds them all up'. Neither do we have a precise formula to weigh the importance of how likely the answer is to be true, against how favourable the information is which it provides. But I doubt the value of attempting to combine all these aspects into a single-valued measurement.

In any case, we are not finished. For there are relations among answers that go beyond the comparison of how well they do with respect to the criteria considered so far. A famous case, again related to Simpson's Paradox, goes as follows (also discussed in Cartwright's paper): at a certain university it was found that the admission rate for women was lower than that for men. Thus 'Janet is a woman' appears to tell for 'Janet

[18] Nancy Cartwright, 'Causal Laws and Effective Strategies', in her *How the Laws of Physics Lie* (Oxford and New York: Oxford University Press, 1986), 21–43.

was not admitted' as against 'Janet was admitted'. However, this was not a case of sexual bias. The admission rates for men and women for each department in the university were approximately the same. The appearance of bias was created because women tended to apply to departments with lower admission rates. Suppose Janet applied for admission in history; the statement 'Janet applied in history' *screens off* the statement 'Janet is a woman' from the topic 'Janet was not admitted' (in the Reichenbach–Salmon sense of 'screens off': P screens off A from B exactly if the probability of B given P and A is just the probability of B given P alone). It is clear then that the information that Janet applied in history (or whatever other department) is a much more telling answer than the earlier reply, in that it makes that reply irrelevant.

We must be careful in the application of this criterion. First, it is not important if some proposition P screens off A from B if P is not the core of an answer to the question. Thus if the why-question is a request for information about the mechanical processes leading up to the event, the answer is no worse if it is statistically screened off by other sorts of information. Consider 'Why is Peter dead?' answered by 'He received a heavy blow on the head,' while we know already that Paul has just murdered Peter in some way. Secondly, a screened-off answer may be good but partial rather than irrelevant. (In the same example, we know that there must be some true proposition of the form 'Peter received a blow on the head with impact x,' but that does not disqualify the answer, it only means that some more informative answer is possible.) Finally, in the case of a deterministic process in which state A_i, and no other state, is followed by state A_{i+1}, the best answers to the question 'Why is the system in state A_n at time t_n?' may all have the form 'Because the system was in state A_i at time t_i', but each such answer is screened off from the event described in the topic by some other, equally good answer. The most accurate conclusion is probably no more than that if one answer is screened off by another, and not conversely, then the latter is better in some respect.

When it comes to the evaluation of answers to why-questions, therefore, the account I am able to offer is neither as complete nor as precise as one might wish. Its shortcomings, however, are shared with the other philosophical theories of explanation I know (for I have drawn shamelessly on those other theories to marshal these criteria for answers). And the traditional main problems of the theory of explanation are solved not by seeing what these criteria are, but by the general theory that explanations are answers to why-questions, which are themselves contextually determined in certain ways.

4.5. Presupposition and Relevance Elaborated

Consider the question 'Why does the hydrogen atom emit photons with frequencies in the general Balmer series (only)?' This question presupposes that the hydrogen atom emits photons with these frequencies. So how can I even ask that question unless I believe that theoretical presupposition to be true? Will my account of why-questions not automatically make scientific realists of us all?

But recall that we must distinguish carefully what a theory *says* from what we believe when we accept that theory (or rely on it to predict the weather or build a bridge, for that matter). The epistemic commitment involved in accepting a scientific theory, I have argued, is not belief that it is true but only the weaker belief that it is empirically adequate. In just the same way we must distinguish what the question says (i.e. *presupposes*) from what we believe when we ask that question. The example I gave above is a question which arises (as I have defined that term) in any context in which those hypotheses about hydrogen, and the atomic theory in question, are *accepted*. Now, when I ask the question, if I ask it seriously and in my own person, I imply that I believe that this question arises. But that means then only that my epistemic commitment indicated by, or involved in, the asking of this question, is exactly—no more and no less than—the epistemic commitment involved in my acceptance of these theories.

Of course, the discussants in this context, in which those theories are accepted, are conceptually immersed in the theoretical world-picture. They talk the language of the theory. The phenomenological distinction between objective or real, and not objective or unreal, is a distinction between what there is and what there is not which is drawn in that theoretical picture. Hence the questions asked are asked in the theoretical language—how could it be otherwise? But the epistemic commitment of the discussants is not to be read off from their language.

Relevance, perhaps the other main peculiarity of the why-question, raises another ticklish point, but for logical theory. Suppose, for instance, that I ask a question about a sodium sample, and my background theory includes present atomic physics. In that case the answer to the question may well be something like: because this material has such-and-such an atomic structure. Recalling this answer from one of the main examples I have used to illustrate the asymmetries of explanation, it will be noted that, *relative to* this background theory, my answer is a proposition necessarily equivalent to: because this material has such-and-such a characteristic spectrum. The reason is that the spectrum is unique—it identifies the material as having that atomic structure. But, here is the asymmetry, I could not well have answered the question by saying that this material has that characteristic spectrum.

These two propositions, one of them relevant and the other not, are equivalent relative to the theory. Hence they are true in exactly the same possible worlds allowed by the theory (less metaphysically put: true in exactly the same models of that theory). So now we have come to a place where there is a conflict with the simplifying hypothesis generally used in formal semantics, to the effect that propositions which are true in exactly the same possible worlds are identical. If one proposition is relevant and the other not, they cannot be identical.

We could avoid the conflict by saying that of course there are possible worlds which are not allowed by the background theory. This means that when we single out one proposition as relevant, in this context, and the other as not relevant and hence distinct from the first, we do so in part by thinking in terms of worlds (or models) regarded in this context as impossible.

I have no completely telling objection to this, but I am inclined to turn, in our semantics, to a different modelling of the language, and reject the simplifying hypothesis. Happily there are several sorts of models of language, not surprisingly ones that were constructed in response to other reflections on relevance, in which propositions can be individuated more finely. One particular sort of model, which provides a semantics for Anderson and Belnap's logic of tautological entailment, uses the notion of *fact*.[19] There one can say that

It is either raining or not raining
It is either snowing or not snowing

although true in exactly the same possible situations (namely, in all) are yet distinguishable through the consideration that today, for example, the first is *made true* by the fact that it is raining, and the second is made true by quite a different fact, namely, that it is not snowing. In another sort of modelling, developed by Alasdair Urquhart, this individuating function is played not by facts but by bodies of information.[20] And still further approaches, not necessarily tied to logics of the Anderson–Belnap stripe, are available.

---

[19] See my 'Facts and Tautological Entailment', *Journal of Philosophy*, 66 (1969), 477–87 and reprinted in A. R. Anderson and N. D. Belnap, jun., *Entailment* (Princeton, NJ: Princeton University Press, 1975); and 'Extension, Intension, and Comprehension', in M. Munitz (ed.), *Logic and Ontology* (New York: New York University Press, 1973).

[20] For this and other approaches to the semantics of relevance see Anderson and Belnap, *Entailment*.

In each case, the relevance relation among propositions will derive from a deeper relevance relation. If we use facts, for example, the relation R will derive from a request to the effect that the answer must provide a proposition which describes (is made true by) facts of a certain sort: for example, facts about atomic structure, or facts about this person's medical and physical history, or whatever.

5. CONCLUSION

Let us take stock. Traditionally, theories are said to bear two sorts of relation to the observable phenomena: *description* and *explanation*. Description can be more or less accurate, more or less informative; as a minimum, the facts must 'be allowed by' the theory (fit some of its models), as a maximum the theory actually implies the facts in question. But in addition to a (more or less informative) description, the theory may provide an explanation. This is something 'over and above' description; for example, Boyle's law describes the relationship between the pressure, temperature, and volume of a contained gas, but does not explain it—kinetic theory explains it. The conclusion was drawn, correctly I think, that even if two theories are strictly empirically equivalent they may differ in that one can be used to answer a given request for explanation while the other cannot.

Many attempts were made to account for such 'explanatory power' purely in terms of those features and resources of a theory that make it informative (that is, allow it to give better descriptions). On Hempel's view, Boyle's law does explain these empirical facts about gases, but minimally. The kinetic theory is perhaps better *qua* explanation simply because it gives so much more information about the behaviour of gases, relates the three quantities in question to other observable quantities, has a beautiful simplicity, unifies our over-all picture of the world, and so on. The use of more sophisticated statistical relationships by Wesley Salmon and James Greeno (as well as by I. J. Good, whose theory of such concepts as weight of evidence, corroboration, explanatory power, and so on deserves more attention from philosophers), are all efforts along this line.[21] If they had succeeded, an empiricist could rest easy with the subject of explanation.

[21] I. J. Good, 'Weight of Evidence, Corroboration, Explanatory Power, and the Utility of Experiments', *Journal of the Royal Statistical Society*, ser. B, 22 (1960), 319–31; and 'A Causal Calculus', *British Journal for the Philosophy of Science*, 11 (1960/1), 305–18, and 12 (1961/2), 43–51. For discussion, see W. Salmon, 'Probabilistic Causality', *Pacific Philosophical Quarterly*, 1980.

But these attempts ran into seemingly insuperable difficulties. The conviction grew that explanatory power is something quite irreducible, a special feature differing in kind from empirical adequacy and strength. An inspection of examples defeats any attempt to identify the ability to explain with any complex of those more familiar and down-to-earth virtues that are used to evaluate the theory *qua* description. Simultaneously it was argued that what science is really after is understanding, that this consists in being in a position to explain, hence what science is really after goes well beyond empirical adequacy and strength. Finally, since the theory's ability to explain provides a clear reason for accepting it, it was argued that explanatory power is evidence for the *truth* of the theory, special evidence that goes beyond any evidence we may have for the theory's empirical adequacy.

Around the turn of the century, Pierre Duhem had already tried to debunk this view of science by arguing that explanation is not an aim of science. In retrospect, he fostered that explanation-mysticism which he attacked. For he was at pains to grant that explanatory power does not consist in resources for description. He argued that only metaphysical theories explain, and that metaphysics is an enterprise foreign to science. But fifty years later, Quine having argued that there is no demarcation between science and philosophy, and the difficulties of the ametaphysical stance of the positivist-oriented philosophies having made a return to metaphysics tempting, one noticed that scientific activity does involve explanation, and Duhem's argument was deftly reversed.

Once you decide that explanation is something irreducible and special, the door is opened to elaboration by means of further concepts pertaining thereto, all equally irreducible and special. The premisses of an explanation have to include lawlike statements; a statement is lawlike exactly if it implies some non-trivial counterfactual conditional statement; but it can do so only by asserting relationships of necessity in nature. Not all classes correspond to genuine properties; properties and propensities figure in explanation. Not everyone has joined this return to essentialism or neo-Aristotelian realism, but some eminent realists have publicly explored or advocated it.

Even more moderate elaborations of the concept of explanation make mysterious distinctions. Not every explanation is a scientific explanation. Well then, that irreducible explanation-relationship comes in several distinct types, one of them being scientific. A scientific explanation has a special form, and adduces only special sorts of information to explain— information about causal connections and causal processes. Of course, a causal relationship is just what 'because' must denote; and since the *summum bonum* of science is explanation, science must be attempting even to describe something beyond the observable phenomena, namely causal relationships and processes.

These last two paragraphs describe the flights of fancy that become appropriate if explanation is a relationship *sui generis* between theory and fact. But there is no direct evidence for them at all, because if you ask a scientist to explain something to you, the information he gives you is not different in kind (and does not sound or look different) from the information he gives you when you ask for a description. Similarly in 'ordinary' explanations: the information I adduce to explain the rise in oil prices, is information I would have given you to a battery of requests for description of oil supplies, oil producers, and oil consumption. To call an explanation scientific, is to say nothing about its form or the sort of information adduced, but only that the explanation draws on science to get this information (at least to some extent) and, more importantly, that the criteria of evaluation of how good an explanation it is, are being applied using a scientific theory (in the manner I have tried to describe in Section 4 above).

The discussion of explanation went wrong at the very beginning when explanation was conceived of as a relationship like description: a relation between theory and fact. Really it is a three-term relation, between theory, fact, and context. No wonder that no single relation between theory and fact ever managed to fit more than a few examples! Being an explanation is essentially relative, for an explanation is an *answer*. (In just that sense, being a daughter is something relative: every woman is a daughter, and every daughter is a woman, yet being a daughter is not the same as being a woman.) Since an explanation is an answer, it is evaluated *vis-à-vis* a question, which is a request for information. But exactly what is requested, by means of the question 'Why is it the case that *P*?', differs from context to context. In addition, the background theory plus data relative to which the question is evaluated, as arising or not arising, depends on the context. And even what part of that background information is to be used to evaluate how good the answer is, *qua* answer to that question, is a contextually determined factor. So to say that a given theory can be used to explain a certain fact, is always elliptic for: there is a proposition which is a telling answer, relative to this theory, to the request for information about certain facts (those counted as relevant for *this* question) that bear on a comparison between this fact which is the case, and certain (contextually specified) alternatives which are not the case.

So scientific explanation is not (pure) science but an application of science. It is a use of science to satisfy certain of our desires; and these desires are quite specific in a specific context, but they are always desires for descriptive information. (Recall: every daughter is a woman.) The exact content of the desire, and the evaluation of how well it is satisfied, varies from context to context. It is not a single desire, the same in all cases, for a very special sort of thing, but rather, in each case, a different desire for something of a quite familiar sort.

Hence there can be no question at all of explanatory power as such (just as it would be silly to speak of the 'control power' of a theory, although of course we rely on theories to gain control over nature and circumstances). Nor can there be any question of explanatory success as providing evidence for the truth of a theory that goes beyond any evidence we have for its providing an adequate description of the phenomena. For in each case, a success of explanation is a success of adequate and informative description. And while it is true that we seek for explanation, the value of this search for science is that the search for explanation is *ipso facto* a search for empirically adequate, empirically strong theories.

VAN FRAASSEN ON EXPLANATION*

PHILIP KITCHER AND WESLEY C. SALMON

There should be no doubt about the fact that Bas van Fraassen has made substantial contributions to our current understanding of scientific explanation. But we believe that there is reason for doubt as to exactly what the contributions are. Chapter 5 of *The Scientific Image*, 'The Pragmatics of Explanation', offers the most detailed account of van Fraassen's view of explanation. We find both the title and the view ambiguous. The purpose of the present discussion is to underscore the difference between a theory of the pragmatics of explanation and a pragmatic theory of explanation. We believe that van Fraassen has offered the best theory of the pragmatics of explanation to date, but we shall argue that, if his proposal is seen as a pragmatic theory of explanation, then it faces serious difficulties.

1

Before we turn to van Fraassen's positive views, we want to consider his response to the tradition of theorizing about explanation. According to van Fraassen, there are two main problems 'of the philosophical theory of explanation'. These are 'to account for legitimate rejections of explanation requests, and for the asymmetries of explanation' (146). Van Fraassen's solution to the former problem seems to us to be ingenious and substantially correct (see below). His treatment of the asymmetries of explanation we find deeply puzzling.

Within the mainstream of philosophical reflection about explanation, the problem of asymmetries arises because there are arguments which

P. Kitcher and Wesley C. Salmon, 'Van Fraassen on Explanation', *Journal of Philosophy*, 84 (1987), 315–30. Copyright © 1987, The Journal of Philosophy, Inc. Reprinted by permission of the authors and the journal.

* The present paper grew out of discussions between us at the Minnesota Center for the Philosophy of Science during the autumn of 1985. We would like to acknowledge the support of the National Endowment for the Humanities for a grant that made our collaboration possible. We are also grateful to Bas van Fraassen for some helpful clarifications of his ideas, made in response to an earlier draft. Parenthetical page references are to *The Scientific Image* (New York: Oxford, 1980).

are closely related, which accord equally well with the conditions set down by models of explanation, and which differ dramatically in their explanatory worth. For present purposes assume either that some explanations (including the examples to be considered) are arguments or that some arguments (including those to be considered) provide a basis for acts of explanation.[1] Then the challenge is to differentiate between the argument that derives the length of a shadow from the height of a tower, the elevation of the sun, and the principles of optics and the argument that derives the height of the tower from the length of the shadow, the elevation of the sun, and the principles of optics. The former seems to be (a potential basis for) an explanation, whereas the latter does not.

One line of solution, hinted at by Carl Hempel in discussion of an analogous case,[2] is to propose that there is no real difference between the two arguments and that the feeling of difference arises from anthropomorphic ideas from which we ought to liberate ourselves. This is not very convincing, and van Fraassen appears to adopt a more satisfactory method of dissolving the problem. *One* way to understand his fable, 'The Tower and the Shadow', is as an attempt to show that the claim of explanatory difference is shortsighted. Failing to appreciate that arguments are explanations (the basis of explanations) only relative to context, we assess the explanatory merits of the derivations by tacitly supposing contexts that occur in everyday life. With a little imagination, we can see that there are alternative contexts in which the argument we dismiss would count as explanatory.

In van Fraassen's story, a character offers the following explanation of the height of a tower:

That tower marks the spot where [the Chevalier] killed the maid with whom he had been in love to the point of madness. And the height of the tower? He vowed that shadow would cover the terrace where he first proclaimed his love, with every setting sun—that is why the tower had to be so high (133–4).

Now we grant that van Fraassen's story describes a context in which the utterance of these words constitutes an explanation for the position and height of the tower. But this will solve the *traditional* problem of the asymmetries of explanation only if one can claim that the argument

---

[1] The idea that arguments might provide the basis for acts of explanation is suggested in Philip Kitcher, 'Explanatory Unification', *Philosophy of Science*, 48 (Dec. 1981), 507–31; and is articulated in some detail by Peter Railton, 'Probability, Explanation, and Information', *Synthese*, 48 (August 1981), 233–56. Although one of us (W.C.S.) rejects the thesis that explanations are arguments, or involve arguments in any essential way, this issue does not affect the present discussion of van Fraassen's views in any significant fashion.

[2] 'Aspects of Scientific Explanation', in *Aspects of Scientific Explanation* (New York: The Free Press, 1965). See pp. 352–3.

underlying the quoted passage is the argument that the unimaginative have dismissed as non-explanatory.

It seems obvious that this is not so. For the (D-N; deductive-nomological) argument that provides the basis for the act of explanation van Fraassen relates does not take the form of deducing the height of the tower from the length of the shadow (with the elevation of the sun and the principles of optics as the only other premises). Rather, we begin with some initial conditions about the psychological characteristics of the Chevalier—he wanted to build a tower with certain properties; he knew certain physical facts. Using general principles of rationality, we infer a statement to the effect that the Chevalier came to believe that, if he built a tower of the appropriate height on the appropriate spot, it would meet his desiderata. Using yet another principle of rationality, we infer that the Chevalier built the tower to these specifications, and, using background principles about the stability of the height and position of such large physical objects, we conclude that the tower has the height and position it has.

It appears that an obvious way to interpret van Fraassen is mistaken: his story does not provide a context in which an argument wrongly dismissed as explanatory shows its explanatory worth. Moreover, since van Fraassen points out, quite explicitly, the dependency on desires (132), we take him to appreciate that his story does not solve the traditional problem of the asymmetries of explanation. Instead, we construe him as claiming that the problem as we have posed it—a problem that talks about arguments and their merits as explanations (the bases of explanations)—is *mis*posed. Once the topic is approached in terms of van Fraassen's favoured pragmatic machinery, we are to see that an *answer* that we might have considered inappropriate can have explanatory worth in the right context.

But this leaves us with puzzles. If we cannot formulate the traditional problem of the asymmetries of explanation in terms of arguments, then how is the problem to be formulated? Does an analogous problem arise within van Fraassen's own theory? Is it resolved by that theory? We shall return to these questions below.

2

According to van Fraassen, an explanation is an answer to a question Q of the form: Why P_k? where P_k states the fact to be explained—i.e. the explanandum (phenomenon). Any such question can be identified as an ordered triple $\langle P_k, X, R \rangle$, where P_k is called 'the topic' of the question, $X = \{ P_1, \ldots, P_k, \ldots \}$ is its contrast class, and R is its relevance relation. Such a question is posed in a context that includes a body of background knowledge K. Q also has a *presupposition*, namely

P_k is true	(a)
each P_j in X is false if $j \neq k$	(b)
there is at least one true proposition A that bears relation R to $\langle P_k, X \rangle$	(c)

(a) and (b) together constitute the *central presupposition* of Q. The why-question Q *arises* in the given context if K entails the central presupposition of Q and does not entail the falsity of (c). That is, it is altogether appropriate to raise Q even if we do not know whether there is a direct answer or not, provided the central presupposition is fulfilled.

If the question does not arise in the context, it should be rejected rather than answered directly. This can be done by offering a corrective answer, i.e. a denial of one or more parts of the presupposition. If the central presupposition is satisfied but (c) is in doubt, a corrective answer to the effect that (c) is false may be suitable.

If the question arises in the given context, it is normally appropriate to provide a *direct answer*. The canonical form of a direct answer to Q is

(∗) P_k in contrast to the rest of X because A

The following conditions must be met:

(i) A is true
(ii) P_k is true
(iii) No member of X other than P_k is true
(iv) A bears R to $\langle P_k, X \rangle$

A is the *core* of the answer, for the answer can be abbreviated 'Because A'.

Since, typically, the person S_q who asks the question Q might be someone with a different body of knowledge from the respondent S_r, we might be tempted to say that two different contexts are involved. It seems more in keeping with van Fraassen's approach, however, to understand that S_q and S_r are operating in a common context with a common body of background knowledge K determined roughly by the state of science at the time. Thus, K may contain many propositions that neither the questioner nor the respondent knows. Moreover, S_q may have false beliefs that are in conflict with propositions in K. S_r may therefore offer corrective answers to flawed questions by pointing to items in K.

Whether A, the core of the answer to Q, is relevant depends solely on the relevance relation R. If A bears R to $\langle P_k, X \rangle$ then A is, by definition, the core of a relevant answer to Q. This way of stating the matter raises a difficulty. In his informal remarks, van Fraassen repeatedly refers to R as a 'relevance relation', but he incorporates no relevance requirement on R in the formal characterization. Indeed, he points to the absence of any problematic constraint that would seek to capture 'the inextricably modal or counterfactual element' (143). Now, if R happens to be a relevance relation, then it is indeed correct to say that A is relevant to $\langle P_k, X \rangle$. But, as we shall now show, the lack of any constraints on 'relevance' relations allows just about anything to count as the answer to just about any question.

3

Let P_k be any true proposition. Let X be any set of propositions such that P_k belongs to X and every member of X apart from P_k is false. Let A be any true proposition. Let R be $\{ \langle A, \langle P_k, X \rangle \rangle \} \cup S$, where S is any set of ordered pairs $\langle Y, Z \rangle$ such that Y is a proposition and Z is $\langle V, W \rangle$ where V is a proposition and W a set of propositions, one of whose members is V. Then there is a why-question $Q = \langle P_k, X, R \rangle$, and A is the core of a direct answer to Q. Moreover, it is easy to see that, with suitable restrictions on S (i.e. that S contain no $\langle Y, Z \rangle$ such that Y is true and Z is $\langle P_k, X \rangle$), A may be the core of the only direct answer to Q. Hence, for any true propositions P_k and A, there is a why-question with topic P_k such that A is the core of the only direct answer to that question. If explanations are answers to why-questions, then it follows that, for any pair of true propositions, there is a context in which the first is the (core of the) only explanation of the second.

We take it that this result is counterintuitive. Indeed, we would view it as a *reductio* of van Fraassen's account of explanation. How can it be avoided?

One way of blocking the trivialization we have outlined would be to impose restrictions on relevance relations. We shall consider this possibility below. First, let us note that van Fraassen's theory of explanation comes in two parts: there is a thesis about what answers to why-questions are, and there is a thesis about how to evaluate answers to why-questions. Perhaps we can use the latter part of the story to defend against the trivialization that threatens the former.

According to van Fraassen, we evaluate answers to why-questions on three different grounds. We ask whether those answers are probable in light of our knowledge, we ask whether they *favour* the topic against the other members of the contrast class, and we ask whether they are made wholly or partially irrelevant by other answers that could be given. Using a notion that van Fraassen often employs in his informal remarks but does not define, let us say that an answer is *telling* if it scores well according to these criteria. More exactly, let us propose that an answer is more or less telling according to its performance on the three criteria. We shall be most interested in maximally telling answers. We shall call them *perfect* answers.

Notice that the theory of evaluating answers to why-questions allows us to compare different answers to the same questions. It does not enable us to assess the degree to which an answer to one question is more telling than an answer to another question. If the questions are of the contrived kind that we introduced at the beginning of this section, then there will be no more telling answer to them than the contrived answer. However, we may easily introduce a grading of questions by considering whether they admit of answers that favour their topic.

Let us say that questions are more or less *well-founded* to the extent that they admit of telling answers. Thus, a question will be *maximally well-founded* if it admits of a perfect answer. Suppose now that P_k is any true proposition, A any proposition, and X any set of two or more propositions such that P_k is its only true member. Let K be a set of propositions that includes both P_k and A, as well as the negations of all the other propositions in X. Then, we claim, there is a why-question whose topic is P_k, whose contrast class is X, such that A is an essential part of a perfect answer to that why-question.

To demonstrate this we need to examine in somewhat more detail van Fraassen's criteria for evaluating answers. On the first criterion, we award high marks to answers if they receive high probability in light of our background knowledge. A corollary of this is that, if the answer belongs to our background knowledge—as is often the case when we give scientific explanations—then it does as well as possible according to this criterion.

The second criterion (favouring) is less straightforward. Van Fraassen's idea is that the answer, to score well, should increase the distance between the probability of the topic and the probabilities of the other members of the contrast class. Typically, the answer *alone* will not redistribute probabilities in this way. Rather, the answer, taken together with certain auxiliary information, will redistribute the probabilities. However, we cannot suggest that the answer plus the total background knowledge K achieves this result; for, in cases where the topic and the negations of the other members of the contrast class belong to K, the suggestion would lead to immediate trivialization. Van Fraassen there-fore suggests that the redistribution of probabilities be achieved by the answer in conjunction with 'a certain part $K(Q)$ of K', where $K(Q)$ is supposed to be contextually determined.

We need not delve into the problems of deciding exactly what counts as singling out the topic within the contrast class, since we shall use a case in which van Fraassen explicitly recognizes that an answer is maxi-mally successful. He writes: 'If $K(Q)$ plus A implies B and implies the falsity of $C, \ldots, N$ then A receives in this context the highest marks for favouring the topic B' (147). [There is a switch in notation here; the topic is B, the contrast class is $\{B, C, \ldots, N\}$].

Van Fraassen's third criterion concerns the availability of superior answers. The answer A loses marks if it has a rival that fares better, perhaps because the rival receives higher probability in the light of background knowledge K, perhaps because the rival favours the topic more than A does, perhaps because the rival screens off A from the topic. Now A does not have to fear any rival if A belongs to K and if A plus $K(Q)$ implies the topic and the negations of the other members of the contrast class. For, under these circumstances, no rival can be more probable in the light of K, no rival can do better at favouring the topic, and no rival can screen A off from the topic.

We conclude that any A belonging to K which, in conjunction with $K(Q)$, implies the topic P_k is a perfect answer to the question $\langle P_k, X, R \rangle$ (provided, of course, that it is an answer to this question).

Let us therefore define a 'relevance relation' R as follows: we stipulate that R holds between B and $\langle P_k, X \rangle$ just in case P_k is a logical con-sequence of B. Let Z be the disjunction of all the propositions in X apart from P_k, and let B be the proposition

$$A \cdot (A \supset P_k) \cdot \sim Z$$

This proposition bears R to $\langle P_k, X \rangle$, and hence it counts as the core of a direct answer to the why-question $\langle P_k, X, R \rangle$. Moreover, by our earlier assumption, A, P_k, and $\sim Z$ belong to K. Thus, B will be completely successful according to van Fraassen's first criterion for evaluating answers. Because $P_k \cdot \sim Z$ is a logical consequence of B, B maximally favours P_k—and we do not need to worry about how $K(Q)$ is selected since $P_k \cdot \sim Z$ is a consequence of B alone. Finally, because of this implication, there is no reason to fear that B will be screened off by some rival answer. Therefore, B is a perfect answer to $\langle P_k, X, R \rangle$.

We have devised one way of finding, for any pair of true propositions A, P_k, a why-question with P_k as topic to which there is a perfect answer with A as an essential part of its core. Moreover, once we see how the construction we have given is possible, it is easy to generate variations on the same theme. For example, if van Fraassen's account does not contain context-independent principles that preclude the possibility of assigning $(A \supset P_k) \cdot \sim Z$ to $K(Q)$, then it will be possible to claim that A is the core of a direct perfect answer to some question with P_k as topic.

We conclude that the machinery that van Fraassen introduces in his discussion of the evaluation of answers does not avail in protecting him against the kind of trivialization we presented at the beginning of this section. The moral is that, unless he imposes some conditions on relevance relations, his theory is committed to the result that almost anything can explain almost anything. Some kinds of relations R are silly, and why-questions that embody them are silly questions. If we pose silly questions, we should not be surprised to get silly answers.

4

Let us now consider a concrete example. Suppose S_q asks why John F. Kennedy died on 22 November 1963, where

P_k = JFK died 22 Nov. 63, X = {JFK died 1 Jan. 63, JFK died 2 Jan. 63,...,
JFK died 31 Dec. 63, JFK survived 1963}

and R is a relation of astral influence. (One way to define R is to consider ordered pairs of descriptions of the positions of stars and planets at the time of a person's birth and propositions about that person's fate.) An answer with core A might consist of a *true* description of the positions of the stars and planets at the time of JFK's birth. Moreover, using astrological theory as background, one might be able to infer (at least with high probability) that JFK would die on 22 Nov. 1963.

We suggest that, in the context of twentieth-century science, the appropriate response to the question is rejection. According to our present lights, astral influence is not a relevance relation. We believe that the positions of the stars and planets on JFK's birthday have no effect on the probability of death on any particular day. Adding the knowledge of those positions does nothing to redistribute the probabilities of death among the members of the contrast class. The moral we draw here—as in the last section—is that van Fraassen's conditions (a)–(c) on answers to why-questions need to be supplemented by adding

R is a relevance relation (d)

Moreover, we claim that (d) cannot be analysed simply in terms of demanding that, if A bears R to $\langle P_k, X \rangle$, then A must redistribute the probabilities on X. For we can meet that demand by considering the proposition

$B = A \cdot$ (If A, then JFK died on 22 Nov. 63) $\cdot$ JFK did not die on 1 Jan.
  63; $\cdot$ . . . . JFK did not die on 21 Nov. 63 $\cdot$ JFK did not die on 23
  Nov. 63; $\cdot$ . . . . JFK did not survive 1963)

and defining the relation of astral influence R so that R contains $\langle B, \langle P_k, X \rangle \rangle$.

Once again, let us consider the question from the perspective of van Fraassen's account of evaluating answers. We note, first, that the true description of the positions of stars and planets at JFK's birth accords with our current scientific knowledge. So the answer gets high marks on this score. Second, we ask to what extent A favours P_k *vis-à-vis* the other members of X. On this criterion A fares poorly (although B, of course, does not). Perhaps an answer that negatively favoured the topic might get still lower marks—though it is not clear to us that it should, since discovering a relevant factor seems better than offering an irrelevancy. Third, we must compare A with other answers to Q. This criterion has three parts. (i) Since A is true and since it belongs to our body of knowledge, no other answer can be more probable. (ii) Since no astrological answer is relevant, all astrological answers equally fail to favour the topic. (iii) Since every astrological answer is irrelevant, screening off is beside the point.

The result is that A is not telling. There is no telling answer to our original question. If we amend the question, we can produce a relative to which B is a maximally telling answer. In our view, both the questions ought to be rejected, and van Fraassen needs to supplement his theory of explanation with an account of relevance relations.

The astrological answer has a further twist, however. As van Fraassen explains, our general background knowledge K—suitably restricted to $K(Q)$ to avoid trivialization—furnishes a *prior distribution* of probabilities over the contrast class X. (Note that, in discussing favouring with respect to the contrived example of Section 3, and answer B of this section, we were entitled to take $K(Q)$ to be any subset of K because we had no need of any additional premisses in generating the most extreme distribution of probabilities over the contrast class.) Given A, we have a *posterior distribution* of probabilities over X. (It should *not* be assumed that the prior distribution assigns equal probabilities to all members of X; surely, survival beyond 1963 was antecedently more likely than death on any given day, and surely some days are more dangerous than others in the life of a US President.) A is the core of a relevant answer to Q only if addition of A to $K(Q)$ would yield a posterior distribution different from the prior distribution. But what sorts of probabilities are these? If they are S_q's personal probabilities, then, given that S_q is a believer in astrology, we might well expect that knowledge of A would lead to a different distribution. So A would be relevant after all.

Van Fraassen might reply that astrological answers are debarred by his (frequently repeated)[3] remarks that explanations make use of accepted scientific theories. The astrological answers are precluded by the fact that they contain statements that are inconsistent with the background knowledge K. But this seems to mistake the purport of our examples. The statement A *belongs* to the background corpus; it is simply a report of the positions of the heavenly bodies at the time of JFK's birth, and we can assume that this report derives from the best current science. Of course, if the answer includes further bits of astrological theory designed to connect A with the statement that JFK died on 22 Nov. 1963, then van Fraassen will have grounds for ruling it out. But, if the favouring of the topic is achieved solely through S_q's personal probabilities, then there is nothing in the answer to which van Fraassen can point as defective. Similarly, there is nothing in B that would be debarred on the basis of an appeal to background knowledge, for all the statements in B belong to the background corpus.

It should now be clear that these examples work by exploiting the laxity of the conditions on the relevance relation in order to reintroduce 'explanations' that van Fraassen hopes to debar by emphasizing the idea that good explanation must use good science. Unless there are constraints on genuine relevance relations, we can mimic the appeal to deviant beliefs in giving pseudoexplanations by employing deviant relevance relations. Hence, if van Fraassen is serious in his idea that genuine explanations must not make appeal to 'old wives' tales', then he ought to be equally serious about showing that relevance is not completely determined by subjective factors. If we are talking about distributions

---

[3] At the outset of his exposition of his theory of why-questions, van Fraassen remarks, 'This evaluation [of answers] proceeds with reference to the part of science accepted as "background theory" in that context' (141). Earlier, he had remarked that, 'To ask that . . . explanations be scientific is only to ask that they rely on scientific theories and experimentation, not on old wives' tales' (129), and 'To sum up: no factor is explanatorily relevant unless it is scientifically relevant; and among the scientifically relevant factors, context determines explanatorily relevant ones' (126). In conclusion, he says, 'To call an explanation scientific is to say nothing about its form or the sort of information adduced, but only that the explanation draws on science to get this information (at least to some extent) and, more importantly, that the criteria of evaluation of how good an explanation it is, are being employed using a scientific theory' (155–6).

and redistributions of personal probabilities, they must be subject to some kinds of standards or criteria. Coherence is one such criterion, but it cannot be sufficient. To be scientifically acceptable, the redistribution of probabilities must involve differences in objective probabilities (frequencies or propensities) in some fashion.

5

When van Fraassen explicitly discusses kinds of relevance relations, the kinds he picks out are fairly familiar from the literature on scientific explanation: we discover such relations as physical necessitation, being etiologically relevant, fulfilling a function, statistical relevance, and, in the fable of 'The Tower and the Shadow', a relation of intentional relevance. We have been arguing that there are some relations that ought not to be allowed in *any* context as genuine relevance relations. Thus, there appears to be a distinction to be drawn between the relations that can serve, in some context or another, as relevance relations (paradigmatically those relations which figure in van Fraassen's discussions) and those which cannot (such as the contrived relations of the last two sections).

How the distinction should be drawn depends on a very general issue about scientific explanation. Is there a set of genuine relevance relations that underlie the genuine why-questions for all sciences and for all times? Those who give an affirmative answer will see a full theory of explanation as offering a specification of the kinds of relevance relations that may underlie genuine why-questions. That specification would be strongly context-independent in that it would pick out the candidates for any given context of posing a why-question, and the candidates would always be the same.

But perhaps there is no such invariant set of genuine relevance relations. The set of genuine relevance relations may itself be a function of the branch of science and of the stage of its development. Consider the abandonment of teleological explanations in physics after the scientific revolution. This can be viewed as a modification of the set of relevance relations: in the context of Aristotelian physics, the notion of teleological relevance was a genuine relevance relation, in the context of Newtonian physics it lost this status. Uniformitarians (those sympathetic to the view of the last paragraph) will deplore this relativism, contending that the notion of teleological relevance never was a genuine relevance relation and that its status was exposed during the scientific revolution. They will accuse relativists of confusing the variation in beliefs about relevance with the relativity of relevance itself.

We do not need to settle this dispute because, on both accounts, there is a non-trivial task of distinguishing genuine relevance relations from the contrived relations of the last two sections. Just as pluralists about literary works will insist that there are many interpretations of *Hamlet* while denying that any reader's fancy counts as an interpretation, so too, relativists should concede that there are some relations that are not genuine relevance relations at any historical stage of any science. The most thorough version of a relativist account of explanation would consist in specifying those principles which determine, for each historical stage of each science, the selection of certain relations as genuine relevance relations. A more modest (and more sensible) approach would be to consider some particular science (or sciences) throughout some particular period and to identify the pertinent relevance relations. Thus, one might focus on contemporary physics and try to distinguish the associated genuine relevance relations from the residue of relations—the contrived, the discarded, and so forth.

Uniformitarians, ambitious relativists, and modest relativists all face the same kind of task. Although we do not know which version of the task he would wish to undertake, van Fraassen has remarked to us (in conversation) that he recognizes the importance of distinguishing genuine relevance relations and that he takes Aristotle's list of types of causes to be a promising start on drawing the distinction. We now want to suggest that completion of the task will require that van Fraassen solve most (if not all) of the traditional problems that have beset theories of explanation. For, depending on one's commitment on the large issue we have left unresolved, these problems take the form of showing why certain relations do not belong to the single set of genuine relevance relations which is associated with all sciences at all times, or of showing why certain relations do not belong to any of the sets of genuine relevance relations associated with different sciences at different times, or of showing, for some particular science(s) and period of interest, why certain relations do not belong to the associated set of genuine relevance relations. Henceforth, we intend that our presentations of problems should be systematically ambiguous among these forms.[4]

To simplify matters, we shall confine our attention to difficulties that arise in what Hempel would have viewed as deductive explanation. Con-

---

[4] We are very grateful to an editor of this *Journal* who raised a question which prompted us explicitly to distinguish these three ways of pursuing the theory of explanation and, thus, substantially to improve on the formulations of an earlier draft.

sider the simple relation of derivation. This relation holds between A and $\langle P_k, X \rangle$ just in case there is a (first-order) derivation of P_k from A plus additional premises in $K(Q)$. We can define any number of relations by imposing constraints on the kinds of statements that should figure in the premises. Thus, to recall a famous Hempelian example, let P_k be the proposition that Horace is bald, and R be the relation of Greenbury-School-Board-derivation that holds between A and P_k just in case A is a conjunction of propositions one of whose conjuncts is the proposition that Horace belongs to the Greenbury School Board, P_k is derivable from A, and there is no conjunct in A that could be deleted while still enabling P_k to be derivable from the result. Suppose that X includes the propositions that Horace is bald and that Horace is not bald. Let A be the proposition that Horace is a member of the Greenbury School Board and that all members of the Greenbury School Board are bald. $\langle P_k, X, R \rangle$ is a van Fraassen why-question to which A is a direct answer, and a perfect answer, to boot.

We claim that the question we have just artificially constructed is not a genuine why-question and that A is no explanation of Horace's baldness. Moreover, we suggest that most (if not all) of the examples of non-explanatory arguments that Hempel hoped to exclude—both those he succeeded in debarring and those which have caused persistent problems for the theory of D-N explanation—give rise to corresponding 'relevance' relations that van Fraassen ought to exclude. As an illustration, let us return to his solution to the problem of the asymmetries of explanation in the light of what we have discovered about his treatment of why-questions.

The proposition that the tower was built on the spot where the Chevalier killed his beloved and that it was built to such a height that its shadow would fall across the terrace where he first vowed his love is relevant to the topic of the question, 'Why is the height of the tower h?' if we construe the relevance relation to be that of intentional relevance. That is just another way of putting the point that there is a perfectly good Hempelian argument that derives the height of the tower from premises about the Chevalier's attitudes and from psychological laws. But, if we are moved by the traditional problem of the asymmetries of explanation, what we want to know is whether there is a context in which the statement 'The length of the shadow is l' answers the question 'Why is the height of the tower h?' in virtue of the fact that the assertion about shadow length, together with premises about the angle of elevation of the sun and the propagation of light [which may be relegated to the background $K(Q)$] favour the topic as against other propositions ascribing different heights. For that (or something very like it) is the translation into van Fraassen's idiom of the asymmetry problem that has bedevilled Hempel and his successors.

Now, unless we impose very delicate constraints on relevance relations, it is easy to contrive a maximally well-founded question $\langle P_k, X, R \rangle$ such that the proposition ascribing shadow length will be the core of a perfect answer. The trick should be apparent by now: take P_k to be the proposition that ascribes the actual height to the tower, let X be a collection of propositions ascribing different heights, let R be the relation of *censored Hempelian derivation*—a relation that holds between A and $\langle P_k, X \rangle$ just in case there is a D-N argument that derives P_k from A plus additional premisses in $K(Q)$. (Quite evidently, we could impose additional constraints so as to rule out the use of the psychological principles on which van Fraassen's account turns, and thus to ensure that the only available D-N arguments are those which invert the usual order of explanatory derivation). We take $K(Q)$ to be fixed in such a way as to include the proposition ascribing the elevation of the sun and the laws of propagation of light. This is surely quite reasonable, for some such $K(Q)$ will have to be allowed if we are to countenance the proposition that the height of the tower is h as the core of an answer to the question, Why is the length of the shadow l? So van Fraassen's theory allows explanations which correspond to those D-N explanations which intuitively 'run the wrong way'.

We suggest that this is a mistake. Just as the contrived questions of sections 3 and 4 should be eliminated by the imposition of constraints on relevance relations, so too the question of the last paragraph and its accompanying perfect answer ought to be banished. For otherwise, van Fraassen's account of explanation will be deficient in exactly the way that Hempel's own treatment was. Every kind of asymmetry that arises for the D-N model can be generated within van Fraassen's framework. This means that, far from solving the problem of the asymmetries of explanation, van Fraassen presupposes a solution to that problem. Thus, if we are right, van Fraassen has offered a beautiful treatment of the pragmatics of explanation which should be viewed as a supplement, rather than a rival, to the traditional approaches to explanation.

6

As we have remarked (see note 3), there are many suggestions in van Fraassen's text that he does not intend to offer an 'anything goes' account of explanation. In the last section, we have attempted to show that this intention ought to commit him to solving most (if not all) of the traditional problems of the theory of explanation. We want to conclude by considering an obvious question. If we interpret van Fraassen as supposing that there are constraints on why-questions and their answers, how does this affect the general argumentative strategy of *The Scientific Image*?

Van Fraassen's discussion of scientific explanation is part of an effort to show that theoretical virtues beyond the saving of the phenomena are pragmatic. That argument eliminates a certain strategy for defending theoretical realism. If the realist proposes that (1) there is an objective criterion of explanatory power that distinguishes among empirically equivalent theories and (2) theories with greater explanatory power have a stronger title to belief, then the doctrine of *The Scientific Image* appears to oppose the proposal by denying (1).[5] If we are correct in our assessment of van Fraassen's position, then it seems that the realist can get at least as far as (1). For, if it is once granted that we can produce statements which favour the topic of a why-question but which do not stand in any objective relevance relation to that topic (or, more exactly, to the ordered pair of topic and contrast class), then it appears that a theory may save the phenomena without generating answers to why-questions founded on genuine relevance relations.

We have argued that, if he is to avoid the 'anything goes' theory of explanation, van Fraassen must offer a characterization of objective relevance relations which, in effect, overcomes the traditional problems of the theory of explanation. Now, within the traditional theories, there is ample room for prediction without explanations: we can have deductive arguments that fail to explain their conclusions, assemblages of statistical relevance relations that bestow high probability on a statement without explaining it. Once van Fraassen has introduced analogous distinctions within the theory of why-questions and their answers, through the provision of constraints on genuine relevance relations that separate mere favouring from the adducing of relevant information, there can be theories yielding statements that favour the set of topics in a given class (or even imply those topics) without generating answers to any genuine why-question with any of those topics. We would thus have the basis for claiming that such theories are objectively inferior to their rivals that do furnish explanatory answers.

---

[5] It is clear from a subsequent paper, in which he discusses Clark Glymour's views about explanation, that van Fraassen would also object to (2). See 'Glymour on Evidence and Explanation', in John Earman (ed.,), *Testing Scientific Theories* (Minneapolis: University of Minnesota Press), 165–76.

The consequence would be that van Fraassen would have to revise his account of what it is to accept a scientific theory by adding the idea that acceptance involves believing that the theory has explanatory power as well as believing that it saves the phenomena (or, perhaps, believing that the theory offers the best trade-off between saving the phenomena and having explanatory power)—indeed, he seems to take just this tack in the article cited in n. 5. Since van Fraassen can still avail himself of a (different) distinction between acceptance and belief, this consequence should be seen as providing only the entering wedge for an argument for realism.

We conclude that, if van Fraassen avoids the Scylla of the 'anything goes' theory of explanation, then he is plunged into what he would view as the Charybdis of supposing that there is an objective virture of theories distinct from their salvation of the phenomena. From our perspective, Scylla is (to say the least) uninviting, but Charybdis feels like the beginning of the way home.

THE PRAGMATIC CHARACTER OF
EXPLANATION*

PETER ACHINSTEIN

Some of those, including the present writer, who criticize standard models of explanation, such as Hempel's D-N model or Salmon's S-R model, do so on the grounds that explanation is a 'pragmatic' or 'contextual' concept—an idea which the standard models seem to reject. Yet the sense in which explanation is, or is not, pragmatic is not always made clear by the critics or champions of the models. Indeed, some critics and some champions may even mean different things by 'pragmatic' or 'contextual'. In this chapter I want to try to clarify a sense in which explanations might reasonably be considered pragmatic, discuss a couple of theories that are or are not pragmatic in this sense, argue the advantages of a pragmatic account, and briefly note some consequences of this for those seeking models of explanation.

1. HEMPEL'S CHARACTERIZATION OF 'PRAGMATIC'

Hempel certainly acknowledges that there is a pragmatic aspect of explanation. He writes:

Very broadly speaking, to explain something to a person is to make it plain and intelligible to him, to make him understand it. Thus construed, the word 'explanation' and its cognates are *pragmatic* terms: their use requires reference to the persons involved in the process of explaining. In a pragmatic context we might say, for example, that a given account A explains fact X to person P_1. We will then have to bear in mind that the same account may well not constitute an explanation of X for another person P_2, who might not even regard X as requiring an explanation, or who might find the account A unintelligible, or unilluminating, or irrelevant to what puzzles him about X.

Peter Achinstein, 'The Pragmatic Character of Explanation', *PSA 1984*, ii. 275–92. Copyright © 1985 by the Philosophy of Science Association. Reprinted by permission of the Association and the author.

* I am indebted to the National Science Foundation for support.

Explanation in this pragmatic sense is thus a relative notion: something can be significantly said to constitute an explanation in this sense only for this or that individual (1965: 425).

Now although Hempel recognizes a pragmatic use, or sense, or concept, of explanation, he sees his own task as one of 'constructing a non-pragmatic concept of scientific explanation—a concept which is abstracted, as it were, from the pragmatic one, and which does not require relativization with respect to questioning individuals' (1965: 426). I take Hempel to be saying something like this. There are sentences, such as ones of the form

(1) Account A explains fact X to person P,

which make essential reference to some person or type of person who is explaining or being explained to. Such sentences are examples of a *pragmatic* use or concept of explanation. By contrast, there are other sentences, such as ones of the form

(2) Account A explains fact X,

which make no reference to any (type of) explainer or audience. These sentences are examples of a *non-pragmatic* use or concept of explanation. Hempel's D-N and I-S models are meant to provide truth-conditions for certain sentences of this type.

Let me use the term 'explanation-sentence' to refer to any sentence containing the terms 'explains' or 'explanation'. I shall say that the terms for persons replacing S and P in sentences with forms such as the following are terms for explainers or audiences:

S explains fact X to P
The explanation of X given by S to P is . . .
S gave account A to P as an explanation of . . .

S and P may be terms for particular explainers and audiences or for types. For example, we might have 'Achinstein explained his theory to philosophers at the 1984 PSA meetings' for a particular explainer and audience, and 'the contemporary physicist explains the structure of matter by invoking quarks' for a type of explainer.

Now I shall broaden what I take to be Hempel's characterization by saying that an explanation-sentence is 'pragmatic' if (a) it contains terms for a (particular or type of) explainer or audience, or if (b) its truth-conditions contain such terms or others defined using such terms. Clause (b) will take into account a view which says that although some explanation-sentences are not explicitly pragmatic they are implicitly so. For example, one might hold the view that an explanation-sentence of the form 'Account A explains fact X' is true iff some (type of) explainer S

explains (or could explain) fact X to an audience (of type) Y by citing A. On this conception, the explanation-sentence in question would be pragmatic.

Whether this characterization of 'pragmatic' captures what Hempel has in mind I will take up later. For the present let us accept it as a sufficient condition.

Hempel's claim can now be put like this. Admittedly, there are pragmatic explanation-sentences, e.g. ones of the form

Account A explains fact X to person P

Explainer S explains X to person P by giving account A

But there are also non-pragmatic explanation-sentences. Most important for our purposes (Hempel will claim) an explanation-sentence of the following form is non-pragmatic:

(2) Account A explains fact X

I shall say that someone holds a pragmatic theory of explanation with respect to explanation-sentences of a given form if he maintains that explanation-sentences of that form are pragmatic. Someone holds a non-pragmatic theory with respect to explanation-sentences of a given form if he maintains that explanation-sentences of that form are not pragmatic. Hempel holds a pragmatic theory with respect to sentences of form (1) but not of form (2).

I want to raise some questions about non-pragmatic theories of sentences of form (2) and others like it. But before doing so let me turn to someone who claims to be an arch-pragmatist, namely, Bas van Fraassen.

2. VAN FRAASSEN'S PRAGMATISM

In the first part of Chapter XI, 'The Pragmatics of Explanation', van Fraassen seems to be arguing in direct opposition to Hempel's non-pragmatic theory of explanation-sentences of form

(2) Account A explains fact X

Van Fraassen writes:

The description of some account as an explanation of a given fact or event is incomplete. It can only be an explanation with respect to a certain *relevance relation* and a certain *contrast-class*. These are contextual factors, in that they are determined neither by the totality of accepted scientific theories, nor by the event or fact for which an explanation is requested (Section 2.9).

I shall briefly characterize van Fraassen's position by using as an example some of my home-town lore. By the dawn's early light Francis Scott Key is able to see the flag atop Fort McHenry. And he asks:

Q: Why is our flag still there?

This interrogative, van Fraassen will say, can be used to pose different questions depending on the contrast intended. For example, Key might be asking:

Why is *our* flag (rather than some other flag) still there?
Why is our flag *still there* (rather than somewhere else)?
Why is our *flag* (rather than something else) still there?

And so forth. The contrast-class includes what is presupposed by the question (our flag being there) together with the alternatives (there being some other flag there, our flag being somewhere else, etc.). More generally, van Fraassen claims, in the case of any why-question there is a contrast-class that is usually implicit in the context: 'In general, the contrast is not explicitly described because, *in context*, it is clear to all discussants what the intended alternatives are.' (ibid.) For Key the context will tell us that the likely contrast is between our flag being there and the British flag being there.

Now let us turn to the second contextual concept van Fraassen mentions, the relevance relation. Francis Scott Key's interrogative

Q: Why is our flag still there?

might be construed (in van Fraassen's terms) as a request for the 'events leading up to its being still there'. If so, we might answer by appeal to the battle raging throughout the night and the failure of the British to capture Fort McHenry. However, there is another possible (though perhaps less likely) interpretation of this interrogative, namely, as a request for the function or purpose of our flag's being there. What we need to know, says van Fraassen, is what 'relevance relation' is being requested—'events leading up to', 'function', or something else. And this, as in the case of the contrast-class, is to be determined by looking to the context. 'Looking to the context' in our example means invoking the intentions, beliefs, and puzzlements of Francis Scott Key. And this is pragmatic.

Now let us apply this to explanation-sentences of the Hempelian type (2). For our example consider:

(3) The hypothesis that the British failed to capture Fort McHenry during the night's battle explains the fact that our flag is still there

Recall van Fraassen's words, 'The description of some account as an explanation of a given fact or event is incomplete. It can only be an explanation with respect to a certain *relevance relation* and a certain *contrast-class*.' And the latter are contextual, requiring reference to some

particular person. Well, if (3) is incomplete, let us complete it by specifying some relevance relation and contrast-class. We can do so, van
Fraassen tells us, by understanding the question being raised as having
three components: the topic (in this case 'our flag is still there'), the
contrast-class (in this case let us say: 'our flag is still there', 'the British
flag is there'), and the relevance relation (in this case let us say: 'events
leading up to'). Although van Fraassen does not do it quite this way we
might now reformulate (3) above by writing:

(4) The hypothesis that the British failed to capture Fort McHenry
during the night's battle explains (by citing 'events leading up to')
why our flag is still there (rather than the British flag being there)

We now have an explanation-sentence which provides the sort of information van Fraassen wants. Is it pragmatic?

It is not explicitly pragmatic, since it contains no terms for an explainer
or audience. Is it implicitly so? Do its truth-conditions contain terms for
an explainer or audience or others defined by reference to these? Van
Fraassen points out, correctly I think, that to determine what relevance
relation and contrast-class are being requested appeal is made to the
context. We look to the explainer, Francis Scott Key, and what intentions
and beliefs he had. But this is not sufficient to show that the truth-conditions for (4) must contain terms for an explainer or audience.

Indeed, Hempel—presumably van Fraassen's arch-foe—could agree
that in order to determine what question someone wants to answer, or
what event someone wants to explain, essential reference to the intentions and beliefs of the questioner will need to be made. This is no
damaging admission for the non-pragmatist, Hempel will say. The important issue is whether once the question being asked has been identified,
it can be determined whether the explanation explains without invoking
any term for an explainer or audience. So far van Fraassen has offered
no reason why this cannot be done. All he has said is that (3) is incomplete. By analogy, Hempel might say, the following sentence is incomplete:

The hypothesis that the British failed to capture Fort McHenry during
the night's battle explains . . .

Suppose we find this incomplete sentence in a history book. To complete
it appeal is made, let us say, to the historian's likely intentions and
beliefs, and/or perhaps to those of Francis Scott Key. That will not make
the resulting completed sentence 'pragmatic' in what I have so far taken
to be Hempel's sense. Suppose we complete the sentence by identifying
the explanandum as

why our flag (rather than the British) is still there

Just because we have appealed to pragmatic considerations in identifying the explanandum, Hempel will ask, how does that show that the truth-conditions for the completed explanation-sentence must contain terms for an explainer or audience? Indeed, Hempel will urge us to accept his own truth-conditions for the completed sentence—say those of the D-N or I-S model—which contain no terms for an explainer or audience.

What about van Fraassen's truth-conditions? I find his intentions a bit cloudy at this point. He seems to present two sets of conditions, one set (perhaps) for the concept of a (merely, or minimally) correct explanation, and another for the concept of a good explanation. To give the first set of conditions we have a question Q determined by the topic P, the contrast-class X, and the relevance relation R. And we have an answer of the form

P in contrast to X because A

Van Fraassen asks: what is claimed in this answer (section 4.3)? He gives four conditions. First, that P is true. Second, that the other members of the contrast-class are not true. Third, that A is true. And fourth, that A does bear the relevance relation R to P and X—e.g. that the answer A does give the events 'leading up to' the event in P. I am not sure if this is supposed to be a set of sufficient conditions, or only necessary ones, or, indeed, if it is supposed to be a set of conditions for the *truth* of sentences of the above form (the latter is suggested by van Fraassen's question 'What is claimed in this answer?').[1] In any case, these conditions, let it be noted, contain no terms for an explainer or audience. Nor does their application to sentences of the form 'P in contrast to X because A' require any reference to explainers or audiences once the question Q is given. Nor do the definitions of van Fraassen's technical terms in these conditions ('topic', 'contrast-class', and 'relevance relation') appear to require the concept of an explainer or audience.

What about van Fraassen's second set of conditions for (as he puts it) 'evaluating' answers? Again, we have a question Q determined by the topic P, the contrast-class X, and the relevance relation R. How good is the answer

P in contrast to X because A?

[1] In conversation van Fraassen suggests that the answer 'P in contrast to X because A' should be understood as relativized to some particular set of assumptions B made in the context. If so his conditions might be construed as truth-conditions for sentences of the form 'P in contrast to X because A, given B'.

Van Fraassen proposes three things that must be determined (section 4.4):

1. We must determine whether proposition A is 'acceptable' or 'likely to be true'.
2. We must determine whether A shifts the probability towards P more than towards other members of the contrast-class X.
3. We must compare 'because A' with other possible answers to the explanatory question in three respects:
   a. Is A more probable than other answers given the background information K?
   b. Does A shift the probability towards P more than other answers do?
   c. Does some other answer probabilistically 'screen off' A from P? (Is there an answer A' such that $p(P/A' \& A) = p(P/A')$?)

This evaluation of explanations introduces two important new factors: a set of background beliefs K relative to which probabilities are to be determined, and a set of answers to the question Q with which the answer A is being compared. Both factors might be deemed pragmatic or contextual. To determine what background beliefs should be used, and what alternative answers proposition A should be compared with, reference will be made to intentions and beliefs of the explainer or perhaps of the evaluator of the explanation. (Indeed, van Fraassen insists that only part of the background information K is to be used in the evaluation, and that which this is 'must be a further contextual factor' (Section 4.4).)

I do not propose here to assess van Fraassen's conditions. (For criticisms see my (1983), ch. 4.) I will simply note what I believe the non-pragmatist's response is likely to be. Just as van Fraassen earlier accused the non-pragmatists of focusing on an imcomplete explanation-sentence, so the non-pragmatists will retort '*tu quoque*' to van Fraassen. All van Fraassen is arguing, the non-pragmatist will say, is that sentences of the following form are incomplete:

'P in contrast to X because A' is a good explanation of q

The (more) complete form of such explanation-sentences is

(5) 'P, in contrast to X, because A' is a good explanation of q relative to alternatives $A_1, \ldots, A_n$, and relative to background information K (or relative to such-and-such a subset of K)

Now that we have completed the explanation-sentence by relativization to a specific set of alternative hypotheses and to background information we are in a position to use the three conditions van Fraassen presents. These conditions invoke no terms for an explainer or audience. Nor will their application to sentences of form (5) require any such terms. Indeed Hempel himself insists on relativizing inductive-statistical (I-S) explanations to a set of background beliefs K, which, of course, can be different from one explanatory context to the next. This does not suffice to make Hempel believe that he is analysing a pragmatic concept of explanation when he offers his inductive-statistical model.

I conclude that van Fraassen ought not to view his position as a pragmatic one—at least with reference to complete explanation-sentences such as those of forms (4) and (5). To be sure, to obtain such complete sentences to begin with reference may have to be made to explainers. With this Hempel could agree. But once the sentences are complete no reference to any (particular or type of) explainer or audience needs to be made to understand what they mean, or to determine whether or not they are true.

3. THE ORDERED PAIR THEORY

Let me turn from van Fraassen's theory to one that I elaborate in my recent book *The Nature of Explanation*. Here I do not plan to present the theory in detail but only to say enough about it to show that it is pragmatic and to argue the advantages of a pragmatic account.

As did Sylvain Bromberger in a seminal essay in 1965, I begin with the concept of an explaining act. The explanation-sentences of concern to me are ones of the form

(6) S explains q by uttering u

where q is the indirect form of a question Q. Simplifying my view, such sentences are true iff S utters u with the intention of rendering q understandable by producing the knowledge that u expresses a correct answer to the question Q. To develop this one needs to talk about the concept of understanding. I do so in the book, but the discussion is complex, and I will not attempt to summarize it here. In any case there is no need to do so, for explanation-sentences of the form 'S explains q by uttering u' are clearly pragmatic in the Hempelian sense. Such sentences make essential reference to an explainer.

The second stage in my theory consists in an attempt to provide a definition of an explanation itself—i.e., the product of an act of explaining or at least of a potential act of explaining. For certain reasons which we need not explore here I say that an explanation of q can be construed as an ordered pair whose first member is a proposition or set of propositions that constitutes an answer to Q, and whose second member is a type of explaining act, namely, explaining *q*. So, e.g., if Newton explains why the tides occur by saying that they occur because of the gravitational pull of the moon, then his explanation—whether good or bad, right or wrong—can be construed as the ordered pair

(7) (The tides occur because of the gravitational pull of the moon; explaining why the tides occur).

The second member of this pair invokes the concept of a type of explaining act, to which the account briefly summarized above is applicable. The first member of the pair is a proposition that constitutes an answer to the question cited in the second member. Unlike usual accounts, an explanation need not be restricted to why-questions. There can be an explanation of what event is now occurring in the bubble chamber, of what significance the American election has for Europe, and so forth. The view I develop attempts to characterize in a general way the kinds of questions (which I call content-questions) that can appear in explanations, and also to characterize in a general way what constitutes an answer to a content-question. The present manner of viewing explanations allows us easily to distinguish explanations from other products, whose second members will not be types of explaining acts, but something else. Furthermore, although this account defines explanation by reference to the concept of an explaining act, for something to *be* an explanation it is not required that it be the product of some particular explaining act. The ordered pair above would be an explanation, on my account, even if neither Newton nor anyone else expressed the proposition that is its first member (i.e., even if no one ever explained why the tides occur by uttering any sentence expressing that proposition).

The latter point is important for the issue of the pragmatic character of explanation, so let me take it just a bit further. Let us consider explanation-sentences of the form

E is an explanation of q

where there is no implication regarding E's goodness or correctness. On the ordered pair theory, the following is a set of truth-conditions for sentences of this form:

(i) Q is (what I call) a content-question;
(ii) E is an ordered pair whose first member is (what I call) a complete content-giving proposition with respect to Q and whose second member is the act-type *explaining q*

Do these truth-conditions contain terms for an explainer or audience or any terms defined by reference to these? They do not do so explicitly.

Nor do the definitions of 'content-question' and 'complete content-giving proposition'. This leaves the act-type 'explaining q', which I take to be definable as a type of act whose instances are acts in which explainers explain q. (a is a type of act 'explaining q' iff (S) (S performs an act of type $a \equiv (\exists u)$ (S explains q by uttering u).) If so then a term for an explainer is invoked in defining one of the concepts in the truth-conditions. And by our previous criterion of 'pragmatic', this suffices to make sentences of the form 'E is an explanation of q' pragmatic.

Yet there is something different about this case and the ones Hempel may have in mind. For although a term for an explainer is invoked, the truth-value of sentences of the form 'E is an explanation of q' will not vary with who, if anyone, is giving or receiving the explanation E mentioned in the explanation-sentence. Earlier I characterized an explanation-sentence as pragmatic if it contains terms for a (particular or type of) explainer or audience or if its truth-conditions contain such terms or others defined using such terms. We might now introduce a second condition, and say that the truth-value of explanation-sentences of that form can vary with a change in the person giving or receiving the explanation mentioned or referred to in the explanation sentence. If both of these conditions are satisfied, let us say that the explanation-sentence is *strongly pragmatic*. If only the first is satisfied, the explanation-sentence is *weakly pragmatic*. By this criterion, sentences of the form

S explains q by uttering u

are strongly pragmatic. (Such sentences contain a term for an explainer, and their truth-value can vary with a change in explainer.) On the ordered pair theory, sentences of the form

E is an explanation of q

are only weakly pragmatic. Truth-conditions for sentences of this form (according to the ordered pair theory) invoke a term for a type of explainer, one who explains q; but the truth-value of sentences of this form does not vary with any change in who is giving E as an explanation of q, or to whom. On the ordered pair theory the concept of an explanation is defined by reference to the concept of an act in which an explainer explains something (thus making 'E is an explanation of q' weakly pragmatic). But whether some particular sentence of the form 'E is an explanation of q' is true will not depend upon who, if anyone, gives the explanation (thus preventing such sentences from being strongly pragmatic). By contrast, according to Hempel's models of explanation, sentences of the form 'E is an explanation of q' are neither strongly nor weakly pragmatic.[2]

I am inclined to think that when Hempel uses the term 'pragmatic' he has in mind 'strongly pragmatic', and that he would not object too strenuously to a 'weakly pragmatic' concept of explanation, since the latter can be 'objective'. But this is speculation on my part.

Let me turn to another, perhaps more important, concept for which the ordered pair theory offers an account, namely, that of a 'good explanation'. Are sentences of the form 'E is a good explanation of q' pragmatic in either sense?

Different evaluations of explanations are possible depending on what ends are to be achieved. The ends might be purely universal ones, e.g. the achievement of truth, empirical adequacy, simplicity, unification, etc. Or they might be more contextual. The end I am particularly concerned with is one that, by the definition given in the first part of the theory, an explainer has when he performs an act of explaining q, namely, rendering q understandable (in some appropriate way) by producing the knowledge of the answer one gives that it is a correct answer to Q. An evaluation with this end in view will take into account both universal and contextual criteria. Very roughly, E will be a good explanation for an explainer to give in explaining q to an audience if E is capable of rendering q understandable in an appropriate way to that audience by producing the knowledge of the answer to Q that it supplies that it is correct; or if it is reasonable for the explainer to believe that this obtains. The appropriateness of the understanding will depend on what the audience already knows and is interested in finding out. It will also depend on what it would be valuable for the audience to know—which, especially in the sciences, can bring in universal criteria. (For details see Achinstein (1983: 107–17).)

In the case of such evaluations, which I call 'illocutionary', sentences of the form 'E is a good explanation of q' will be construed as elliptical for 'E is a good explanation for an explainer to give in explaining q to an audience'. Explanation-sentences of the latter form are strongly pragmatic. They contain terms for an explainer and audience, and the truth-

---

[2] The two conditions are independent. We have already seen an example satisfying the first but not the second. Here is something satisfying the second but not the first: 'The fact that I was delayed in traffic is the correct explanation of why I am late.' This sentence contains no terms for an explainer or audience (in the sense indicated earlier: it contains nothing of the form 'S explains q to P' or 'the explanation of q given by S to P is . . .'). Yet its truth-value will vary with a change in the person giving the explanation mentioned. By the definitions above, this explanation-sentence is neither strongly nor weakly pragmatic. (To transform it into a strongly pragmatic explanation-sentence satisfying both conditions we could write: 'The fact that he was delayed in traffic is the correct explanation given by Danny Dawdle of why he is late.')

value of sentences of this form can vary with a change in explainer or audience.

Now I am not claiming that illocutionary evaluations are the only possible ones. I do insist that they are important, that they are frequently given, and that using them, by contrast to non-illocutionary, non-pragmatic evaluations, will enable us to see why certain scientific explanations are generally judged better than others. Let me illustrate this by invoking a simple example, Rutherford's 1911 explanation of the results of scattering experiments involving alpha particles.

In experiments published in 1909 Geiger and Marsden showed that when alpha particles are directed at a thin metal foil most of them go through the foil with small angles of deflection, but some are scattered through an angle of more than $90°$, thus emerging on the side of incidence. In order to explain these surprising results Ernest Rutherford proposed a new theory of the structure of the atom. He assumed that an atom contains a positive charge that is not evenly distributed but is concentrated in a nucleus whose volume is small compared to that of the atom. He also assumed that the positively charged nucleus is surrounded by a compensating charge of moving electrons. Finally, he assumed that each scattering was the result of a single encounter between an alpha particle and a foil atom. Since most alpha particles penetrate the foil without being appreciably scattered, the foil atoms are mostly empty of matter. An alpha particle that is scattered at a wide angle is not scattered by a much less massive electron, but by a positive charge concentrated in the nucleus. From these assumptions, together with classical principles including conservation of energy and momentum, Rutherford derived a formula which gives the number of alpha particles falling on unit area deflected through an angle θ as a function of several other quantities. From this formula it is possible to calculate the number of alpha particles scattered at wide angles such as $150°$ or $135°$.

Is Rutherford's explanation of the scattering results a good one? If we evaluate it in a non-illocutionary way using only criteria that are non-pragmatic, it would, I suppose, get a mixed review. True, it derives the wide scattering angles in a precise way from lawlike, quantitative assumptions; it appeals to micro-entities; and it offers a cause of the scattering—all of which physicists and philosophers of science tend to regard with favour. But, as later developments in physics show, it is only a crude approximation to what actually occurs in the foil atoms. And it introduces a conception of the atom as involving moving electrons that is incompatible with classical electrodynamics. (Moving electrons should radiate energy and collapse into the nucleus, which clearly they do not.) Furthermore, if we use only non-pragmatic criteria in our evaluation, we will have a difficult time seeing why Rutherford's explanation is better than certain others we might construct that are clearly inferior.

Consider, e.g., the following quantitative hypothesis that Geiger and Marsden could have constructed from their experiments without appeal to Rutherford's theory. (I shall call it the G–M hypothesis.)

The G–M hypothesis: When alpha particles are directed at thin metal foils the atoms comprising the foils cause the alpha particles to be scattered at various angles in accordance with the formula

$$N = Qnt(Ze)^2E^2/4r^2(MV^2)^2\sin^4\theta/2$$

(N is the number of alpha particles deflected at angle θ, Q is the total number of alpha particles incident on the foil, n the number of atoms per unit volume in the foil, t the thickness of the foil, Z the atomic number of the metal of the foil, e the elementary unit of charge, E the charge of the alpha particle, r the distance from the foil to the detection screen, and M and V the mass and velocity of the incident alpha particle.)

From the G–M hypothesis, together with information about a particular experimental set-up indicating the number of alpha particles, the thickness and the atomic number of the foil material, and so forth, the number of alpha particles scattered at various angles, including large ones, can be described in a precise way, using lawlike, quantitative assumptions. Moreover, this explanation is unifying in the sense that it permits the derivation of several different results obtained in the experiments of Geiger and Marsden. (For example, it permits a derivation of the fact that the number of alpha particles scattered through a given angle is directly proportional to the thickness of the scattering foil, and that the number is inversely proportional to the square of the energy of the alpha particles.) The explanation is causal in the sense that the G–M hypothesis contains a description of something that causes the scattering, viz., the presence of the atoms in the metal foil. And in so doing it appeals to micro-entities. Yet I think it would be regarded as vastly inferior to Rutherford's explanation. But objective, non-pragmatic values such as derivability from quantitative laws, unification, causation, and micro-entities will not *by themselves* tell us why Rutherford's explanation is a good one by contrast with the G–M hypothesis. Rutherford's explanation is good, or is as good as it is, not simply because it answers a causal question about the scattering in a quantitative way at a unifying micro-level, but because it does so at the *subatomic level of matter in a way that physicists at the time were interested in understanding the scattering.*

By 1911, the time of Rutherford's paper, the atomic theory of matter was widely accepted in physics, as was the idea that the atom itself is not atomic but has an internal structure. The latter idea emerged from the discovery of radioactivity and the electron, and the results of scattering of beta particles by atoms. It was also thought reasonable to suppose that alpha particle scattering was produced by events at the subatomic level. The question was how to work this out quantitatively using some theory about the structure of the atom. About five years before Rutherford's paper, J. J. Thomson had proposed the 'plum pudding' model of the atom according to which the positive electricity in the atom is uniformly distributed throughout the atom and the electrons are held stationary in equilibrium positions by the positive charges surrounding them and the repulsion of other electrons. However, it was impossible to derive the wide scattering angles of alpha particles from the Thomson model.

One of the reasons Rutherford's explanation is highly regarded is that it does derive these angles from a model of the internal structure of the atom—which physicists at the time were seeking. And I think that the major reason the G–M explanation would not be so highly regarded—despite the fact that it derives the wide scattering angles from quantitative hypotheses—is that it does not give an explanation by appeal to subatomic structure. (It simply says that the scattering is produced by atoms, and it provides an empirical formula for the scattering.) But to assess Rutherford's explanation in the manner suggested is to offer an illocutionary evaluation. In the present case we are considering whether Rutherford's explanation (by contrast say to G–M) is a good one for Rutherford to have given. To determine this we need to look at the situation of Rutherford and other physicists in 1911. What did they know, and what did they seek to know? Doing this means treating the explanation-sentence 'Rutherford's explanation of the alpha scattering is a good one' as strongly pragmatic. We need not, of course, treat it this way only with reference to Rutherford as explainer or a 1911 audience. The explanation-sentence might have a different truth-value if construed as elliptical for one making reference to a contemporary explainer and audience.

Now let me offer a conjecture. Suppose, following in the footsteps of Hempel and Salmon, you formulate a set of objective, non-pragmatic criteria that you think all scientific explanations must satisfy to be evaluated highly. These criteria will be universal in the sense that they are not to vary from one explanation to the next, but are to be ones applicable to all scientific explanations. They are also universal in the sense that they are not to incorporate specific empirical assumptions or presuppositions that might be made by scientists in one field or context but not another. So they might include the use of laws, causal factors, and quantitative hypotheses, the satisfaction of some criterion of unification or simplicity, and so forth. My conjecture is that whatever set of objective, non-pragmatic, universal criteria you propose you will be able to find or construct counter-examples to it, both as a set of necessary conditions and as a set of sufficient conditions. You will be able to find explanations that you will want to evaluate highly, despite the fact that they violate one or more of your favorite criteria. (Although this is not something I have illustrated here, you will evaluate them highly because they satisfy pragmatic criteria that are appropriate to use in the context of evaluation.) And you will be able to find, or at least construct, explanations (as I tried to do with the G–M explanation) that satisfy your criteria yet would not be highly regarded. You can emphasize criteria such as the introduction of laws, causal factors, and unification. But unless you say something more specific about the kinds of laws and causal factors to be used, or what is to be unified, you won't find your criteria sufficient to exclude examples you want excluded. But this 'something more', as I tried to illustrate in the Rutherford case, will involve fairly specific empirical assumptions that may be made by certain scientists at certain times but not by others at other times: You want to derive the scattering angles not just from any laws that will do the job, or from any causes no matter how described, but (e.g.) from ones that invoke events occurring within the atom. You desire an explanation that provides unification, but not just any sort of unification. (One that unifies only various results obtained in scattering experiments, as does the G–M hypothesis, may not be of sufficient interest to you.) To determine what this 'something more' is requires pragmatic assumptions about the explanatory context.

Now let me consider one major objection the non-pragmatist may offer. It is the one mentioned earlier that he might make against van Fraassen. Even if you accept the importance of illocutionary evaluations, the non-pragmatist may say, all this shows is that sentences of the form 'E is a good explanation of q' are incomplete. In the case of illocutionary evaluations the view I have espoused completes such sentences by writing 'E is a good explanation for an explainer to give in explaining q to an audience'—which makes them strongly pragmatic. But there may be ways to complete such sentences that yield the same evaluations but that are not pragmatic.

Let me use the term 'instructions' to refer to a set of rules or guidelines an explainer may be following when he explains q to an audience, or that an audience may want followed when q is explained to it. Instructions impose conditions on the answer to the explanatory question. They may incorporate very specific empirical conditions assumed by the explainer or audience. (For example, 'Describe the structure of the atom in such a way that the interaction between alpha particles and either positively or negatively charged constituents of the atom produces the scattering.') They may also incorporate some very general conditions ('Derive the scattering angles from quantitative laws'). Suppose that by appeal to a particular explanatory context—by appeal to the knowledge, beliefs, desires, and values of the explainer and audience—we determine that some set of instructions I is an appropriate one for that explainer to follow in explaining q to that audience. (The instructions themselves will not include reference to any explainer or audience.) We can now take the (allegedly) incomplete sentence 'E is a good explanation of q' and complete it by relativizing it to the instructions I (and perhaps also to some set of beliefs K of explainer and/or audience):

(8) E is a good explanation of q relative to instructions I (and K)

We then supply truth-conditions for sentences of this form which are 'objective' and are not relativized to explainer or audience. Here is one possibility:

(9) A sentence of form (8) is true iff
  (a) E satisfies instructions I, and E provides a correct answer to question Q; or
  (b) Given K, it is probable that (a) obtains

I do not wish to defend these conditions but only to use them as an example. By our earlier definition, sentences of the form (8) should be neither strongly nor weakly pragmatic. Such sentences contain no terms for an explainer or audience; their truth-conditions (9) do not contain such terms; and their truth-values will not vary with a change in who is explaining or to whom (as long as instructions I are kept the same). So, the non-pragmatist will admit, just as you need to appeal to the context to determine what question Q is being raised, and what beliefs K can be assumed, so you need to appeal to the context to determine what instructions I are to be followed. But once all these things are determined, then the issue of whether E is a good explanation of q relative to I and K is settleable in an objective, non-pragmatic way (by determining, e.g., whether (a) or (b) of (9) is satisfied).

This reply, I suggest, trivializes the non-pragmatist's position with regard to the evaluation of explanations. The aim of non-pragmatists such as Hempel and Salmon is to provide non-pragmatic criteria of evaluation—criteria whose applicability does not depend on, or vary with, who is explaining or to whom. What I have called 'instructions' are rules that incorporate criteria to be used in evaluating explanations. And the non-pragmatist is now agreeing with me that the applicability or appropriateness of some set of instructions will depend upon, and vary with, explainer and audience. But this is too much of an admission. When it comes to evaluating explanations I take the non-pragmatist to be seeking a set of instructions whose appropriateness is not affected by context.

Let me put this in another way. The non-pragmatist should not transform a sentence of the form 'E is a good explanation of q' into 'E is a good explanation of q relative to instructions I', but into 'E is a good explanation of q relative to *appropriate* instructions I'. Or better, he should say that sentences of the form 'E is a good explanation of q' are true only if there is some set of *appropriate* instructions that E satisfies. In either case the instructions are to be appropriate ones. And if, as above, the non-pragmatist admits that appropriateness always depends, in part, on context, he is in agreement with the pragmatist. If the very definition of 'appropriateness' with regard to instructions requires reference to an explainer and audience (see Achinstein (1983: 112 ff.)), and if the truth-conditions for 'E is a good explanation of q' require the satisfaction of appropriate instructions, then 'E is a good explanation of q' is strongly pragmatic.

In sum, the situation here is different from that of van Fraassen, who appeals to the context to determine only the question being raised, a set of alternative hypotheses, and the background information. By contrast, the instructions he formulates for evaluating explanations are not pragmatic. Their applicability does not depend on, or vary with, explainer or audience.

4. IMPLICATIONS

Let me comment briefly on the implications of a pragmatic theory of explanation for two recently contested issues in the philosophy of science.

(*a*) *Realism* v. *Anti-Realism*. Of course, a good deal depends on how you define 'realism' and 'anti-realism'. According to van Fraassen's formulation, the realist aims to give 'a literally true story of what the world is like', while the anti-realist aims to give 'theories that are empirically adequate' (1980: 9, 12), i.e., theories that yield truths about 'observables'.

The first point I want to make is that, contrary to what might be thought, a pragmatic theory of explanation does not commit one to anti-realism. Consider a theory of the sort I offer for pragmatic explanation-sentences of the form

E is a good explanation for an explainer to give in explaining q to an audience

The theory proposes several truth-conditions for sentences of this form, but the important one for the present issue is that E provides a correct answer to question Q or that it is reasonable for the explainer to believe it does. The fact that E provides a correct answer to Q is not by itself sufficient to make E a good explanation of q; further contextual conditions need to be satisfied. But these contextual conditions in no way prevent a realist construal of 'correct answer to Q' as one that, among other things, provides a 'literally true story'. The contextual conditions do not require that we construe a 'correct answer' to be one that simply 'saves the phenomena'. By reference to the context of Rutherford's 1911 explanation, we can determine the need to provide an explanation of the scattering that appeals to the inner structure of the atom. We may evaluate Rutherford's explanation highly, in part because it satisfies such contextually determined instructions. But this need to appeal to context does not mean that we must construe Rutherford's explanation non-realistically.

Indeed, so far as I can see, even van Fraassen's own evaluative theory—which earlier I argued is not pragmatic—does not require an anti-realist position of the sort he himself urges. We are supposed to evaluate the goodness of the explanation 'P in contrast with X because A' by determining whether proposition A is 'acceptable' or 'likely to be true', and by determining certain probabilistic relationships between A, the contrast-class X, and the other answers being considered. None of this would seem to require anti-realism. And the fact that the contrast class and alternative answers are determined contextually in no way precludes a realistic construal of answer A.

Conversely, pragmatism with regard to explanation does not commit one to realism. A 'correct answer to question Q' might be construed anti-realistically as one that 'saves the phenomena'. Or, perhaps better, one might drop the condition that the explanation provides a correct answer to Q in favour of the condition that the explanation provides an answer to Q that saves the phenomena. This modification is in no way precluded by the need to appeal to contextual facts about an explainer or audience. (There are other more compelling reasons to resist anti-realism having to do with the concept of understanding that I will not explore here. My point is only that the need to invoke explainers and audience is not a compelling reason.)

(b) *Relativism* v. *Absolutism*. Pragmatism with regard to explanation, particularly strong pragmatism, is a form of relativism. The truth-value of a strongly pragmatic explanation-sentence will vary with explainer and/or audience. But this relativism does not necessarily commit one to particularly virulent forms such as subjectivism or (Feyerabendian) anarchism. For example, it will not be the case that an explanation will be a good one for an explainer to give an audience if it simply satisfies any criteria set by the explainer or audience. For one thing, the explanation must satisfy some truth or confirmation requirement. For another, there may be certain criteria the satisfaction of which by the explanation is valuable for the explainer or audience, despite their own beliefs about these criteria. The form of relativism I would support could agree that the introduction of laws, causes, unification, and so forth, are general methodological criteria valued in science. They are '*prima-facie*' virtues. But in giving assessments of explanations of the sort I have been describing—in giving illocutionary evaluations—they cannot be treated as necessary or sufficient conditions. They are relevant, but they must be combined in appropriate ways with pragmatic information.

REFERENCES

Achinstein, Peter (1983), *The Nature of Explanation* (New York: Oxford University Press).
Bromberger, Sylvain (1965), 'An Approach to Explanation', in R. J. Butler (ed.), *Analytic Philosophy*, 2nd ser. (Oxford: Basil Blackwell), 72–105.
Geiger, H., and Marsden, E. (1909), 'On a Diffuse Reflection of the Alpha-Particles', *Proceedings of the Royal Society* (A) 82, 495–500.
Hempel, Carl G. (1965), *Aspects of Scientific Explanation and Other Essays in the Philosophy of Science* (New York: The Free Press).
Rutherford, E. (1911), 'The Scattering of Alpha and Beta Particles by Matter and the Structure of the Atom', *Philosophical Magazine*, 21, 669–88.
Thomson, J. J. (1904), 'On the Structure of the Atom: An Investigation of the Stability and periods of Oscillation of a Number of Corpuscles Arranged at Equal Intervals around the Circumference of a Circle; with Application of the Results to the Theory of Atomic Structure', *Philosophical Magazine*, 7, 237–65.
van Fraassen, Bas (1980), *The Scientific Image* (Oxford: Clarendon Press).

XII

EXPLAINING AND EXPLANATION*

ROBERT J. MATTHEWS

It is commonplace in science to speak of one thing as explaining another. The sorts of things that are said to explain or to be explained are various: theories, hypotheses, laws, facts, events, phenomena, etc. Carl Hempel has attempted to explicate this way of speaking by proposing an analysis of sentences of the form given by (1):

(1) E explains x

where E is a set of sentences (or propositions) constituting the explanation (the 'explanans') and x is the phenomenon to be explained (the 'explanandum-phenomenon').[1] That analysis, the deductive-nomological model of explanation, has been subjected to persistent criticism on a number of different grounds. Of the criticisms that have been adduced, one is especially relevant here: the claim that E explains x surely has something to do with the (potential) role of E in explaining episodes described by sentences of the form given by (2):

(2) A explained x (to B) by citing (by means of, etc.) E

where A is the explainer and B the intended audience. Yet there is nothing in Hempel's analysis that would suggest that the senses of 'explain' in (1) and (2) are related. Indeed, it is not clear why we speak of E as *explaining* x, since on his analysis the relation of E to x turns out to be one of the logical deducibility from E of a sentence describing x—something not obviously related to the activity of explaining x to someone. But instances of (1) and (2) *are* related: to say, for example, that Einstein's general theory of relativity explains the precession of the perihelion of

Robert J. Matthews, 'Explaining and Explanation', *American Philosophical Quarterly*, 18 (1981), 71–7. Reprinted by permission.

* Work on this paper was begun while I was an Andrew W. Mellon Faculty Fellow at Harvard University, 1977–8; I should like to thank Sylvian Bromberger, Richard Henson, Hilary Putnam, and George Smith for their helpful discussion and criticisms of earlier drafts of this paper.

[1] *Aspects of Scientific Explanation* (New York, 1965), 245–58, 333–76. (For simplicity of exposition I shall neglect those explanations subsumed by Hempel's inductive-statistical model; the special problems that they pose are not relevant here.)

Mercury is to say something about the theory's explanatory powers, i.e. about the potential role of that theory in explaining episodes.

In this paper I shall sketch an account of explanation that manifests the relation between (1) and (2); in particular, I shall propose a causal theory of explanation (hereafter referred to as the CTE) which bears important similarities to recently proposed causal theories of perception, knowing, and evidence. On this theory, explanations are characterized in terms of their role in a causal process that eventuates in an audience's coming to understand the phenomenon being explained.

1

Substitution instances of (1) and (2), exemplified by (3*a*) and (3*b*), are importantly similar to sentences of the sort exemplified by (4*a*) and (4*b*) respectively:

(3*a*) Einstein's general theory of relativity explains the precession of the perihelion of Mercury.

(3*b*) Jones explained the precession of the perihelion of Mercury by means of Einstein's general theory of relativity

(4*a*) The key opens the lock

(4*b*) Jones opened the lock using the key

In particular, substitution instances of (1) and (2) are instances of the more general schemata (5) and (6), respectively:

$$(5) \qquad\qquad NP_{instr.} + Verb_{causative} + NP_{obj.}$$
$$\text{[present]}$$

$$(6) \qquad\qquad NP_{agentive} + Verb_{causative} + NP_{obj.}$$
$$+ \begin{Bmatrix} using \\ by\ means\ of \\ etc. \end{Bmatrix} + NP_{instr.}$$

where in (5) the instrumental noun-phrase from (6) replaces the deleted agentive noun-phrase in the subject position. Arthur Collins was apparently the first to note this similarity,[2] but he mistakenly takes sentences exemplified by (3*a*) and (4*a*) to instance a common figure of speech, given by (7):

(7) If appeal to or use of *X* is commonly a means by which a person can get *Y* done, then speaking figuratively we can say, '*X* gets *Y* done'

---

[2] 'Explanation and Causality', *Mind*, 75 (1966), 482–5.

Collins seemingly overlooks the fact that sentences instantiating (5) can be (and most often are) used in a non-occurrent sense. Thus, Collins's claim to the contrary notwithstanding, a particular key opens a particular lock, if it does, even though it is not commonly used for that purpose. Indeed, it does so, if it does, even if it has never been used for that purpose. Similarly, one thing explains another, if it does, even if no one has ever actually explained the latter in terms of the former. The claim that X gets Y done is not always a claim that anyone has actually used X for this purpose, much less is it always a claim that the use of X is a common means by which a person can get Y done.

Sentences instantiating (5) are typically dispositional claims to the effect that the particular referred to by the instrumental noun-phrase can (i.e. is able to) bring about the state of affairs described by the sentence formed by taking the object noun-phrase as subject and the past participle of the causative verb as predicate. In other words, substitution instances of (5), and hence of (1), are typically power ascriptions of the form 'i brings about ϕj' that ascribe to some particular i the power of bringing about the state of affairs that j is ϕed. Thus, for example, to say that STP reduces engine friction is to say of STP that it has the power of bringing it about that engine friction is reduced. Similarly, I wish to argue, to say that the kinetic theory of gases explains Boyle's law is to say of that theory that it has the power to explain Boyle's law, i.e. to bring it about that Boyle's law is explained.

If, as I shall argue, sentences of the form 'E explains x' are power ascriptions, then the relation between (1) and (2) is straightforward: sentences of sort (2) are assertions to the effect that the explanatory powers ascribed to an explanans E by a sentence of sort (1) have been (might be, are being, etc.) exercised successfully by some explainer A with respect to some audience B. Of course, the explanatory powers of an explanans E must like the powers of any other particular be suitably activated, initiated, released, or otherwise brought into play by some external stimulus or agent. An explanans E that explains x will be efficacious in explaining x to B only if A cites E. Because the explainer's citing of E constitutes the immediate cause of the manifestation of E's powers, we typically speak of the *explainer* as having explained x, just as we speak of the doctor who administered penicillin as having cured a patient's pneumococcal infection.

The suggestion that we take talk of 'explanatory powers' at face value must meet two immediate objections. First, there are the usual sceptical worries about causal powers: despite their ubiquity in science and every-day life, power ascriptions are thought by some to involve the ascription of mysterious or occult qualities to particulars. Second, 'abstract' entities such as theories, hypotheses, and the like are seemingly not of a sort that can be causally efficacious; hence, the ascription of causal powers to these entities seems inappropriate.

There are several different ways of responding to the second objection depending on one's ontological proclivities. My own predilection is to take theories, hypotheses, and the like as universals, more specifically as 'norm kinds',[3] that have as correctly formulated instances physical objects or events possessing certain causal powers. On this view, theories *virtually* possess the causal powers that they do in virtue of their correctly formulated instances actually possessing those powers. Inasmuch as explanantia are causally efficacious only when suitably cited by some explainer, some may prefer to avoid this issue by taking the citing of E rather than E itself as causally efficacious; my subsequent analysis of ascriptions of explanatory powers can be readily reconstrued in these terms.

The sceptical worries about causal powers voiced in the first objection are probably best answered by providing an analysis of power ascriptions that specifies the sort of property ascribed to a particular i by power ascriptions of the form 'i brings about ϕj'. On the analysis that I shall propose here, power ascriptions are true of a particular, if they are, in virtue of properties actually possessed by that particular. Such ascriptions are peculiar not in the sort of property that they ascribe to particulars, but rather in the sort of predicates that are used to pick out the properties ascribed: they ascribe to the particular the property of being of such a nature that the particular would bring about a certain state of affairs if that particular were suitably activated, initiated, released, etc. More specifically, an analysis sufficient for present purposes is given by (8):

(8) 'i brings about ϕj' means 'i is of such a nature that if i were suitably[4] activated, initiated, released, etc., then j would be thereby ϕed'

The proposed analysis has just the consequences that one should expect: power ascriptions of the form 'i brings about ϕj' entail subjunctive conditionals of the form 'if i were suitably activated, initiated, released, etc., then j would be thereby ϕed'; furthermore, these ascriptions entail that the entailed subjunctive conditional is true of i in virtue of i's having the intrinsic nature that it does. The first entailment is uncontroversial, even though there is no general agreement as to the proper analysis of

---

[3] Cf. N. Wolterstorff, 'Towards an Ontology of Artworks', *Nous*, 9 (1965), 115–42.

[4] Appeal to the notion of 'suitable' antecedent conditions is innocuous so long as they can be specified non-circularly.

subjunctive conditionals. The second entailment is evidenced by the fact that we withdraw the power ascription if we discover that the behaviour thought to manifest the power is not to be explained causally in terms of the intrinsic nature of the particular. Thus, for example, we withdraw claims for the curative powers of a drug if we find that a placebo is equally efficacious.

There is nothing occult or mysterious about the notion of the 'nature' of a particular. The expression 'of such a nature that' functions in (8) as a placeholder for an unprovided characterization of the power in terms of the particular's fundamental properties, i.e. those properties in terms of which the true theory of the particular would explain manifestations of its powers. Power ascriptions of the form 'i brings about ϕj' ascribe to i a second-order property, namely, the second-order property of having certain unspecified first-order (i.e. fundamental) properties such that if i were suitably activated, initiated, released, etc., then j would be thereby ϕed.[5] Thus, (8) should be construed as follows: 'i brings about ϕj' means 'there exist certain unspecified fundamental properties $F_1, \ldots, F_n$ such that in virtue of having these properties, if i were suitably activated, initiated, released, etc., then j would be thereby ϕed'. Power ascriptions, it should be noted, do not provide a specification of the nature of the particular in virtue of which it has the powers that it does; nor do they preclude the possibility that two particulars having different natures might each have the same power of bringing about a certain state of affairs. (Poisons are a case in point.)

2

The task of accounting for the explanatory powers of explanantia is considerably more difficult than that of accounting for the more ordinary powers of keys, medicine, and the like, because one has the added problem of explaining just what the power to explain comes to. As a first step we must ascertain what someone succeeds in doing when he (or she) succeeds in explaining x to someone. There are two different but closely related senses of 'explain' that need to be distinguished here: when we say of some explainer that he successfully explained x to someone by citing E, we may mean that he succeeded in bringing his audience to understand x by citing E. Or we may mean only that he cited E with the intention of achieving this effect. I shall refer to these two senses as

---

[5] Cf. Hilary Putnam, 'On Properties', *Mathematics, Matter, and Method* (Cambridge, 1975), 305–22.

respectively the perlocutionary and illocutionary senses of 'explain'. Thus, when we ask what someone succeeds in doing when he succeeds in explaining x to someone by citing E, we must distinguish two different sorts of success: (1) the explainer's success in doing something that constitutes an instance of explaining in the strong perlocutionary sense of 'bringing to understand by citing E', and (2) his success in doing something that constitutes an instance of explaining in the weak illocutionary sense of 'citing E with the intention of bringing to understand'.

A's explaining x to B by citing E (in the illocutionary sense) involves more than A's merely citing E with the intention of bringing B to understand x. As Peter Achinstein remarks, 'not every way of attempting to render q understandable [where q is the *oratio obliqua* form of a question Q] counts as attempting to explain q'.[6] He offers the following illustration: I know that the sentence 'God is love' is so causally efficacious with you that the mere uttering of it will cause you to understand anything, and in particular why atoms emit only discrete radiation. Knowing this, I utter the sentence with the intention of bringing you to understand why atoms emit only discrete radiation. My utterance achieves the intended effect, but surely I have explained nothing. The point seems clear: A's citing of some word, sentence, theory, etc. may have the effect of bringing an audience to understand q without its being true that A explained q to them. There are further conditions on A's citing that must be met if A is to have explained q by citing E. Minimally, A must cite an appropriate explanans E, and that explanans must bring the intended audience to understand q in the appropriate manner. The following examples illustrate these two further conditions. Suppose that A intends to explain q to an audience B, believes on good evidence that E^* explains q, and therefore attempts to explain q to B by citing E^*. Suppose further that E^* does not explain q, but that the mistake is just of a sort that permits B to recognize E as the correct explanans and hence to come to understand q. A's citing of E^* causes B to come to understand q, but A has not explained q, except in the weak illocutionary sense of having cited E^* with the intention of bringing his audience to understand q. Suppose next that A, having been apprised of his mistake, sets out to explain q to a different audience, this time by citing the correct explanans E. Suppose that A's audience comes to understand q, again as a result of his citing E, but for reasons having solely to do with the manner in which he cites E (e.g., its loudness, eloquence, etc.). A's citing of E causes his audience to come to understand q, but here again A has not explained q.

---

[6] *Law and Explanation* (Oxford, 1971), 63.

There is, I think, a third condition on A's citing of E, namely, that the explanans must itself be produced in the appropriate manner in virtue of A's understanding of q, but I shall not argue for it here.

Achinstein's consideration of various cases in which A's uttering a sentence would not count as an instance of A's explaining q to B leads him to an analysis that effectively accepts the two conditions formulated above. His analysis would satisfy the first condition by requiring that A believe that what he utters by way of explanation expresses a correct answer to Q. The analysis would satisfy the second condition by requiring that A intend that his utterance of u render q understandable to B solely by producing in B the recognition that u expresses a correct answer to Q. This Gricean requirement is intended to rule out cases in which A might intend that his utterance of u bring B to understand q, but in a way that would be inappropriate. Achinstein concludes that A explains q by uttering u (in the illocutionary sense of 'explain') if and only if

(9) A utters u with the intention that his utterance of u render q understandable solely by producing the recognition that u expresses a correct answer to Q[7]

Achinstein requires that A believe that his utterance expresses a correct answer to Q. But this requirement is either wrong or misleading, depending on how one construes the notion of an 'answer' to Q. If the notion is construed narrowly so that u expresses a correct *direct answer* to Q,[8] and nothing more than that, then the requirement is wrong, since (9) cannot then capture the explainer's intentions in many cases. For it is often the case that in explaining q to B one must provide not only a correct (direct) answer to Q, but also such other information as may be necessary in order to bring B to understand q.[9] In addition to providing a correct (direct) answer to Q, A's explanation may have to provide additional background information, correct certain of B's mistaken background beliefs that are inconsonant with the correct answer to Q, show B that certain of his correct background beliefs are in fact consonant with the answer, and so on. If, on the other hand, the notion of an 'answer' to Q is construed broadly so that u provides whatever information may be necessary to bring B to understand q, then Achinstein's requirement is

---

[7] 'What Is an Explanation?', *American Philosophical Quarterly*, 14 (1977), 2.

[8] A direct answer to a question Q is a proposition that completely, but just completely answers Q. Cf. Nuel D. Belnap and Thomas B. Steel, *The Logic of Questions and Answers* (New Haven, Conn., 1976); and Lennart Åqvist, *A New Approach to the Logical Theory of Interrogatives* (Tübingen, 1975).

[9] See Sylvain Bromberger, 'An Approach to Explanation', in R. J. Butler (ed.), *Studies in Analytical Philosophy* (Oxford, 1965), 73–105.

misleading, since it gives the mistaken impression that the eventual explication of the broad notion of a correct answer to Q is a reasonable expectation for erotetic logic. In fact, this expectation is unreasonable, for it is very unlikely that a developed erotetic logic would individuate questions in such a way that a single interrogative construction would express a different question for each possible set of background conditions. Certainly no results to date in the semantics of interrogatives gives any ground for optimism. Achinstein presumably overlooks this difficulty because he fails to distinguish the broad and narrow senses of an 'answer' to Q. Thus, he assumes mistakenly that (9) can be true on a sense of an 'answer' to Q such as might be characterized by an erotetic logic, presumably along lines suggested by his notion of a 'complete answer' to Q (i.e. a proposition formed by replacing the *wh*-pronoun in the interrogative sentence expressing Q by a phrase giving a reason, time, manner, etc).[10]

On the broad sense of an 'answer' to Q for which (9) is misleading, Achinstein's analysis nevertheless suggests a useful first characterization of the causal role that the citing of an appropriate explanans E must play in bringing B to understand q by way of explanation: A's utterance must manifest his intention to bring B to understand q; furthermore, the propositions expressed by that utterance must provide whatever information B must learn in order to come to understand q, and B's coming to understand q must result from his recognition that A's utterance provides this information. Or to put the point more directly:

(10) A explains q to B by citing E (in the illocutionary sense) if and only if A cites E with the intention of (i) presenting to B such information as B must learn in order to bring B to understand q, and (ii) having B come to understand q as a result of his having recognized that E provides this information.

This analysis differs from Achinstein's in two respects. First, explaining is here defined in terms of citing certain propositions rather than in terms of uttering certain sentences. This is done primarily to facilitate the subsequent analysis of sentences of the form 'E explains x' in which E ranges over propositions (theories, hypotheses, etc.), though it also reflects my suspicion that 'explain' is not a genuine illocutionary-act verb. (Explaining is supervenient on the performance of certain illocutionary acts.) Second, and more important, the analysis leaves unspecified the information that must be provided in order to bring B to understand q: A is not required to provide a particular piece of information (e.g. a correct

---

[10] 'What Is an Explanation?', 9.

direct answer to Q); rather A must provide whatever information must, in A's opinion, be provided in order to bring B from his presumed state of not understanding q to a state of understanding q. Because (10) does not require that A provides a correct direct answer to some question Q for which q (the *oratio obliqua* form of Q) is the object of explanation, (10) provides a general analysis of sentences of the form 'A explains x to B by citing E' (in the illocutionary sense of 'explain'), where x is any suitable object of explanation. In other words, the class of permissible grammatical objects of 'explain' in (10) is not restricted to interrogative constructions of the sort 'why such-and-such is the case', 'how such-and-such occurs', and so on. In the argument that follows, I shall therefore construe all occurrences of 'q' in (10) as having been replaced by 'x'.

It is a crucial tenet of the CTE that if A is to explain x to B by citing E, then A must intend that his citing of E play the appropriate causal role in bringing B to understand x. A need not know the causal role in fact played by the citing of explanantia in successful explaining episodes in order that he be able to intend that his citing of E play that role. He may simply intend that his citing of E play the role in fact played by the citing of explanantia in successful explaining episodes, whatever it may be. The requirement in (10) that B come to understand x as a result of his having recognized that E provides whatever information he needs in order to achieve this understanding attempts to capture this condition on explaining. It has been suggested to me that this Gricean formulation of the condition, adopted from Achinstein, may not have the intended effect of ruling out cases in which the etiology of B's coming to understand x is inappropriate. Specifically, B might come to understand x as a result of his having recognized that E provides the requisite information without his actually having understood E. Now it does seem that B's understanding E should be part of the etiology of his coming to understand x, so if there could be such cases of recognition without understanding, then the Gricean formulation of (10) would need strengthening. I am undecided whether such strengthening is in fact needed; however, if it is, the direction of the needed refinements should be clear.

We are now in a position to say what the power to explain comes to. From (10) we can infer:

(11) A explains x to B by citing E (in the perlocutionary sense) if and only if A's citing of E presents to B such information as B must learn in order to come to understand x, and in so doing A intentionally brings B to understand x as a result of his having recognized that E provides this information

From (11) and our earlier analysis of power ascriptions, given by (8), we can infer:

> (12) *E* explains *x* to *B* when cited by *A* if and only if *E* is of such a nature that if *E* were suitably[11] cited to *B* by *A*, then the citing of *E* by *A* would present to *B* such information as *B* must learn in order to come to understand *x*, and in so doing *A* would bring *B* to understand *x* as a result of his having recognized that *E* provides this information

(12) provides an analysis not of categorical power ascriptions of the form '*E* explains *x*' but of ascriptions of explanatory powers that are relativized to both explainer *A* and audience *B*. The proper analysis of categorical ascriptions is obtained from (12) by generalization:

> (13) *E* explains *x* if and only if *E* is of such a nature that for any explainer *y* and any audience *z*, if *E* were suitably cited to *z* by *y*, then the citing of *E* by *y* would present to *z* such information as *z* must learn in order to come to understand *x*, and in so doing *y* would bring *z* to understand *x* as a result of *z*'s having recognized that *E* provides this information

At first blush (13) seems too strong. On the analysis provided, virtually all power ascriptions of the form '*E* explains *x*' will turn out to be false, since for any explanans *E* there will surely be at least one audience *B* for which the citing of *E* would not be efficacious in bringing that audience to understand *x* in the prescribed manner. We speak, for example, of the kinetic theory of gases as explaining Boyle's law, but we can easily imagine audiences for which the citing of that theory would not be efficacious in bringing them to understand the regularity in question. The audience may be distracted, fail to understand the theory, not be able to follow the derivation, etc. On the analysis provided by (13), the claim that the kinetic theory of gases explains Boyle's law is therefore false. Critics of Hempel's deductive-nomological model have attempted to establish the falsity of sentences of the form '*E* explains *x*' by adducing similar examples. Because *E* cannot be used successfully to explain *x* to just any audience, they conclude that *E* explains *x* only for certain audiences.

I endorse their conclusions which support the seemingly untoward consequence of (13) that sentences of the form '*E* explains *x*' are, strictly speaking, false. The assertion that *E* explains *x* is categorical: the unquali-

---

[11] By 'suitable' here I mean that *E* is cited in a language familiar to *B*, at a level audible to him, etc.

fied power of explaining x is ascribed to E; no explicit mention is made of particular audiences for which the citing of E will be efficacious, even though the citing of E will be efficacious only for certain audiences. Nor is there any compelling reason, so far as I can see, for thinking limitations on E's powers to be implicit in the ascription. One may be tempted to suppose, as some have, that an assertion to the effect that E explains x is in fact elliptical for an ascription that is suitably relativized to an audience intended by the speaker who asserts that E explains x. But I know of no principled grounds of a sort provided by either semantic or pragmatic theories for thinking this so. Indeed, if this supposition were correct, then there should be no objection to concluding from the fact that A successfully explained x to B by citing E that E explains x. But surely there is. E may explain x only for B. It would also be a consequence of the supposition that categorical ascriptions are elliptical for such audience-relativized ascriptions that some E that now explains x may fail to explain x at some later time, if the knowledge, interests, and beliefs in terms of which the intended audience is defined changes. But surely E explains x, if it does, irrespective of the vicissitudes of human knowledge, interests, and beliefs.

A more plausibly relativized version of (13) might analyse power ascriptions of the form 'E explains x' in terms of E's power, when suitably cited, to bring a *standard* audience to understand x in the appropriate manner. Such an analysis would aim to capture the intuition that the categorical ascription is warrantedly assertable just in case the citing of E would normally be efficacious in explaining x to the sort of audience that would normally receive explanations of x. Yet even if we set aside difficulties of spelling out what is to count as a 'standard' audience, I see little to recommend such a weakened version of (13). The only ground that I can see for thinking that ascriptions of the form 'E explains x' are not in fact categorical in the manner suggested by (13) has to do with the dubious assumption that assertions that have a prominent place in scientific discourse cannot be admittedly false. Thus, it becomes a desideratum, if not a requirement, of any adequate analysis that on that analysis power ascriptions turn out to be true. In fact, falsehoods enjoy a very prominent place in scientific discourse. Approximation and idealization, which is to say, something less than strict truth, are the threads from which the fabric of science is woven. Categorical power ascriptions are no exception: they are advanced as idealizations, i.e., as approximate truths recognized to hold only in certain unspecified (and perhaps unspecifiable) situations. Thus, for example, when penicillin fails to save a dying pneumonia victim, we acknowledge the idealization present in the claim that penicillin cures pneumonia, usually by mentioning certain extenuating circumstances that set this case apart from the large number of standard cases for which penicillin is efficacious. If we are unsuccessful in explaining away the failure, we withdraw the categorical ascription or qualify it appropriately. Of course, our ability to explain away failures is closely tied to our ability to explain successes, since explanations of the former sort will appeal to our best explanation of what goes on in the successful cases. But what must be emphasized here is that we are presently able to explain away many potential counter-instances to power ascriptions of the form 'E explains x'. We are able to do this because we already know much about the situations in which the explanatory powers of an explanans E that explains x can be successfully exercised. We already possess something like a tacit theory of explanation, and within this theory categorical power ascriptions of the form 'E explains x' are construed as idealizations, i.e. as approximate truths recognized to hold only in certain unspecified situations. These idealizations find special application in the sciences because the audiences to whom particular scientific explanations are directed are typically quite uniform: audiences that find themselves on the receiving end of an explanation of the precession of the perihelion of Mercury, for example, typically share the knowledge, beliefs, and interests required for the success of an explanation that appeals to the general theory of relativity. Categorical ascriptions of the form 'E explains x' find very little application outside the sciences, precisely because such uniformity of audience is lacking. This difference bears emphasis because many philosophers of science have attributed the prevalence of categorical ascriptions in the sciences to a special character of scientific explanations rather than to the special character of the audiences to which they are addressed.

3

The causal theory of explanation sketched in this paper vindicates the commonsensical notion that E's *explaining* x has something to do with actual (or potential) explaining episodes involving the citing of E, since on the analysis provided by (13), E explains x if and only if the citing of E is capable of playing the appropriate causal role in a particular causal process—the 'explanatory process'—that eventuates in an audience's coming to understand x. The CTE also vindicates the intuition that there is something common to explanatory endeavours across disparate domains of inquiry that warrants the use of a single term to designate this activity. Hempel would seemingly have us believe that the senses of 'explain' involved in talk of explaining the rules of a contest, explaining the meaning of a cuneiform inscription or a complex legal clause or a passage of a symbolist poem, or explaining how to bake Sacher torte or how to repair a radio, are no more related to the senses of 'explain' in talk of explaining Boyle's law or explaining the precession of the perihelion of Mercury than are the senses of 'proof' in 'mathematical proof' and '86 proof Scotch'—which is to say, not related at all.[12] Yet the CTE makes clear that in each case we are talking about an attempt, not necessarily successful, to bring someone to understand something. What Hempel takes to be different sorts of explanation can be seen to be explanations that bring about different sorts of understanding, but explanations none the less. Hempel readily concedes that there is a pragmatic sense of 'explain' on which explaining is a matter of bringing an audience to understand something. But this, he insists, is not the sense of 'explain' that interests him. He is concerned only with what he calls the 'logical aspect' of explanation. The account provided by the CTE challenges the notion that there is a non-pragmatic sense of 'explain' that is instanced by sentences of the form 'E explains x'. The sense of 'explain' instanced by such sentences is irrevocably pragmatic inasmuch as their proper analysis, given by (13), mentions the potential role of an explanans E that explains x in explaining episodes in which someone brings an audience to understand x by citing E.

An obvious difficulty for an account of explanation such as mine that relies on the notion of understanding is that there are decisive objections to all the received accounts of understanding.[13] But while the CTE's reliance on this as yet unexplicated notion signals a piece of unfinished business, and a big piece at that, it seems to me a virtue of the theory that it locates the difficulty correctly. It was surely a fundamental mistake on Hempel's part to have proposed the deductive-nomological model as a model of scientific explanation, for in doing so he encouraged the persistent conflation of pragmatic with non-pragmatic criticisms of the model. That model, I believe, is more properly construed as a model of scientific understanding. The 'complete explanations' of the model stand to actual explanations (the 'sketches' of the model) in just the relation that understanding some phenomenon stands to an explanation of that

---

[12] *Aspects of Scientific Explanation*, 413.

[13] See Michael Friedman, 'Explanation and Scientific Understanding', *The Journal of Philosophy*, 71 (1974), 5–19. Serious difficulties with Friedman's own account of understanding are identified by P. Kitcher, 'Explanation, Conjunction, and Unification', *Journal of Philosophy*, 73 (1976), 207–12.

same phenomenon: in each case the latter is a more or less incomplete expression of the former, tailored to the needs of the explanatory situation. So construed, the deductive-nomological model can be saved from most pragmatic criticisms, including those raised by the CTE, for the model will then have no more (or less) to do with explanation than does any account of understanding.

NOTES ON THE CONTRIBUTORS

PETER ACHINSTEIN is Professor at Johns Hopkins University. He is the author of *Concepts of Science, Law and Explanation*, and *The Nature of Explanation*.

BARUCH A. BRODY is Leon Jaworski Professor of Biomedical Ethics at Baylor College of Medicine. He is the author of *Abortion and the Sanctity of Human Life*, *Identity and Essence*, and *Life and Death Decision-Making*. He is currently working on two books, one on the development of new drugs, and one on international health expenditures.

J. ALBERTO COFFA was Professor of the History and Philosophy of Science at Indiana University, until his untimely death in 1984. His doctoral dissertation, *The Foundations of Inductive Explanation*, and his subsequent seminal articles, have been widely influential.

CARL G. HEMPEL was Stuart Professor of Philosophy at Princeton University until his retirement in 1973, and subsequently taught at the University of Pittsburgh. He is the author of *Aspects of Scientific Explanation*, *Fundamentals of Concept Formation in Empirical Science*, *Philosophy of Natural Science*, and of many influential articles in the philosophy of science.

JAEGWON KIM is Professor at Brown University. His many articles span such topics as causation, explanation, mind and body, and supervenience. Some of his essays have appeared in a collection, *Supervenience and Mind*.

PHILIP KITCHER teaches at the University of California, San Diego. He is the author of *The Nature of Mathematical Knowledge*.

DAVID LEWIS is Professor at Princeton University. He has written many books and articles, including *Convention*, *Counterfactuals*, and *Philosophical Papers*, volumes i and ii.

PETER LIPTON is in the Department of the History and Philosophy of Science at Cambridge University. He is the author of *Inference to the Best Explanation*.

TIMOTHY MCCARTHY teaches at the University of Illinois at Urbana-Champaign. He has written extensively on logic, philosophy of mathematics, and philosophy of language. He is currently completing a book on the indeterminacy of reference.

ROBERT J. MATTHEWS is Professor of Philosophy at Rutgers University. He is the author of articles in the philosophy of psychology and in theoretical psycholinguistics; he is co-editor of *Learnability and Linguistic Theory*.

PETER RAILTON is Professor at the University of Michigan, Ann Arbor. In addition to his work in the philosophy of science, he has made important contributions to moral philosophy. His work on explanation, realism, normative ethics, and metaethics has appeared in some of the leading journals of philosophy.

DAVID-HILLEL RUBEN is Professor of Philosophy at the London School of Economics, University of London. He is author of *The Metaphysics of the Social World* and *Explaining Explanation*.

WESLEY C. SALMON, who teaches at the University of Pittsburgh, has written several books and many important articles on explanation, including *Scientific Explanation and the Causal Structure of the World*, and on other topics of interest in the philosophy of science.

BAS C. VAN FRAASSEN is Professor at Princeton University. He has written *The Scientific Image*, *Laws and Symmetry*, and *Quantum Mechanics*.

JAMES WOODWARD is Professor of Philosophy at the California Institute of Technology. He works mainly in philosophy of science, and has written a number of papers on explanation. He is currently working on a book on explanation.

SELECT BIBLIOGRAPHY

Philip Kitcher and Wesley Salmon (eds.), *Scientific Explanation, Minnesota Studies in the Philosophy of Science*, 13 (Minneapolis: University of Minnesota Press, 1989), includes a comprehensive chronological bibliography on explanation, which aims for near completeness for the period 1947–88. The reader who wishes a fuller bibliography than that which I have provided below is advised to consult their list. As in the collection itself, this bibliography ignores items which focus primarily on functional explanation, action explanation, and explanation in history and social science.

ANTHOLOGIES

GRÜNBAUM, ADOLF, and SALMON, WESLEY (eds.), *The Limitations of Deductivism* (Berkeley and Los Angeles: University of California Press, 1988).

KITCHER, PHILIP, and SALMON, WESLEY (eds.), *Scientific Explanation, Minnesota Studies in the Philosophy of Science*, 13 (Minneapolis: University of Minnesota Press, 1989).

KNOWLES, DUDLEY (ed.), *Explanation and Its Limits* (Cambridge: Cambridge University Press, 1990).

KÖRNER, STEPHAN (ed.), *Explanation* (Oxford: Blackwell, 1975).

MCLAUGHLIN, ROBERT (ed.), *What? Where? When? Why?* (Dordrecht: Reidel, 1982).

PITT, JOSEPH (ed.), *Theories of Explanation* (New York: Oxford University Press, 1988).

BOOKS

ACHINSTEIN, PETER, *The Nature of Explanation* (New York: Oxford University Press, 1983).

ARISTOTLE, *The Posterior Analytics*. Any edition, but especially Jonathan Barnes (trans.) (Oxford: Oxford University Press, 1975).

BRAITHWAITE, RICHARD, *Scientific Explanation* (Cambridge: Cambridge University Press, 1953).

GARFINKEL, ALAN, *Forms of Explanation* (New Haven, Conn.: Yale University Press, 1981).

HEMPEL, CARL, *Aspects of Scientific Explanation* (New York: The Free Press, 1965).

—— *Philosophy of Natural Science* (Englewood Cliffs, NJ: Prentice-Hall, 1966).

HUMPHREYS, PAUL, *The Chances of Explanation* (Princeton, NJ: Princeton University Press, 1989).

LIPTON, PETER, *Inference to the Best Explanation* (London: Routledge, 1991).

MILL, JOHN STUART, *A System of Logic*, book 3, any edition.

RUBEN, DAVID-HILLEL, *Explaining Explanation* (London: Routledge, 1990).

SALMON, WESLEY, *Four Decades of Scientific Explanation* (Minneapolis: University of Minnesota Press, 1989), and also as an article in Kitcher and Salmon (eds.), above.

—— *Scientific Explanation and the Causal Structure of the World* (Princeton, NJ: Princeton University Press, 1984).

—— JEFFREY, RICHARD, and GREENO, JAMES, *Statistical Explanation and Statistical Relevance* (Pittsburgh: University of Pittsburgh Press, 1971).

VAN FRAASSEN, BAS, *The Scientific Image* (Oxford: Oxford University Press, 1980).

VON WRIGHT, GEORG HENRIK, *Explanation and Understanding* (London: Routledge & Kegan Paul, 1971).

ESSAYS AND ARTICLES

ACHINSTEIN, PETER, 'A Type of Non-Causal Explanation', in Peter French, Theodore Uehling, Jr., and Howard Wettstein (eds.), *Causation and Causal Theories, Midwest Studies in Philosophy* (Minneapolis: University of Minnesota Press, 1984), 221–44.

ACKERMAN, ROBERT, 'Discussion: Deductive Scientific Explanation', *Philosophy of Science*, 32 (1965), 155–67.

ARONSON, JERROLD, 'Explanations Without Laws', *Journal of Philosophy*, 66 (1969), 541–57.

BROMBERGER, SYLVAIN, 'An Approach to Explanation', in R. J. Butler (ed.), *Analytic Philosophy*, 2nd ser. (Oxford: Blackwell, 1965), 72–105.

—— 'Why-Questions', in *Mind and·Cosmos: Essays in Contemporary Science and Philosophy*, ed. Robert Colodny, (Pittsburgh: University of Pittsburgh Press, 1966), 86–111.

COLLINS, ARTHUR, 'Explanation and Causality', *Mind*, 75 (1966), 482–500.

DAVIDSON, DONALD, 'Causal Relations', *Journal of Philosophy*, 64 (1967), 691–703.

FETZER, JAMES, 'Probability and Explanation', *Synthese*, 48 (1981), 371–408.

FORGE, JOHN, 'The Instance Theory of Explanation', *Australasian Journal of Philosophy*, 64 (1986), 127–42.

—— 'Theoretical Explanation in Physical Science', *Erkenntnis*, 23 (1985), 269–94.

FRIEDMAN, MICHAEL, 'Explanation and Scientific Understanding', *Journal of Philosophy*, 71 (1974), 5–19.

—— 'Theoretical Explanation', in Richard Healey (ed.), *Reduction, Time, and Reality* (Cambridge: Cambridge University Press, 1981), 1–16.

GÄRDENFÖRS, PETER, 'A Pragmatic Approach to Explanations', *Philosophy of Science*, 47 (1980), 404–23.

HANNA, JOSEPH, 'On Transmitted Information as a Measure of Explanatory Power', *Philosophy of Science*, 45 (1978), 531–62.

HARMAN, GILBERT, 'The Inference to the Best Explanation', *Philosophical Review*, 74 (1965), 88–95.

HUMPHREYS, PAUL, 'Aleatory Explanations', *Synthese*, 48 (1981), 225–32.

JOBE, EVAN, 'A Puzzle Concerning D-N Explanation', *Philosophy of Science*, 43 (1976), 542–9.

KIM, JAEGWON, 'Explanatory Exclusion and the Problem of Mental Causation', in E. Villanueva (ed.), *Information, Semantics, and Epistemology* (Oxford: Blackwell, 1990), 36–56.

—— 'Noncausal Connections', *Nous*, 8 (1974), 41–52.

KINCAID, HAROLD, 'Supervenience and Explanation', *Synthese*, 77 (1988), 251–81.

KITCHER, PHILIP, 'Explanatory Unification', *Philosophy of Science*, 48 (1981), 507–31.

LYON, ARDON, 'The Relevance of Wisdom's Work for the Philosophy of Science', in Renford Bambrough (ed.), *Wisdom: Twelve Essays* (Oxford: Blackwell, 1974), 218–48.

MACKIE, J. L., 'Extensionality—Two Kinds of Cause', ch. 10 of his *The Cement of the Universe* (Oxford: Oxford University Press, 1974), 248–69.

McMULLIN, ERNAN, 'Structural Explanation', *American Philosophical Quarterly*, 15 (1978), 139–47.

MEIXNER, JOHN, 'Homogeneity and Explanatory Depth', *Philosophy of Science*, 46 (1979), 366–81.

MELLOR, D. H., 'Probable Explanation', *Australasian Journal of Philosophy*, 54 (1976), 231–41.

MORAVCSIK, JULIUS, 'Aristotle on Adequate Explanations', *Synthese*, 28 (1974), 3–17.

NICKLES, THOMAS, 'Davidson on Explanation', *Philosophical Studies*, 31 (1977), 141–5.

NOZICK, ROBERT, 'Why Is There Something Rather Than Nothing?', in his *Philosophical Explanations* (Oxford: Oxford University Press, 1984), 115–64.

OPPENHEIM, PAUL, and PUTNAM, HILARY, 'Unity of Science as a Working Hypothesis', in Herbert Feigl, Michael Scriven, and Grover Maxwell (eds.), *Minnesota Studies in the Philosophy of Science* 2 (Minneapolis: University of Minnesota Press, 1958), 3–36.

RAILTON, PETER, 'A Deductive-Nomological Model of Probabilistic Explanation', *Philosophy of Science*, 45 (1978), 206–26.

RYLE, GILBERT, ' "If", "So", and "Because" ', in Max Black (ed.), *Philosophical Analysis: A Collection of Essays* (Englewood Cliffs, NJ: Prentice-Hall, 1963), 302–18.

SALMON, WESLEY, 'A Third Dogma of Empiricism', in Robert Butts and Jaakko Hintikka (eds.), *Basic Problems in Methodology and Linguistics* (Dordrecht: Reidel, 1977), 149–66.

SCRIVEN, MICHAEL, 'Causation as Explanation', *Nous*, 9 (1975), 3–16.

—— 'Explanations, Predictions, and Laws', in Herbert Feigl and Grover Maxwell (eds.), *Minnesota Studies in the Philosophy of Science*, 3 (Minneapolis: University of Minnesota Press, 1962), 170–230.

—— 'Truisms as Grounds for Historical Explanation', in Patrick Gardiner (ed.), *Theories of History* (New York: Free Press, 1959), 443–75.

SOBER, ELLIOTT, 'Common Cause Explanation', *Philosophy of Science*, 51 (1984), 212–41.

—— 'Explanatory Presupposition', *Australasian Journal of Philosophy*, 64 (1986), 143–9.

STEGMÜLLER, Wolfgang, 'Two Successor Concepts to the Notion of Statistical Explanation', in G. H. von Wright (ed.), *Logic and Philosophy* (The Hague: Nijhoff, 1980), 37–52.

STRAWSON, PETER, 'Causation and Explanation', in B. Vermazen and J. Hintikka (eds.), *Essays on Davidson* (Oxford: Oxford University Press, 1985), 115–35.

TELLER, PAUL, 'On Why-Questions', *Nous*, 8 (1974), 371–80.

THAGARD, Paul, 'The Best Explanation: Criteria for Theory Choice', *Journal of Philosophy*, 75 (1978), 76–92.

TUOMELA, RAIMO, 'Explaining Explaining', *Erkenntnis*, 15 (1980), 211–43.

—— 'Inductive Explanation', *Synthese*, 48 (1981), 257–94.

WOODWARD, JAMES, 'Are Singular Causal Explanations Implicit Covering-Law Explanations?', *Canadian Journal of Philosophy*, 16 (1986), 253–80.

—— 'Developmental Explanation', *Synthese*, 44 (1980), 443–66.

INDEX OF NAMES

I have omitted indexing references to Carl Hempel. His importance to the subject of explanation is indicated by the fact that, if he were to have been included in the Index, it could only have made sense to list the pages which did not refer to him or to his ideas.